What on Earth Happened?

Christopher Lloyd

Illustrations by Andy Forshaw

BLOOMSBURY

Part 1 Mother Nature

(13.7 billion – 7 million BC)

How the universe was formed, and the unfolding of life on earth before mankind.

Part 2 Homo Sapiens
(7 million–5000 BC)
How humans evolved as hunter-gatherers, living within the state of nature.

Part 3 Settling In
(5000 BC – c.570 AD)

*How farm animals and crops led to the growth of a range
of different human civilizations.*

Part 4 Going Global
(c.570 AD–present day)

How the fates of human civilizations and the natural world fused into a global whole.

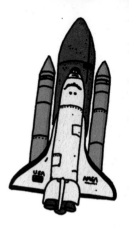

30 What a Revelation!

How a series of visions appeared to Mohammed, a man from Mecca, giving birth to Islam, a new way of life that promised to perfect the errors of mankind.
page 231

31 Paper, Printing & Powder

How Chinese scientific discoveries gradually spread westwards via Islam to Europe, supercharged by a Mongol chief who created the world's largest empire.
page 243

32 Medieval Misery

How plague, invasions and famine impoverished Christian Europe, which found itself surrounded by Islamic civilizations, impenetrable desert and endless blue seas.
page 256

33 Treasure Hunt

How each of the world's settled human societies sought out their own various fortunes through a mixture of trade, toil and theft.
page 268

34 Moules Marinière

How a few maritime explorers accidentally discovered a New World. Their arrival proved fatal to ancient civilizations and provoked a fierce contest between the rival nations of Europe.
page 278

35 Fancy a Beer?

How European merchants pioneered a new way of life overseas that fuelled a taste for growing lucrative crops, making some very rich and many more poor.
page 293

36 New Pangaea

How crops and creatures were farmed, harnessed, transported and exploited for the fancies of a single, global and mostly civilized species: man.
page 304

37 Mixed Response

How different human civilizations reacted to the arrival of European businessmen-cum-soldiers eager to trade for profit.
page 313

38 Free Reign

How extreme inequality between people ignited rebellions in the name of freedom, and how armies were conscripted for the sake of a feeling, flag or song.
page 322

39 Monkey Business

How the human species freed itself from nature's limitations by mastering its own source of transportable power and how its populations increased beyond all reasonable measure.
page 334

40 White Man's Race

How people from the West became convinced they were superior to all living things, believing it was their duty to subjugate the globe to their way of life.
page 345

41 Back to the Future

How some people tried to resist the advance of Western civilization, wishing instead to return to a more natural order, but whose attempts often met with catastrophic consequences.
page 359

42 Witch Way?

How the whole world was bound into a single system of global finance, trade and commerce, sustained by relentless scientific endeavour. Can the earth and its living systems sustain humanity's ever increasing demands?
page 370

Acknowledgements

The year I have spent researching and writing this book would have been entirely impossible were it not for the unwavering support of my wonderful wife. Apart from being the best companion anyone could ever hope for, her no-nonsense proof reading stopped me drifting down many narrative cul-de-sacs. The constant inspiration of our two lovely daughters has been just as important. A big thank you also to our dog, Flossie. It would have been quite impossible to work out where on earth to begin and end each chapter without our many country walks.

I have also greatly benefited from the generosity, enthusiasm and support of my parents, family, friends and colleagues for which I am eternally grateful. To mention everyone would be unfair to forests. But I would especially like to thank Richard Balkwill, Satish Kumar, Andrew Lownie and Felipe Fernandez-Armesto for their advice and guidance. I owe a huge debt of gratitude to Mike Jones, who had the extraordinary courage and confidence to commission this book. The publishing team at Bloomsbury has been a complete delight to work with at every stage of this project. Their professionalism and sheer passion for publishing beautiful books is second to none. In particular I would like to thank Richard Atkinson, Robert Lacey, Will Webb, Anne-Marie Ehrlich, Andy Forshaw, Polly Napper, Penelope Beech, Anya Rosenberg, Ruth Logan, Katie Mitchell, Sarah Barlow and Vicki Robinson. Sincere thanks also to Ludger Ikas, Malte Ritte and Sebastian Vogel at Berlin Verlag, and to Colin Dickerman at Bloomsbury USA. However, the biggest debt of gratitude I owe is to Natalie Hunt, whose boundless professionalism, patience and sheer hard work saw this huge project safely through from early manuscript to final polish.

Other thanks go to all those unknown trees that have been sacrificed in the course of telling this extraordinary story. There's a good 300 million years of evolutionary wizardry just in the paper between these covers.

Finally, I dedicate this book to the memories of two special people. Dodo, my grandmother, would have taken such a keen interest in this whole project. My other dedication is to dear Christo, my namesake and late great uncle, whose generosity of spirit and passionate interest in other people knew no bounds.

Comments and suggestions about this book can be posted on the discussion forum at www.whatonearthhappened.com. All contributions and feedback are most welcome.

Prologue

HISTORY IS IN TROUBLE. It has been splintered by experts into discrete topics and chopped up by governments to suit their educational fancies. Worst of all, it is almost never presented chronologically. How old is the universe? When did life on earth begin? Who was mankind's oldest ancestor? How did ancient Chinese science shape the modern world? Why did democracy begin in Greece? Are humans really superior to other living things? Many people today are understandably confused about the answers to such a broad but fundamental range of questions, when all they learned at school was a few facts about kings and queens, a world war or two and maybe something about creatures called dinosaurs.

Knowledge about the past is now scattered among so many separate subject disciplines and buried within so many different books, it is easy to see why people may find it hard to enjoy what history has to offer without getting hopelessly lost or tied up in knots.

What on Earth Happened? is a journey that begins at the beginning, 13.7 billion years ago. Using the metaphor of a twenty-four-hour clock, it tells the story of everything in four parts: the origins of the universe and life on earth before mankind (Part 1: Mother Nature); the evolution of humans within the natural world (Part 2: Homo Sapiens); the development of a range of different human civilizations (Part 3: Settling In); and the fusion of those civilizations and the natural world into a global whole (Part 4: Going Global).

While writing the story, I have tried to draw on the widest possible range of sources and have used the most up-to-date knowledge about the universe, life on earth and humanity. I have made every effort to avoid mistakes, but in a history as wide-ranging as this, who knows what gremlins may have occasionally sneaked in? If and where they are to be unearthed, of course, the fault is entirely mine. My hope is that what emerges is a uniquely connected story that dovetails the growth of human civilizations with evolutionary biology, modern science with prehistoric art, and the rise of world religions with the irrepressible forces of mother nature.

This has been an exhilarating, life-changing project – back in time billions of years and several times all the way around the world. It has convinced me more than ever that all history should begin as a single, chronological tale that's written in an accessible way for everyone. That has been my primary purpose. I hope, as you read this text, at least some scales will fall from your eyes, in the way so many have from mine.

Christopher Lloyd
June 2008

Part 1
Mother Nature
(13.7 billion–7 million BC)

ECHOES OF A COLOSSAL explosion that triggered the beginning of our universe still reverberate today, 13.7 billion years after the Big Bang. Microseconds later the universe inflated to become billions of miles wide. New stars were born, old ones burned out. Some 9.2 billion years later our sun ignited, made up of the leftovers of burned-out stars.

Giant balls of hot dust and gas, pulled together by the gravity of the sun, jostled for position in the newly formed solar system. A collision between the early earth and another planet, Theia, scattered so much debris that it created our moon. A great bombardment of comets, kicked across the solar system by the powerful force of Jupiter, fell to the earth, evaporating into rain. Hot gases trapped inside the earth's core escaped out of volcanoes, forming the earth's first atmosphere.

A few hundred million years after the earth's fiery birth, inert chemicals began to duplicate into simple single-celled forms of life that we now call bacteria. Sometimes errors crept into their copying systems, creating variety. One type used the sunlight to make food, giving off oxygen as waste. Over the next 2.5 billion years these simple life forms filled the air with surplus supplies of this vital energy-rich gas, giving birth to a new atmosphere. Teamwork between the earth, its environment and early bacteria improved conditions for more sophisticated forms of life. Bacteria merged to create more complex cells which themselves began to link together into the earth's first multi-cellular beings.

Eventually the seas were filled with exotic creatures – some with stalky eyes, grasping arms and other bizarre appendages – thanks to the advent of sexual reproduction. A cast of sponges, jellyfish and corals were joined by bony fish, scorpions and trilobites. Spores germinated on land, where mosses evolved over millions of years into herby plants and leafy trees that could live miles away from the water's edge. As oxygen levels rose, sea creatures came ashore to explore new sources of food and shelter. Giant insects and dragonflies provided food for amphibians who ruled on land. Thanks to the processes of life, the earth's land was smothered in a nourishing blanket of nutrient-rich soil.

As the earth's crusts collided into a single super-continent, Pangaea, hard-shelled eggs provided reptiles with a way of reproducing inland. Enormous lizards called dinosaurs dominated the land, emerging alongside the earth's first flowers, birds and new civilizations of insects. But then, 65.5 million years ago, the world was traumatized by the strike of an enormous meteorite, causing a massive extinction. A family of small nocturnal creatures filled the void, radiating into many creatures great and small that filled the earth's continents as they slowly drifted apart.

Crunch, Bang, Ouch!

How an invisible speck of infinite energy exploded our universe into existence, with its galaxies of stars and constant laws of physics.

TAKE A GOOD look around. Put everything you can see inside an imaginary but enormously powerful crushing machine. Plants, animals, trees, buildings, your entire house (including contents), your home town as well as the country where you live. See it all get pulverized into a tiny ball.

Now put the rest of the world in there, too. Add the other planets in our solar system and the sun, which is about 1,000 times bigger than all the planets put together. Then put in our galaxy, the Milky Way, which includes about 200 billion other suns, and finally all the other galaxies in the universe, many of which are bigger than ours – there are about another 125 billion of these. See all this stuff squeezed together, reduced to the size of a brick, then a tennis ball, now a pea – finally, see it crushed even smaller than the dot on top of this letter i.

Then it disappears. All those stars, moons and planets vanish into a single, invisible spec of nothing. That was it – the universe began as an invisible dot, a 'singularity', as scientists like to call it.

This invisible, heavy and very dense dot was so hot and under such enormous pressure from all the energy trapped inside it that about 13.7 billion years ago something monumental happened.

It burst.

This was no ordinary explosion. It was an almighty explosion, the biggest of all time – it was what we now call the Big Bang. What happened next is even more dazzling. It didn't just make a bit

The Milky Way galaxy contains about 200 billion stars. Younger stars, such as our sun, are located in its spiral arms.

of a mess; it made a huge mess, billions of miles wide. In a fraction of a second the universe expanded from being an invisible speck of nothing to something so enormous that it includes everything we can see, including all the matter needed to make the earth, the sun, the moon and the stars.[1] And there's also a whole lot more that we can't see yet, because our telescopes can't peer that far. In fact, the universe is so big that no one really knows how wide or deep it actually is.

Why do experts think such an incredible event took place, especially when apparently it happened so long ago, and no one was there to witness it? Many people to this day are understandably suspicious of the whole Big Bang idea. But scientists are in broad agreement about what happened, because, they say, the evidence is all around us.

Frenchman Georges Lemaître was so shocked by the slaughter he witnessed on the battlefields of the First World War that he devoted most of his life to studying the stars. He first became interested in space in 1923, during a trip to Cambridge University, where there was an observatory that contained some of the biggest telescopes in the world. By 1927 he had become recognized as a great mathematician, and he developed a new theory of an expanding universe which all started with a very big bang.

Just two years after Lemaître published his ideas, another scientist called Edwin Hubble claimed that through a powerful telescope he could see other galaxies moving away from the earth, and that the further away they were, the faster they were moving. Here was visible evidence that the universe was still expanding. Hubble reasoned that a long time ago, something must have forced the stars and galaxies outwards – something like Lemaître's Big Bang.

Echoes from thunderstorms can bounce around mountains and valleys for a long time, sometimes more than a minute. The Big Bang was such an enormous explosion that scientists reckoned it should still be possible to detect its echo.[2]

It was first picked up in 1964 by two American engineers in New Jersey: Arno Penzias and Robert Wilson. At the time they were trying to find ways to improve the design of radio telescopes. But their new telescope kept picking up a mysterious noise. Regardless of which direction they pointed it, the irritating interference would not go away. The two engineers went off to investigate a nearby radio transmitter in New York City, thinking it could be the source of the trouble. When they found it was full of pigeons, they thought what they were hearing was perhaps the sound of the birds being amplified by the radio mast. The pigeons were removed, the transmitter was cleaned up, but still the mysterious noise remained.

Just thirty miles away another team of scientists, led by cosmologist Robert Dickie, was trying to perfect a highly sensitive space microphone which they hoped could detect the echo of the Big Bang. By chance, Penzias and Wilson telephoned Dickie to see if he, or anyone in his team, had any ideas on how they could get rid of the background noise on their new telescope. Almost immediately Dickie suspected that what Penzias and Wilson were hearing might be the echo of the Big Bang. Today you needn't just take their word for it. Think of those fuzzy black and white dots that appear on a television screen when it isn't tuned properly. One in every hundred of those dots is caused by the background echo of the Big Bang.[3]

Even if we accept the idea that an invisible speck exploded our universe into existence, why do scientists believe it happened 13.7 billion years ago? Using modern telescopes, scientists have been able to build on Hubble's observations and calculate the actual speed at which galaxies are spreading outwards. With this data they can project backwards in time to work out how long ago these objects were all together in one place.

Just after the Big Bang, more mysterious things started to happen. An enormous blast of energy was released. First it was transformed into the force of gravity, a kind of invisible glue that makes everything in the universe want to stick together. Then the massive surge of energy created countless billions of tiny building bricks – like microscopic pieces of Lego. Everything that exists today is made of billions of particles that originated a fraction of a second after the Big Bang.

About 300,000 years later things had cooled down enough so that these particles – the most common of which are electrons, protons and neutrons – could start to stick together into tiny blobs which we call atoms. With the help of that glue – gravity – and the passing of a little time, these atoms gathered together to make enormous clouds of very hot dust. Out of these clouds came the first stars, massive balls of hot fire supercharged with energy left over from the Big Bang. Gravity made the fiery stars gather into groups of many different shapes and sizes – some in spinning spirals, some in the shape of spinning plates. We call these clusters of stars galaxies. Our galaxy, the Milky Way, was formed about 100 million years after the Big Bang – that's 13.6 billion years ago.[4] It's in the shape of a large disk – like two back-to-back fried eggs – that spins round at a dizzying speed of about 500,000 miles per hour.

New information about the origins of our universe was gathered by an American spaceship called the Wilkinson Probe, launched in 2001. This has enabled scientists to make the most accurate measurements yet of the Big Bang's echo and of everything else that makes up the universe.[5] The probe also confirmed that, as Hubble saw through his telescope, the universe is still expanding. But many mysteries remain.

For instance, no one knows whether the universe's expansion is slowing down or not. If it is slowing, then perhaps gravity will one day start to pull all the stars and galaxies backwards again – as if they are on an immense, invisible elastic band. This means that the universe might one day get squashed back into a

The deepest ever view of the universe pictured by the Hubble space telescope in 2004. Each dot of light is a separate galaxy, some dating back more than 13 billion years.

00:00:00

tiny invisible speck, and as the pressure inside the speck builds, that could lead to another Big Bang. Some scientists even believe that there may have been millions of previous Big Bangs, and that our current universe is just the latest in line before it gets crunched up and a new one takes its place.

Another mystery is whether or not our universe is the only universe in existence. Recently a growing number of physicists have proposed that our universe may in fact be one of many different universes – perhaps an infinite number – each one exploding like bubbles from numerous Big Bangs. What makes each universe unique are their laws of physics (such as the strength of gravity or the forces between particles in an atom).

Such a 'multiverse' theory would help explain why our universe seems to have just the right physical laws to support the emergence of life. The probability of such laws appearing by accident are so infinitesimally small that the existence of other universes with different physical laws makes more rational sense to atheistic scientists than the idea of an intelligent creator or God. [6]

Extra-terrestrial reincarnation: star-dust, the leftovers of supernovae explosions from burned-out stars, is the stuff from which new stars like our sun are continually reborn.

Galaxies are big. Take a chocolate Smartie and place it in the middle of your kitchen table. Let's say this represents our sun. How far away do you think you should place a second Smartie to represent the next nearest star in our galaxy? One metre, perhaps? Maybe ten metres? In fact, on this Smartie scale the nearest star would be *ninety miles away*.[7]

Our sun is quite a new star – scientists believe it was reincarnated out of the gas and dust left over from a previous star or stars that had burned out and exploded after being crushed by their own gravity. These huge explosions, called supernovae, are still fairly common in space. They leave behind them all the matter – mostly gas and star-dust – needed for new stars continually to be reborn.

About 4.6 billion years ago, the left-over gas and dust cloud from previous burned-out stars collapsed and ignited to form our sun. That means that our sun is only about a third as old as the universe itself. It is just as well for us that the sun is a relatively young star, because the first stars could never have produced orbiting planets like earth which could support life. These first-generation stars were made only of simple gases, such as hydrogen and helium. But because supernovae explode with such massive force, they smash atoms together to create heavier, more useful building materials that make up rocky planets like earth which contain elements such as iron, oxygen and carbon – all vital ingredients for building life.

For a long time people believed that the earth was at the centre of the universe. But we now know that our solar system is located in one of the Milky Way's outer spiral arms, called the Orion arm, and is currently travelling through a sparse and lonely part of the galaxy called the Local Bubble. There are only a few other stars in our neighbourhood, a region that astronomers affectionately call the Local Fluff.

The solar system includes everything that circles around our star, the sun. The most important objects are the planets. These large round balls of rock and gas formed at the same time as the sun, out of the same clouds of dust and gas left over from previous burned-out stars. There may have been as many as twenty-five early planets at the beginning of the solar system. The gas-filled balls tended to drift away from the sun, and became the gas giants, Jupiter, Saturn, Uranus and Neptune. The others contained heavier, more useful building materials that could withstand the heat of the sun. These formed the inner 'rocky' planets – Mercury, Venus, Earth and Mars. For several million years these giant smouldering objects orbited the sun, wobbling and straying about as they tried to find a stable path in the newborn system.

The early solar system was an extremely nasty place – and *very* unsuitable for life. An invisible rain of tiny highly charged particles streamed out of the hot, fiery furnace of the sun like a storm of razor-sharp daggers. These could cut through almost anything. They are still being fired out by the sun, about twenty billion tonnes every day. It is known as the 'solar wind' and can penetrate even the toughest space suits and helmets worn by astronauts.[8] Even if some form of life could have handled the terrible heat at that time, the 'solar wind' would have killed it off instantly.

It was hell on earth. A semi-molten crust of sticky volcanic lava burbled across the planet's surface like burning-hot treacle. There was no solid ground, no water and definitely no life. The unstable earth spun so fast on its axis that each day was only about four hours long.

What happened next was a freak. Experts believe two young planets happened to be on the same orbit around the sun, but moving at different speeds. One was the earth, the other another early planet called Theia. About fifty million years after the sun began to glow, these two newborn planets ploughed into each other. With a massive jolt, the ailing earth fell on to its side, out of control, a crippled, hysterical, shaking wreck.

Thousands of volcanoes erupted following the impact. Huge volumes of gas, previously trapped inside the earth's core, now spurted through the surface, giving birth to the earth's early atmosphere.

Theia's outer layers vaporized into billions of tiny particles. Debris flew everywhere, surrounding the earth with an enormously thick blanket of hot dust, rock and granite. Trapped by the earth's gravity, this fog of rubble swirled around in the sky, making everything go dark. For months not even the brightest sunlight could penetrate the thick layers of dust which were once planet Theia. Her heavy, molten-iron core converged into the centre of the earth, causing an almighty shockwave that fused the two planet's cores into a single, tight, metallic ball, thousands of degrees hot, that sank deep into the middle of the stricken globe, crushed by the force of the impact.[9]

It's just as well for life on earth that this almighty collision happened. The earth's metallic core gave birth to a magnetic shield that deflects the most lethal effects of the solar wind away from the planet's surface. The shield also prevents the solar wind from splitting water (H_2O) into separate hydrogen and oxygen atoms, preserving the earth's vital supplies which would otherwise diffuse into space. Without this shield, life on earth might never have evolved. Other planets that do not have an iron core, such as Mars and Venus, appear never to have developed life.

Today there is no physical evidence on earth of the impact of the collision with Theia – no crater – because such was its force that all the outer material vaporized and exploded into space. But visible evidence isn't far away. The dust and granite that wrapped itself around the earth soon stuck together again, thanks to the glue of gravity, and turned into an enormous ball of dust. Only about a year after the giant impact the earth had a new companion – our huge, bright, crystal-like moon.

The moon soon had a very important stabilizing effect. Its own gravitational force helped save the earth from wobbling uncontrollably after the massive impact of Theia. The moon's gravity has helped slow down the rotation of the earth so that, over a very long period of time, the four-hour day has become a twenty-four-hour day. For billions of years our earth and its moon have danced as partners around the sun, like two graceful ice-skaters steadying themselves by holding hands, facing each other as they swirl in circles around the rink.

First Twitches

How collisions, bombardments and volcanoes pummelled the young earth's hot, lifeless crust, and how chemicals began to replicate into microscopic forms of life.

ONE HOT SUNNY afternoon in the autumn of 1951, Professor Harold Urey strolled into his lecture hall at the University of Chicago. The room was filled with students eager to hear this great scientist talk about his pet subject – the theory of the origins of life on earth.

For more than 150 years scientists had been struggling to come up with credible theories of how life began. The problem, as Urey knew, was that so far it had proved impossible to actually *demonstrate* how life could have started from a ragbag of primitive or 'primeval' substances such as those found on the early, hostile earth. As a result, no one could agree on the origins of life.

Nearly a hundred years had passed since the meticulous French scientist Louis Pasteur had proved that substances that contained no living things remain lifeless if left alone, meaning that apparently not even the simplest forms of life could

emerge of their own accord out of a barren, lifeless earth. So what on earth could have happened? Who or what created the first magical spark that led to the beginnings of life?

The debate was a furious one. On the one hand, some people believed that life was created by the hand of a divine architect – by God – perhaps as spelled out in the Bible. They argued that scientific investigation could never reveal the true mysteries of how life began, since God's powers lay beyond the scope of human understanding. Others thought that life came from outer space. After all, the universe is so enormous, there must be at least a chance that other intelligent life forms are out there somewhere. Maybe life on earth was the result of a horticultural experiment seeded by extra-terrestrial beings billions of years ago.

Urey was convinced that science would eventually come up with the answer. He wanted to find a way to show how the chemicals needed for building life, called amino acids, could have been created on the early earth. He believed that, with the passing of a lot of time, simple single-celled life forms may have developed naturally – perhaps inevitably – into the complex and beautiful world as we know it today.

Urey dreamed of concocting a laboratory experiment that would simulate conditions on early earth, and show how life was created out of a lifeless jumble of nothing. One of the students in the audience that day was utterly gripped by what Urey was saying. Stanley Miller had stopped off in Chicago on his way across America. He was in the process of trying to decide on a research project that would complete his training as a scientist.

The more Urey spoke, the more excited the twenty-one-year-old Miller became. At the end of the lecture he went up to the professor and – after a great deal of talking – persuaded Urey to work with him on a project to try to create life in a laboratory out of nothing more than a cocktail of chemical junk.

The two men secretly set to work. They started by designing an elaborate glass apparatus, in the middle of which was a large jar that would contain all the substances which Urey and Miller believed existed at the time of the early earth, including gases – most of which would have erupted out of volcanoes – such as hydrogen, methane and ammonia. Steam was fed into the glass jar through a tube connected to a flask of boiling water. Inside the jar were two metal rods, or electrodes. A powerful electric current would surge through these to make sparks – re-creating smaller versions of the violent lightning strikes that were common on early earth. The whole apparatus was designed to reproduce the earth's early atmosphere, complete with thunder and lightning.

Miller started off the experiment by boiling the water in the flask. Steam climbed up through a tube connected to the large glass chamber, where it mixed with the primeval gases. Next, he flicked the electric switch. Some 60,000 volts of electric current surged into the electrodes, beginning a constant stream of mini lightning strikes.

To his bitter disappointment, nothing happened, and a despondent Miller left the lab that night with nothing to show for his efforts.

But when Miller arrived the next morning, he found that the water in the flask had turned pink, indicating that some kind of chemical reaction must have taken place. After running the experiment for a week, the results he had hoped for were unmistakable: the clear water had turned a definite shade of red. The water now contained amino acids – the vital ingredients for life, used by all plants and animals (including you and me) to construct their living cells. Surely this was the demonstrable proof that Urey had so strongly believed in. Life, Urey and Miller concluded, began by chance on the hell that was the earth some 3.7 billion years ago, because the Goldilocks-like conditions for it to do so happened to be just right.

Miller and Urey's experiment was a turning point in the scientific understanding of how life on earth may have begun. The experiment has since been repeated many times, with slightly different mixtures of substances, because over the years theories of exactly what chemicals existed on early earth, and in what quantities, have changed. Yet Urey and Miller had not actually produced life itself in their test tubes – rather, they had created

Stanley Miller, student of the Nobel Prize-winning Professor Harold Urey, with his scientific apparatus in an experiment that was the first to re-create chemical conditions as they may have been on the early earth.

Gravity from the gas-giant Jupiter is thought to have kicked storms of comets towards the early earth, bringing water and possibly chemicals needed for life.

intergalactic terms, such distances are actually extremely close indeed. In fact, at the right time of year and on a clear night you can see Jupiter shining brightly in the night sky. With a modest telescope at least four of its sixteen orbiting moons are easily visible. Amazing as it sounds, life on earth may have begun right here.

the ingredients. To this day no one has been able to produce an actual living cell out of lifeless chemicals in a laboratory – so the debate about the first moment of life's incarnation still rages on.

Some scientists reckon life began thanks to another dramatic event in the early history of the planet. They think it may have been seeded by what's known as the 'great bombardment' of comets – an episode that started somewhere near the giant red planet Jupiter, about 3.7 billion years ago.

In 1687 Isaac Newton, the famous British scientist, discovered that all objects attract other objects, but that the bigger an object, the greater its gravitational pull. Jupiter, the largest planet in our solar system, is huge – about 1,300 earths could fit inside it. As a consequence it has a very strong gravitational force. Ordinarily such a large force would not affect other planets like earth given that Jupiter is some 391 million miles away. But in

Hundreds of thousands of small rocks were left over after the formation of the planets. About 3.7 billion years ago an especially vast swarm of them ended up orbiting around the sun near Jupiter, trapped by its enormous gravitational pull. At that time the solar system was so unstable that the orbits of large planets such as Jupiter, Saturn and Neptune sometimes strayed off course. During one of these episodes it is thought that their gravity kicked these small comets and asteroids, like footballs, across the solar system. Some of them headed in the direction of the early earth, arriving like a giant barrage of odd-shaped cannonballs.

Evidence for this great bombardment has been gathered by a team of scientists at NASA's Space Flight Centre in Houston, Texas. They have been studying an ancient comet called LINEAR, which recently strayed towards the sun and was vaporized by its heat into a cloud of dust and gas. As it broke up, the NASA team peered through their sophisticated telescopes to see if they could detect what substances were inside the comet. Evidence gathered from their observations suggested that life's amino acids may have come from inside similar comets, meaning that life on earth did not have to start completely from scratch. Perhaps it was delivered in a kind of kit from space?

Further evidence comes from the moon. Its pockmarked appearance, easy to see with a decent pair of binoculars, shows how its surface is a million scars of crater impacts. With little atmosphere to protect it and no living systems to hide the damage, the moon is today's best historical record of the brutal age of huge impacts that took place billions of years ago.

Such life-bearing comets may have brought other precious gifts too. Just seconds before the swarms of comets ploughed into the early earth, they began to melt, their ice heated up by the friction of the earth's atmosphere. It would have been a spectacular sight – millions of comets falling through the sky like enormous snowballs, each one plummeting to the earth followed by a long, curving tail of fiery hot steam. The steam condensed into water, and then, possibly for the first time ever, something happened that we all take completely for granted. It rained. It rained so much that some experts believe that most of the water in our lakes, rivers and oceans may have come from comets that dropped to the earth billions of years ago.

Yet for all the science and technology, and the brilliant academic minds that have tried to come up with an explanation, no one has actually solved the riddle of how those chemical building blocks re-created in the laboratory by Stanley Miller and Harold Urey turned into living cells, the stuff of you and me.

The magic of a living cell is that it can reproduce. It can have buds, offspring, make replicas of itself. Usually single cells made exact copies of themselves – as viruses and bacteria do today, although sometimes a copying error creeps into the system to form a mutant cell. The ability to multiply is what makes life so completely different from anything else in the known universe. Nothing dead can do it.

Put a teaspoonful of salt on the side of your dinner plate and then look away for a few seconds. Next time you look down, imagine that the salt has taken over the whole surface of the plate and is spreading uncontrollably across the table. Sped up a bit, that's what life is like. Salt can't do it, because it's not alive. Even the amino acids, the building blocks that Urey and Miller re-created, can't do it. But yeast can, mould can, bacteria can. Somehow, something triggered these substances into a mode where they started to duplicate. Everything that has happened since is the history of life on earth.

As the rain of comets subsided, the surface of the earth began to cool enough for the molten lava to turn into solid ground. Water fell from the skies, forming the early oceans, which in turn cooled the surface even more. Underground, molten lava and hissing gases trapped beneath the earth's crust needed, literally, to let off steam. Volcanoes erupted *everywhere*. Massive mountains of fire spewed out boiling-hot lava and gas across the entire planet, releasing more trapped gases that formed an early atmosphere containing nitrogen, methane, ammonia, oxygen and carbon dioxide – the essential ingredients used by Miller and Urey in their lab to mix up the stuff of life. When oxygen raced into the earth's atmosphere it combined with another gas left over from the explosion that formed our sun and the solar system – hydrogen. Together these two gases joined to create more water (H_2O), and the floods begun by the comets grew bigger still, until water covered nearly 70 per cent of the planet's surface.

Some experts believe the magical leap from life-giving amino acids to single-celled living organisms may have taken place deep down in these early oceans. Methanogens – one of two basic types of single-celled bacteria – evolved deep within the oceans so they could hide away from the lethal effect of the sun's solar wind. They thrived next to volcanic vents called 'black smokers' that belched out thick, black, acrid fumes from the ocean floor, providing chemicals for food and warmth.

The other type of single-celled life form probably evolved from a copying error when food was scarce. It adapted to live off a completely new energy source – sunlight – which it used to split carbon dioxide (CO_2) and water (H_2O) into food. This simple but ingenious feeding process is what we now call photosynthesis.

Unlike the methanogens, these cyanobacteria needed to live close enough to the surface of the seas to feed off the light of the sun shining through the water. Their photosynthesis transformed the planet's atmosphere, because its waste product is oxygen.[1] Over billions of years cyanobacteria caused surplus supplies of oxygen to build up in the air.[2]

Shark Bay, some 650 kilometres along the coast from Perth in the wilds of western Australia, certainly looks like a prehistoric place. Near the shore live some of the oldest rocks left in the world. Next to them, by the sea, are thousands of mounds or hillocks about half a metre high, resembling over-stuffed cushions. Although they do not look like living things, these weird structures – called stromatolites – contain billions of cyanobacteria. Now only occasionally jutting out of shallow lagoons in places like western Australia, Mexico, Canada and in the depths of the Caribbean sea, stromatolites were once everywhere on the early earth.

To begin with, the oxygen provided by these stromatolites bonded with iron on the sea floor left over from the giant impact with Theia. The iron became iron oxide, a rusty red mineral ore.[3] When all the available substances that oxygen could bond with ran out, it was simply left to hang around in the air – which is where it has remained ever since. Oxygen now accounts for about 21 per cent of all the air we breathe. The rest consists mostly of nitrogen (71 per cent), with a little water vapour and a number of other trace gases in small measures under 1 per cent, including carbon dioxide.

Without oxygen life on earth would have carried on, but it might never have developed beyond columns of sticky, microscopic bacteria. Humans could never have evolved, because oxygen is an energy-rich gas that sustains all forms of advanced animal life. Also, high up in the atmosphere, oxygen provides an essential protective blanket in the form of the ozone layer that protects land life from the sun's powerful ultraviolet rays.[4]

Cyanobacteria helped reduce the boiling-hot temperatures on the early earth by feeding off carbon dioxide in the air. As levels of this heat-retaining 'greenhouse' gas reduced, temperatures fell. Lower temperatures helped life leap to the next level. Single-celled bacteria (prokaryotes) began to merge and to grow into more complicated and sophisticated living things.

About two billion years ago, another mutation managed to creep into nature's systems that allowed a single-celled bacteria to feed off the oxygen-rich atmosphere. Oxygen's enormous energy-giving properties meant that this process – called respiration – could produce up to *ten times* more energy than other ways of life. Soon the oceans were filled with highly energetic microscopic cells that fed off oxygen that was dissolved in the oceans.

So energetic were these microscopic cells that some found they could drill their way *inside* other, larger cells and strike up a mutually beneficial bargain. While the smaller cells fed off the larger cells' waste products, the larger cells used up surplus energy created by the smaller cells' respiration.[5] Through such a collaboration, called endosymbiosis, larger cells were now much better equipped for survival in the increasingly oxygenated world.

Stromatolites, pictured here in Shark Bay, western Australia, once covered the earth's surface and helped give birth to our oxygen-rich atmosphere.

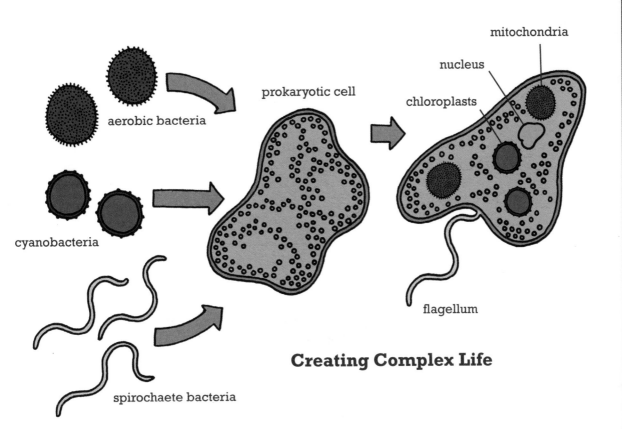

aerobic bacteria

cyanobacteria

spirochaete bacteria

prokaryotic cell

mitochondria

nucleus

chloroplasts

flagellum

Creating Complex Life

By working together, these more complex cells (eukaryotes) developed a range of special skills. Some parts, called mitochondria, turned food into energy. Others, called chloroplasts, became experts at getting rid of the cell's poisonous waste products. Still others turned into something like librarians: their job was to store all the necessary information about how to construct an identical cell from scratch. We call these librarians genes: the word comes from the Greek *genos*, meaning birth. They live in a part of the cell called its nucleus and they are made from a compound called deoxyribonucleic acid, or DNA.

What looks like altruistic teamwork now fell victim to a far more gruesome approach to life. Some of these highly energetic complex cells found that by engulfing another living bacteria whole, they could gain access to a new, more abundant source of food. The world's first mouth was nothing more than a toothless, microscopic hole, yet it triggered a powerful predator/prey relationship amongst living things. Creatures now evolved rapidly in an arms race to protect themselves from being eaten or equip themselves for better attack.

Most found that the best strategy was to link up into teams. Gangs of cells joined together to form the world's first multi-cellular creatures. Some of them became the forefathers of all animal life, while others turned into the ancestors of today's plants and trees.

Already we have travelled more than *three billion* years since the earth was first formed. If we think of the earth's history as a twenty-four-hour clock, we have journeyed from about 05:19, when the first signs of life emerged, to about 16:00, leaving just eight hours to go for all the rest of life to evolve. Although miraculous signs of life have already appeared in the form of complex microscopic bacteria, hundreds of millions of years have yet to pass before fish, animals, plants and trees make their first appearance.

They only ever made it thanks to another piece of extraordinary teamwork – one that prepared the planet for yet more dramatic changes to life on earth.

Rising levels of energy-rich oxygen caused some free-living bacteria to work their way inside other cells making life more complex – a process which is called endosymbiosis.

05:19:48

Tectonic
Teamwork

How conditions for creating new, more complex creatures were improved by the joint forces of the earth's planetary systems and the processes of early life.

DOCTORS IN A MEDICAL emergency always have the same top priority: to protect the body's vital life-support systems. If a patient's internal transport system – the blood – cannot carry oxygen from the lungs and nutrients from the stomach to the cells in the body, then the victim quickly starves. And if waste products such as carbon dioxide and toxic acids are left to fester because they cannot be removed, it is almost as quickly poisoned.

During this period of earth's history a global life-support mechanism emerged that works in a strikingly similar way to the human respiratory system. Without this mechanism microscopic bacteria from two billion years ago would never

have evolved into plants, animals and people. In fact, without the proper working of nature's life-support systems most living things today would die out very quickly indeed.

The first and most simple part of the earth's life-support mechanism is very well known. It is rain – or, to give it its proper term, the water cycle. As the sun beats down on the planet's surface, the seas get warmer, and some of the water evaporates into steam. Once in the air, the steam cools to form clouds, which get blown about by the wind across the planet, eventually to fall elsewhere as rain. Without this automatic fresh water supply, most living things on land and sea would almost certainly perish. No pipes, no pumps, no need for power stations, no people to watch over the machinery – it just happens, every day, the most precious of all free gifts.

But beneath the surface of what seems a very simple process, an important partnership between the earth and her living things developed sometime between 3.7 and two billion years ago.

For rain to fall, clouds need to form. Steam molecules can condense back into water only if there is some kind of surface or 'seed' around which they can cluster. Luckily, waste gases produced by early bacteria provided perfect surfaces around which steam could turn back into water to form rain.[1] In this way bacteria help nature operate one of her most important life-support systems by seeding clouds. Cloud cover also creates a reflective blanket that sends many of the sun's scorching-hot rays shooting back into space. And so they help to cool the planet, greatly improving conditions for life on earth.

This sort of teamwork is just one of a number of partnerships between the earth and living things that help control the climate and prevent excessively hot temperatures from harming life. Scientists are still unclear about exactly how all the earth's life-support systems actually work, but another example of how they have prevented almost certain catastrophe for life on earth is to do with the amount of salt in the seas.

Sea salt comes from rainwater that pounds down on the rocks of the land and dissolves minerals, which are then washed into the sea in streams and rivers. Also, huge quantities of salt trapped beneath the surface of the earth are regularly churned up by volcanic vents at the bottom of the sea. Both these processes are part of what is known as the rock cycle.

Many eras ago the earth and nature evolved a cooperative partnership that helped make sure that salt levels in the sea never became too concentrated. Living creatures risk death by poisoning if the level of salt in the sea rises too high. Without some way of regulating its concentration, life would rapidly have become extinct.

When microscopic sea creatures die, their bodies drop to the bottom of the ocean floor like a torrent of tiny seashells.[2] Over millions of years, thick piles of this dead matter piled up metres high on the ocean floors, forming massive sediments. Eventually the weight of these piles crushed the dead materials into what we now call limestone, or chalk. As all this limestone piled up on the ocean floor, it created a series of huge walls or reefs near the coasts. These barriers trapped sea water into static pools or lagoons, leaving the heat of the sun to evaporate all the water away, revealing a solid white salt powder as a deposit. By this process salt was removed from the seas.[3]

The sun's heat evaporates sea water from lagoons, leaving powdery white salt deposits as found in these Iranian salt flats. This natural process made sure the seas were never too salty for life to thrive.

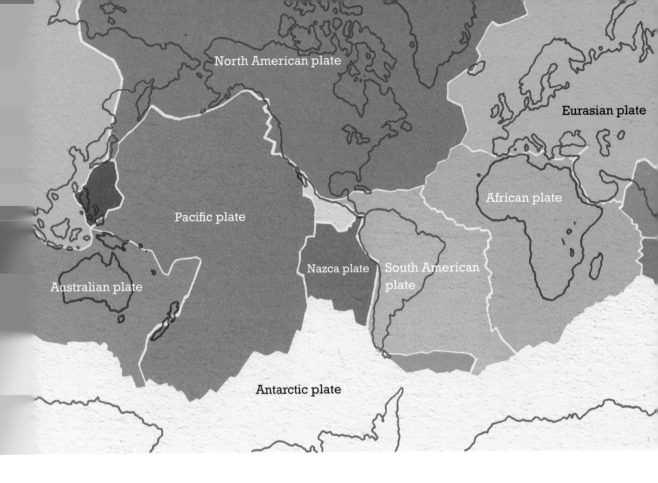

North American plate

Eurasian plate

African plate

Pacific plate

Nazca plate

South American plate

Australian plate

Antarctic plate

The earth's tectonic plates as they are arranged today. Earthquakes, volcanic eruptions and mountain ranges often arise at their boundaries.

It's as if the earth takes away poisons in much the same way as our bloodstream carries away the waste produced by living cells. Of course, the more shallow reefs, shores and beaches there are, the more salt is removed from the sea. Perhaps it was just luck, but between one and two billion years ago the weight of these limestone reefs became so great that they caused the crusts of the earth to start to dip, buckle and eventually collapse into the melting lava beneath the earth's surface, a phenomenon known as subduction.

This may have helped to trigger one of the most bizarre of the earth's life-support processes; one which further reduced the levels of salt in the sea. It is called 'plate tectonics'.

As you read this page, you are sitting on a crust of the earth which is floating like a giant raft on an underground sea of boiling-hot lava. The earth's crust is split up into a number of floating plates that are in constant motion, like enormous, slow-moving bumper-cars. Each plate is either drifting apart

from, or bashing into, another one. When they collide they form massive mountain ranges that soar high into the sky. When they drift apart, huge ocean ridges form in their wake. So much pressure builds up in the rocks of the earth that the movement of the earth's plates causes massive earthquakes and volcanoes, hot geysers and tsunamis. Because the earth's crust has split into separate plates that move around, like giant pieces of a puzzle, the number of shores and beaches available as evaporated salt pans has increased dramatically.

The same process also means that evaporated sea salt gets safely stored deep beneath mountain ranges – millions of tonnes of salt are today buried beneath the European Alps and the Himalayas. As long as the plates continue to move, mountains of salt will always be safely buried under the rocks, leaving the levels of salt in the sea low enough for life to continue to thrive.

Plate tectonics was first discovered by a German scientist called Alfred Wegener, who wrote about it in 1912. He came up with the startling idea after finding identical fossils of early animals on different continents (see page 55). Thousands of miles of ocean separated these ancient animals, yet no one could explain how they could possibly have got to all the world's continents, especially as these were creatures that couldn't swim.

Wegener had the imagination to think that perhaps, once upon a time, the continents weren't separate at all, but joined together like a huge jigsaw. His ideas began to attract attention in the 1920s, when he produced a revised version of his theory, but it wasn't until the late 1960s that scientists began to see that the continents – big land masses such as America, Europe, Asia, Africa and Australia – have actually merged and drifted apart many times in the earth's history. They are still moving today. For example, Europe and America are growing further apart at a rate of about five centimetres a year.

For billions of years the earth's crusty plates have been shuttling this way and that, floating on the surface of the molten rock called the earth's mantle, powered by the enormous heat of the earth's core, which provides the great forces required to make and move mountains. No one knows for sure how often the continents have collided to form one super-continent, and then split apart again into separate land masses. But it seems to have happened at least three times in the earth's history. The first super-continent, called Columbia, is thought to have formed about 1.5 billion years ago (about 4 p.m. on our twenty-four-hour clock).

On another occasion, between 850 million to 630 million years ago, during a period known as the Cryogenian, the earth's continents were arranged as if on a string stretching around the equator. With so much land amassed in the hottest regions of the world, more of the earth's crust was exposed to tropical rainfall than ever before causing huge volumes of atmospheric carbon-dioxide to become dissolved by deluges of rain into carbonic acid and washed into the sea. As levels of the greenhouse gas carbon dioxide in the air reduced, temperatures plummeted, plunging the earth into a deep-freeze, which some scientists have speculated may have caused ice sheets to cover most of the world's surface, turning the globe into a giant snowball. Only once the crusts moved on did volcanic activity replenish the atmosphere's supplies of greenhouse gases to put an end to the millions of years of bitter cold, warming up the planet once again and making it good for a new phase of life on earth.[4]

For billions of years this tectonic cycle has been churning up the surface of the earth in ultra-slow motion, drastically changing weather patterns, burying dangerous salts and minerals, making and destroying super-continents and crumpling crusts as if they were thin pieces of tin foil. Such are the earth's life-support processes that seem to have kept everything – from the composition of atmosphere, global temperatures and the saltiness of the sea – sufficient for life to thrive. Without these systems, the evolution of complex life as we know it would have been impossible.

Fossil Fuss

How life exploded into a variety of new organisms, some of which developed hard shells, bones and teeth that fossilized into a timeless museum of life on earth.

LIFE ON EARTH consisted of only two kinds of living things until about a billion years ago: the original, simple single-celled bacteria that produced methane and oxygen as waste products, and the newer, more complex multi-cellular organisms that fed off the increasingly abundant supplies of oxygen in the air. Deep within these more complex life forms (eukaryotes) – originally created out of the fusion of several simple cells into a complex cell – a small but important revolution was beginning to take place that would prepare the way for an explosion in the variety of living things.

For billions of years – probably until about a billion years ago (about 6.30 p.m. on our twenty-four-hour clock) – these tiny organisms, with their increasingly sophisticated cells, were all the life there was. Then, something triggered a spectacular and dramatic increase in the pace of the evolution. It's difficult to say exactly what happened, because there

is no surviving evidence of what life was like until living things developed shells, bones and teeth, which fossilize to leave impressions in the rocks, a process that only began about 545 million years ago.

During the period before fossils, living things developed a radical new form of duplication called sexual reproduction that revolutionized the way life evolved.

In the beginning when cells made copies of themselves, most of them were just like their parents: they were clones that differed only when a rare copying mistake occurred. The mistake may have been a good one (and the child cell survived) or a bad one (and it probably died off). With sexual reproduction, child cells are always different from their parents'. Normally two parents are involved: one is male, the other is female, each one merging its genes with the other to create a new being that has bits of both.[1]

This means that the new cell always has a unique code, massively increasing the variety of life on earth. Of course, since the child cell's code is basically a mixture of its parents' genes, it can still inherit features from them. Good mistakes from either or both of its parents can still be passed on to the child to help it thrive. It might also inherit bad mistakes or have bad mistakes of its own, in which case it usually dies out, further strengthening the remaining forms of life on earth.

Sex may have originated after the DNA of a bacteria's prey somehow survived the process of digestion and began to coexist with its predator's nucleus. The binding of two strands of DNA – the double helix – has its origins here. Only during the process of reproduction does this double helix peel off into its original parts so that special cells (that evolved into sperm and eggs) could recombine to create a new double-helix that would contain the survival codes of two similar but different beings.

Sexual reproduction helped life shoot forward in complexity, equipping it to survive the challenging conditions on the planet in far fewer generations than it would have taken without it. While it took 2.5 billion years for life to evolve into simple types of microscopic organism, it took less than half that time for life to transform completely into everything we know today – from fish, amphibians and reptiles to plants, trees, birds, mammals and man.

One of the first men who worked all this out was Gregor Mendel, a monk, born in 1822, who lived in what is now the Czech Republic. He spent most of his life absorbed in the natural world, especially in his favourite place for study, the monastic vegetable garden. There he could pursue his particular fascination with growing thousands of different varieties of peas. In fact, Brother Mendel became so interested in these peas that between 1856 and 1863 he studied more than 28,000 different pea plants. What intrigued him was that when these slightly different plants had seedlings, their differences (or characteristics) would often be carried forward to the next generation. The seedlings had inherited features from their parent plants.

The concept of inheritance was first described by Mendel in 1865, in a paper called 'Experiments on Plant Hybridization' which he read to the Natural History Society of Brno. He went on to devise a number of laws that could predict how living characteristics are passed from one generation to another through sexual reproduction. Mendel never became famous for his studies during his lifetime. He died in 1884, after spending the later years of his life fighting disputes with government officials who were demanding that monks pay higher taxes.

Sexual reproduction made a huge difference to the variety of living things. Since creatures had no shells or bones until c.545 million years ago

By cross-breeding pea plants (A and B) Mendel worked out how some characteristics, such as pea colour, were passed down through subsequent generations (C and D). His laws formed the basis of modern genetic theory.

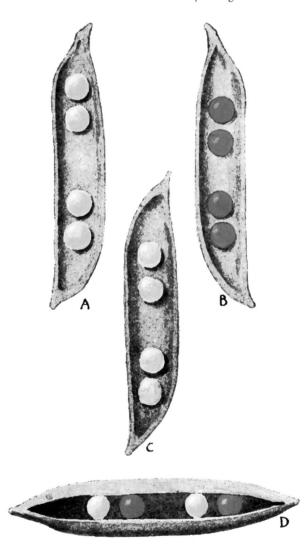

A

B

C

D

20:48:00

generally speaking the imprints of early creatures have not been preserved in ancient rocks. However, as a certain twentieth-century Australian geologist discovered, this was not always the case.

⋈⋈⋈⋈

Like many men of his generation, twenty-one-year-old Reg Sprigg was recruited into the army during the Second World War. Sprigg's charge was to reopen an old uranium mine on behalf of Australia's Commonwealth Scientific and Industrial Research Organization (CSIRO). Uranium is a naturally occurring element found in rocks all over the world. At the time it was thought to be extremely rare, and there was a desperate search under way to find as much as possible as a source of material to make devastating new weapons of mass destruction – nuclear bombs.

One late afternoon in 1946, when Sprigg was studying the ancient ground searching for signs of rocks that might contain uranium, his eye was attracted to a curious-looking rock with shallow indentations on its surface. He rolled it over.

What he found underneath stole his breath away. This was no ordinary rock, but a hoard of geological treasures, an exotic natural stash of very early fossils. An array of bizarre, weird and wonderful shapes had been etched on to these Australian rocks by the forces of time and nature. Sprigg's fossils have been named the Ediacaran fauna, after the Ediacara Hills region in south Australia where he found them.

These are the oldest multi-cellular fossils in the world. They show a wide variety of marine life, ranging from something that looks like a small sea beetle to a flattened, segmented worm-like creature that could grow up to a metre long.[2] Over a hundred different types of visible creature have now been discovered in various parts of the world, including Russia, south-west Africa and north-west Canada, all of which belong to this period of history.

Sprigg's discovery shows that by about 600 million years ago life had evolved way beyond bacteria, smaller than the eye can see, into a host of new creatures, both large and small. Some were no more than living blobs of see-through jelly that sat around on the sea floor, feeding off whatever micro-organisms passed them by. Others had small legs that they used to swim and hunt for food.

So important are these finds that scientists have named this period of the earth's history after the area where they were discovered. By the Ediacaran Period life had transformed from the invisible to the exotic. It is defined as having begun 635 million years ago, and ended some ninety-three million years later.

We are now on the edge of an enormous transformation. The time is just past 9 p.m. on our twenty-four-hour trek across earth time. The rest of history will be played out in the final three hours. There's still no life on land, no plants, no trees, no flowers, no insects, no birds or animals, let alone humans. The earth is very old, but human history is not. Compared to the age of the earth, everything else we will discover is either young, very young, or just hatching out. Mankind is among the youngest of all.

Sprigg's find was spectacular because the fossils he discovered were so old. But what he found was nothing compared to the quantity and variety of life forms coming up just ahead in what geologists call the Cambrian Explosion. The Cambrian Period lasted for fifty-four million years, from 542 to 488 million years ago. Here for the first time a full and clear picture of what life on earth was actually like 500 million years ago finally emerges. When the fossil record begins for real it is rather like a theatre curtain being pulled back to reveal a stage bursting with actors in the middle of a play.

Fossils are wonderful for helping investigators identify what kinds of creatures have lived on the earth. Charles Doolittle Walcott was born near New York in 1850. As a young boy he found school rather boring. It wasn't that he had no interest in things, rather the opposite. He was so curious that he just wanted to get outside and explore the world for himself – in particular he liked to look for minerals, rocks, birds' eggs and fossils.

By 1909 Walcott had become a well-established fossil collector. One day, a freak accident changed the rest of his life. While he was walking high up in the Canadian Rockies, his mule slipped and lost a shoe. In the process it turned over a glistening rock of black shale, a type of rock made out of compacted mud and clay. Usually shale is too dark for any marks on it to be easily visible, but the sun happened to be at just the right angle to reflect a most curious series of outlines. When Walcott stopped to pick up the rock, he saw a row of remarkable, flattened, silvery fossils. These were perfectly preserved creatures from the Cambrian Period.

It turned out that the mountainside had collapsed about 505 million years ago, smothering these creatures, killing them instantly and burying them like a time capsule for posterity. Walcott's discovery became one of the richest hoards of fossils ever. It is known as the Burgess Shale (named after Mount Burgess, near to the site where Walcott found the fossils). Walcott returned to the site many times afterwards, and eventually wrote a library shelf of books about his finds. In all he identified as many as 140 separate species of once-living things, giving us the most intimate view of life in the Cambrian oceans. He eventually collected more than 60,000 fossils. And what a bizarre range of creatures they were.

First, there is the strange-looking Anomalocaris (see the illustration above). This was one of the biggest sea-hunters of its day, and could grow up to a metre long. It used a pair of grasping arms to capture and hold struggling prey. For a long time, the fossils that make up this extraordinary creature were thought to be three separate living things. The body was identified as a sponge, the grasping arms as shrimps, and the circular mouth as a primitive jellyfish.

Another was the remarkable Hallucigenia. This curious worm-like beast also kept fossil-hunters and scientists scratching their heads. It was thought to have walked on stilt-like legs, and to have had a row of soft tentacles on its back which it used for trapping passing food. Thanks to the discovery of similar fossils in other parts of the world (particularly in China), fossil investigators now think they've been looking at it upside down. Instead it walked on paired tentacle-like legs, and

used the spines on its back as a form of body armour to protect itself from being eaten.

But nothing in the wildest imagination of science-fiction writers could conceive of such a beast as Opabinia. This swimming gem had five stalky eyes, a fantail for swimming, and a long, grasping arm for feeding. It was smaller than other predators, being only about four centimetres long, and there's nothing remotely like it alive today.

One of the most common forms of animal life at that time, and the most common of all the fossils unearthed in the Burgess Shale, were trilobites. Fossilized remains of these arthropods, which look like giant woodlice, have been found all over the world. They had a thick, hard external covering – ideal for fossilization. Trilobites were probably the first creatures ever to be able to see. They had eyes, like those of today's flies, which divide into

Fossils of the Anomalocaris were first discovered by Charles Walcott in the Burgess Shale. This metre-long sea creature used two grasping arms to capture its prey.

Trilobites had hard-shelled bodies that fossilized easily, providing a detailed record of these extinct creatures that were once cousins of today's arthropods.

hundreds of different cells, giving them a kind of mosaic view of the world under the seas.

To create a realistic picture of how life on earth actually evolved, it is important to know when each species lived and died, so they can be pieced together in chronological sequence. The genius of one man helped work out how to do this. Charles Darwin (1809–82) worked out that all living things have evolved according to a sequence that is still unfolding today.

Darwin's book, *On the Origin of Species* (published 1859) explained, for the first time, the theory that all living things originally evolved from a single common ancestor. What modern scientists now call the Last Universal Common Ancestor (LUCA), the common ancestor of all living things today, is thought to have lived about 3.5 billion years ago. Since then, life has evolved into a huge number and variety of types and species. Darwin worked this out because the fossil record frequently shows new creatures emerging and others disappearing. His conclusion was that all living things are related to each other, but that only those species best suited to the environment of the day survived. His theory gave scientists the first ever way of arranging fossils in different groups, and eventually into a rough chronological order.

The upshot of Darwin's theory was the inevitable conclusion that even humans must be descended from simpler forms of life, like apes, and before that from mice, reptiles, fish, and ultimately from those bacteria found at the

beginning of life on earth. And how did he work all this out? The answer is that he studied fossils. Lots of them.

Charles Darwin was a scientific super-sleuth who went on an epic adventure around the world on a quest to discover the origins and meaning of life. He set off in 1831 aboard a small Royal Navy surveying boat, HMS *Beagle*, whose captain, Robert FitzRoy, had been tasked with making the first maps of the coast of South America. Darwin went as the captain's companion. During the five-year voyage Darwin collected thousands of fossils, suffered dreadfully from sea-sickness, and experienced an earthquake in Chile.

Darwin always took extensive and detailed notes of all his observations. When he saw giant stepped plains of seashells on the land, raised up high above the sea, he realized that over a long period of time the land must somehow have been pushed above the water. On the Galapagos Islands, off the coast of Chile, he noticed that each island had its own special variety of mockingbird. He also found species of giant tortoises which showed slight differences from one island to another. His observations led him to conclude that these creatures must all have originated from a single ancestor, but had adapted slightly to the specific environments in each island habitat.

Fossil records of now-extinct species became the key for Darwin's understanding of the evolutionary process. By studying fossils and comparing them with living things today, he could see that each

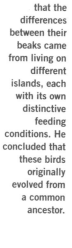

Finches sketched by Charles Darwin. He theorized that the differences between their beaks came from living on different islands, each with its own distinctive feeding conditions. He concluded that these birds originally evolved from a common ancestor.

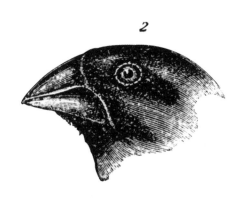

species has adapted itself according to a principle that he called 'natural selection'. Over successive generations those creatures best equipped to live life on earth at that time had survived, flourished and become dominant, while those least well equipped had died off, their species eventually becoming extinct.

Many people were outraged at the implications of Darwin's theory. The suggestion that humans were descended from animals – more specifically apes – threatened the widely held view that mankind was somehow different, superior, to all other living things. Equally as implausible to many was the idea that humans are just another natural species which, like all others, was destined one day to become extinct.

On his five-year voyage Darwin saw at first hand at least two human atrocities: the appalling conditions in which South American slaves were living, and how brutally European settlers treated native Australian and New Zealand Aborigines (see page 346). Such experiences helped him make sense of the evidence of the fossils in the rocks. Man was undoubtedly descended from animals. In one of his most important books, *The Descent of Man* (1871), Darwin concluded that despite mankind's 'noble' qualities, 'Man still bears in his bodily frame the indelible stamp of his lowly origin.'

Then, as now, many religious people believed that humans are different from animals because they have a soul. By saying that humans are descended from animals, did Darwin mean that they don't have souls? If man has no soul, then how can he be saved since he cannot go to heaven or hell? In the minds of some people, Darwin's theories made a mockery of the Christian religion. So worried was he about how his ideas would be received that he refused to publish them for nearly thirty years.

Today, scientists have discovered powerful new ways of dating rocks and fossils which back up Darwin's theories of how creatures have evolved over time. They have found certain types of minerals – including uranium, potassium and a type of carbon called carbon 14 – that act as natural clocks. Over time, these minerals gradually change from one substance into another: a process known as radioactive decay. By working out the rate at which this decay takes place, scientists can use these minerals to date the rocks in which they are found. Using this dating technique, they have been able to construct an accurate picture of how life on earth has changed since the 'Cambrian Explosion', c.530 million years ago. They have also made a map of the past called the geologic column, which is divided into a number of eras.

The Burgess Shale and the fossil record first appear in the Palaeozoic Era (from the Greek meaning 'old life') when animals first developed shells. Each subsequent era is then subdivided into a number of periods. The Palaeozoic Era begins with the Cambrian Period some 542 million years ago. It ends with the Permian Period, about 291 million years later. It was the most dramatic of finales, as we will soon see.

3

4

Geological Time

Millions of years ago

Twenty-four-hour clock

Mya	ERA	PERIOD	EPOCH	SPECIES	EVOLUTIONARY STAGE	Clock
4,600	HADEAN				No life on earth; volcanoes; rain cools the surface; oceans form.	00:00
3,800	ARCHAEAN				Methanogens (prokaryotes); cyanobacteria; stromatolites; oxygen in air.	04:00
2,500	PROTEROZOIC				Complex cells (eukaryotes).	15:00
850		Cryogenian			Snowball earth.	
635		Ediacaran			Multi-cellular creatures.	
542						20:50
488	PALAEOZOIC	Cambrian			Shells, bones and teeth.	
443		Ordovician			Vertebrates.	
416		Silurian			Primitive land plants; worms.	21:50
359		Devonian			Bony fish; tetrapods.	
299		Carboniferous			Amphibians; reptiles; forests; flies.	
251		Permian			Mammal-like reptiles; Pangaea.	
	MESOZOIC	Triassic			First dinosaurs; small mammals; ichthyosaurs.	22:50
199		Jurassic			Dinosaurs dominate land; pterodactyls in the air.	
145		Cretaceous			Last dinosaurs; social insects; flowers; birds; monocots.	
65.5						23:40
55	CENOZOIC	Tertiary	Palaeocene		Mammals grow larger.	
33			Eocene		Whales return to the oceans.	
23			Oligocene		Horses evolve in Americas.	
5			Miocene		Monkey migrations.	23:57
1.8			Pliocene		First bipeds and humans.	
0.11	HISTORIC	Quaternary	Pleistocene		Megafauna extinctions.	23:59
0.02			Holocene		Farming; first human civilizations.	
today			Anthropocene		Globalization; rise in CO_2 levels.	24:00

– 34 –

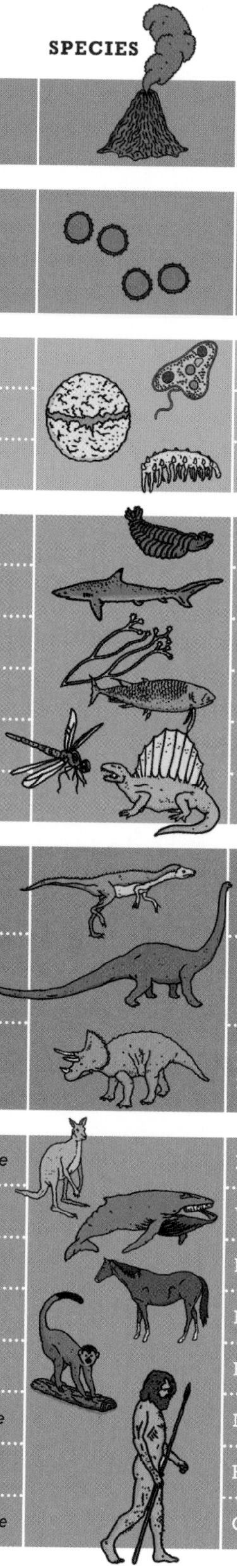

Davy Jones's Locker

How prehistoric life evolved in the seas before living creatures colonized the land, and how some fish developed backbones – becoming humanity's oldest ancestors.

THE BEST WAY to get an idea of what life was like several hundred million years ago is to use our imaginations and dive down to the ocean floor. Our journey will then take us up for a swim with some prehistoric fish before we clamber ashore for a walk through the earth's primeval forests, keeping an eye out for nature's original creepy-crawlies. Next, we'll see the first four-legged animals emerge out of the sea to take charge of the land, ending with the domination of the dinosaurs and a catastrophic event 65.5 million years ago that helped wipe them all out.

On our way we will also try to discover where we, mankind – or *Homo sapiens* to give us our proper scientific name – came from. Darwin worked out that all living things have common ancestors; so what creatures were our prehistoric relatives? Who were our great-grandparents 200 million generations back?

Before we start, here's a quick time-check on our twenty-four-hour clock. On our first stop we will see what life was like between about 9.05 p.m. and 10 p.m.

First, a quick word about how scientists put creatures into categories for easy reference. Agreeing what creatures belong to which groups of living things is still one of science's biggest challenges. Carolus Linnaeus (1707–78) made it his life's work to come up with a proper system for putting living things into different families or groups.

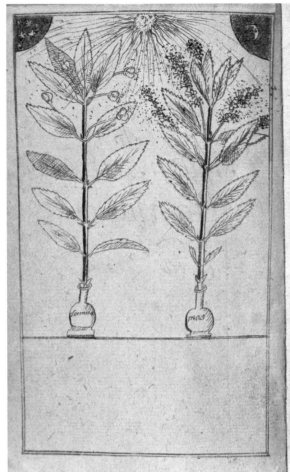

The cover from an early manuscript by the Swedish naturalist Carolus Linnaeus, a man who classified thousands of animals and plants into kingdoms, families and species.

During the 1730s and 1740s Linnaeus travelled around his native Sweden studying nature. On his journeys he devised a system for classifying all the life forms of the natural world.

He published his work in a book called *Systema Naturae* (1735). By 1758 he had classified 4,400 species of animals and more than 7,700 species of plants. His system looks like a tree that, from a few main trunks, branches out in hundreds of different directions. Linnaeus called the main trunks kingdoms, and there are three of them – animal, vegetable and mineral. He then divided these kingdoms into a number of smaller groups, including species defined as a family of living things capable of interbreeding and producing fertile offspring.

Physical appearance was the only method Linnaeus had to help him place living things in different groups. His classification of humans sounds racist today. *Homo Americanus* was, he said, 'reddish, stubborn and angered easily'; *Homo Africanus* was 'black, relaxed and negligent'; *Homo Asiaticus* was 'sallow, avaricious and easily distracted', while he described *Homo Europeanus* as 'white, gentle and inventive…'[1] Such classifications, known as scientific racism, lasted well into the twentieth century (for more on scientific racism, see page 347).

Much of Linnaeus's system has now been revised, especially in the light of Charles Darwin's evolutionary theory, published eighty years after Linnaeus's death. Living things are now ordered not just on the basis of their visible appearance, but in the order in which they are believed to have evolved from one species to another. This takes into account a living thing's internal structure, as well as any information from the code stored in its genes.

New divisions, called phyla, have been added to make sense of our modern understanding of evolution. About thirty-five phyla cover all living things, although most life belongs to just nine big groups. Here are some of the key species that evolved in the prehistoric seas.

Sponges

These were among the simplest of all animals living in the ancient Cambrian seas. There are still many types alive today. About 5,000 different species of them have been discovered so far. They attach themselves to rocky surfaces at the bottom of the sea. The reason we use them for washing ourselves in the bath is that their bodies are full of absorbent holes. Sponges use tiny hairs called flagella to beat sea water through these holes to extract a diet of microscopic nutrients.

For a long time people thought sponges were plants, because they are rooted to the sea floor and they don't seem to move. But actually sponges are in a distant way relatives of mankind. We are much more closely related to a sponge than to, say, a daffodil. Sponge fossils have been found dating back to the earliest Cambrian Period. A famous place for finding them is in the Sponge Gravels of Farringdon, Oxfordshire.

Corals

Most people have heard of coral reefs, but what many probably don't realize is that these enormous constructions were built over hundreds of thousands of years by tiny marine organisms which build their homes on top of the skeletons of their ancestors.

When coral fish die their bones pile up to create vast underwater mountains that provide an ideal marine habitat for future generations of corals and other sea creatures. It is thought that up to 30 per cent of today's marine species camp out in the earth's biggest existing coral reef – the Great Barrier Reef, off the coast of north-east Australia. This colossal structure, composed of more than 1,000 islands, stretches out for more than 1,000 miles.

The Cambrian seas were full of coral reefs, and like today's Great Barrier Reef, they were teeming with life. Reefs are perfect places for sea creatures to live, because they're full of nooks and crannies. Caves, cracks, corners and crevices are ideal for laying eggs, hiding from predators or just taking a break. Coral fish need sunlight in order to live. As each generation of coral dies, the underwater mountain gets taller and taller, meaning that the top of the reef is never far from the sunlit surface. Many reefs have broken through the surface, becoming popular tourist destinations such as the Seychelles and the Maldives in the Indian Ocean, although these are now in peril because of global warming and the increasing acidification of the oceans (see page 376).

Coral reefs are environments that seem to have stimulated an amazing level of trust between different species. For example, small fish are often seen cleaning the larger fish – even entering their mouths to wash their teeth. Communities of these small fish run their own types of 'cleaning station' where larger fish come for rest and relaxation. Corals in the Cambrian seas are a perfect example of natural cooperation and community.

Jellyfish

These are part of the same family as corals, but nowhere near as friendly. The family, or phylum, is called cnidaria. Like sponges, they are primitive creatures, although they can swim using a pumping action of their bell-like heads. Jellyfish have a very simple nervous system, no sensory organs and only one opening – a combined mouth and anus. They were very common in the Cambrian seas, and some could pack a punch worthy of a lion.

The box jellyfish that lives off the coast of Australia is one of the most poisonous creatures ever to have lived. Hanging from its tentacles is a lethal arsenal of harpoons. Inside each is a coiled tube which, when disturbed, fires an arrow-like thread into the body of its victim, injecting it with a paralysing poison. New harpoons are continually being manufactured by this formidable, heavily armed creature of war. [2]

Jellyfish hunt in packs. Great herds of them would have been seen in the Cambrian seas, rising to the surface at night to feed off green algae and falling to the depths by day to avoid being eaten by

21:05:00

fish like squids. Jellyfish are definitely closer relations to humans than sponges, because they are among the first living things to have cell tissues. Over time these evolved into specialized organs and body parts such as hearts and lungs.

Ammonites

Any fossil-hunter would recognize these creatures, even though they have been extinct for many millions of years, having died out, along with so many other species, at the time the dinosaurs became extinct 65.5 million years ago (see page 59). Fossils in this characteristic spiral shape crop up everywhere. Although they look like snails, their closest relatives are actually cephalopods – the family that includes today's octopus and squid.

Ammonites first appeared about 400 million years ago, in the Devonian Period. The animal's living parts were contained in the last and largest of its shell chambers. Shells were ideal protection against sharp-toothed predators. Ammonite fossils have been discovered showing teeth marks, scars from unwelcome attacks.

Ammonites could hide away in their shells for long periods of time. When under attack they had a device like a locking door to close up the entrance to their chambers. They could grow pretty big, too. One fossil found in southern England measures more than two feet across, while another, found in Germany, is over six feet. Their shells make extremely good fossils, which people have been collecting for hundreds of years.

Sea squirts

These look like giant sacks, anchored to the sea floor, which drink massive volumes of water each day, filter out the food and then expel the waste water back into the ocean. At first glance they seem similar to sponges, but actually they're a lot more sophisticated. Not only were squirts a common feature of the prehistoric sea bed, but the way they evolved was important for all kinds of creatures fortunate enough to live on earth in the future – and that includes humans.

Sea squirts have babies that swim about like tadpoles. They propel themselves with a special tail that contains a very primitive form of backbone called a notochord. Descendants of sea squirts developed these notochords into vertebrae – the bones that form our spinal column. All animals that have nerve chords or spines belong to this group, called the chordate, which includes all the

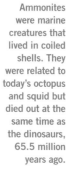

Ammonites were marine creatures that lived in coiled shells. They were related to today's octopus and squid but died out at the same time as the dinosaurs, 65.5 million years ago.

fish, amphibians, reptiles, birds and mammals. Baby sea squirts are the most basic form of chordate that has ever lived, and so they go down in prehistory as the first forefathers of human beings.

They aren't, however, our most intelligent ancestors. As soon as a swimming squirt finds a hard rock it cements itself into place and then eats up the rest of its body that it used for swimming, as it is no longer needed – that includes the precious nerve chord which in other species developed into spines and brains.

Lancelet

Our first fish-like creature may not be big, but it's very old. Something like today's lancelet emerged about 560 million years ago. It seems to have evolved from some copying mistakes in those baby squirts – perhaps one that never glued itself properly to the bottom of the sea.

Like all fish, the lancelet is a distant relative of ours because it has a spinal cord running the length of its body. But that's just about where the similarities stop. Unlike us, it cannot be called a vertebrate, because its cord is not surrounded by bones. The lancelet has no brain, but it does have small gills at the side that breathe sea water in and out. It uses these for feeding by filtering small food particles. These fish also protect themselves from predators by burrowing into the sand on the ocean floor.

Placoderms

Among the most fearsome creatures of the prehistoric seas were the now extinct placoderms. These were among the first fish with jaws and teeth, which were probably adapted from one of their gills. Recent research has shown that some species of placoderm had one of the most powerful bites of any creature ever known. Their teeth could tear a shark in two with a single snap.[3] A placoderm could grow up to ten metres long and weighed over four tonnes. It was built like a tank. Heavy, articulated armour-plating covered its head and throat, and its body was thickly scaled. Even its fins were encased in armour-plated tubes.

Placoderms were some of the world's first true vertebrates. Their spinal cords were protected, like ours, in a series of bony segments. Ugly as they were, they are our cousins nonetheless. They died out in the late Devonian Period, during one of the extinction phases of the early earth (see page 52).

Sea scorpions

Here's a good reason why fish like the placoderms needed to protect themselves with such highly developed body armour. The now-extinct sea scorpion (its scientific name is eurypterid) was formidable. It had a long, spiked tail, equipped with a deadly venomous sting. The creature could grow to more than two metres in length, making it one of the largest arthropods ever to have lived.

Arthropods form another phylum – one that includes the louse-like trilobites. This family is the most numerous of all, and contains all the insects, spiders and crustaceans (like crabs and shrimps). More than 80 per cent of all known living species today are arthropods. They can live everywhere and anywhere – in the sea, on the land and in the air. Arthropods have segmented bodies. They also have a hard outer skeleton that protects them.

Sea scorpions died out along with many other species in what's called the Permian Mass Extinction, 252 million years ago (see page 52). More than 200 fossils of these terrifying creatures have been discovered. In fact, some fossilized tracks made by a 1.6-metre-long sea scorpion were found recently off the coast of Scotland.[4]

Pike

In the prehistoric seas ancestors of today's pike fish developed two remarkable features that proved decisive in their success, and which they passed on to other creatures. The first is that they learned to hunt by stealth – that is to say, unlike sharks, they didn't just rely on speed and brute force. Instead, they would very quietly swim up behind their prey, staying totally motionless in the water, and then

Sea scorpions (eurypterids) could grow more than two metres long and had a venomous tail that made them among the most dangerous creatures dwelling in the prehistoric seas.

suddenly pounce, leaving little chance for their surprised victims to escape.

They were able to do this because, unlike some fish that have to keep swimming to stop themselves from sinking, the pike developed a new system enabling it to stay totally still in the water. To do this it uses an internal bag of air called a swim bladder. When the fish wants to sink, it absorbs some of the air from this sack into its bloodstream. If it wants to rise, it does the reverse, releasing air back into the swim bladder. The result is that these fish can always stay at the same depth in the sea without having to move. It's a bit like the way a submarine works. In fact, many of man's greatest inventions are copies of what already happens in nature.

Ancestors of today's pike fish also developed a pocket of air to help them listen to what was going on in the surrounding water. Pike, and other related fish, called teleosts, were the first animals to hear. Sound waves travel through the water, and cause the air in the swim bladder to vibrate. Tiny bones, very much like those in our ears, send these vibrations to the fish's brain, which interprets them as sound.

Lungfish

Imagine being a medium-sized fish, fighting for survival in the violent, dangerous prehistoric seas. How wonderful it would be to just quit the oceans altogether and wander off onto the beach to start a new life afresh. Forefathers of today's lungfish were among the first creatures to develop the equivalent of an escape hatch from the prehistoric seas by adapting one of their gills into primitive air-breathing apparatus. There are only six species of lungfish alive today, but something closely related to them emerged from the oceans around 417 million years ago.

Lungfish were discovered as recently as 1879 just off the coast of Queensland, in Australia. Near to the site where they were found were fossils of their ancient ancestors, some 200 million years old. These fish looked almost exactly the same as their modern equivalents which is why many scientists today consider the lungfish a living fossil – a creature that almost exactly resembles its ancient ancestors.

Lungfish look like powerful, elongated eels. They burrow into the mud and use their lungs to survive dry periods when water is scarce (this process is called estivation – where animals become dormant during the summer, as opposed to during the winter, which is hibernation). They lived in the estuaries of rivers, and learned to survive in dried-up river mouths by breathing air. They developed other features that helped them live on land, including four highly developed fins, well adapted for 'walking' across hard, dry surfaces. Such devices provided the key to surviving in a dramatically different habitat.

It's time to explore the shore.

6

Friends of
the Earth

*How plants on land eventually evolved into tall trees, and
how the ground was covered with a blanket of nutrient-rich
soil nurtured by insects, worms and fungi.*

FOR MILLIONS OF YEARS heavy rain fell on the earth's barren land masses, wearing down the ground into a lifeless muddy silt. High levels of carbon dioxide in the atmosphere at this time meant the rain was acidic, increasing rock erosion and weathering. The first plants were just squidgy things that looked like small seaweeds and green mosses. They were descendants of the ancient blue-green algae – the oxygen-producing stromatolite cyanobacteria – and they clung close to beaches, rivers and streams.

Transforming these small, soggy clumps of moss into tall, graceful trees that can live thousands of miles from the water's edge presented some awesomely difficult engineering challenges.

Think how hard it would be to design a tall tree that could flourish in the wild. For a start, there is the business of staying upright. Ideally, a forty-metre-high tree should be able to withstand a force ten hurricane without toppling over. Next, a steady supply of water and nutrients are needed to sustain the whole tree. The bits of the tree that make food – the leaves at the top – must be as near to the sun as possible, which in a thick, dark forest means being tall enough to make sure that all the other trees don't block out the sunlight. But being tall means being further away from the main source of water, which is stored somewhere in the ground. Finally, if the tree's family is to flourish in future generations, it must be able to reproduce successfully. That means just dropping seeds willy-nilly on the ground below simply won't do, because young trees can't thrive if they have to compete with their parents for sun, food and water. Seeds have to be spread further afield. How is this to be done, when trees can't walk or swim? (For an exception, see page 64.)

Designing a tree is no simple thing which is probably why the earliest plants – mosses, liverworts and hornworts (the family is called the bryophytes) – stayed exactly where they liked it best: near to the water's edge, thriving only in inlets and bays in river mouths and beside streams. They completely ducked the idea of being tall. Their strategy was to hide from the wind by staying small and keeping close to a good source of easy-to-find water, so as to avoid drying out.

These plants had no proper roots, leaves or internal plumbing system to deliver water and nutrients. But if trees and plants were to colonize the vast tracts of barren land, a half-hearted attempt at escaping from the sea like this was no long-term solution. About 420 million years ago, the first signs of a new approach began to emerge in the form of 'vascular' plants. Ultimately, all the world's trees and forests are derived from them.

The first vasculars weren't anything spectacular to look at. They comprised smallish shoots, only about fifty centimetres high, with thick stems and firm, spiny leaves. We know about these plants thanks to a bizarre discovery in a Scottish village called Rhynie, about forty kilometres north-east of Aberdeen. In 1912 local doctor and amateur geologist William Mackie made an extraordinary find when he was out exploring a nearby piece of ground. After digging out various pieces of earth, he discovered perfectly preserved species of plants fossilized in the rocks.

About 400 million years ago Rhynie was like a steaming cauldron with boiling-hot pools of bubbling mud. Every so often a giant geyser would spout a huge fountain of scorching water filled with silicon from deep inside the earth. Silicon is one of the elements that form sand and rock. When this silicate water landed on nearby vegetation and plants it didn't just kill them instantly: when it cooled, it petrified them into perfect fossils of stone.

The fossils of Rhynie are so well preserved that scientists can see exactly what the plants were made of and how they worked. These vascular plants had ingeniously invented a chemical called lignin, that toughens the walls of plant cells. Plants that have little or no lignin stay small and floppy – like herbs or garden flowers. Although the stems of plants like these can feel rigid, they are held up only by the force of water within them. If the water supply dwindles, the herb or flower wilts.

Plants with lignin in them can survive upright even when a drought sets in. With great precision, lignin-toughened cells are stacked and interwoven in carefully constructed layers to make wood – the magical stuff of trees. Lignin also provides tubes through which minerals and water are transported up and around the tree.

The first evidence of lignin came from plants called rhyniophytes (named after their place of discovery in Scotland). These are now extinct, but their descendants are all round us – indeed, everything woody ultimately comes from these early pioneers of the land. Mind you, it took a while for these small plants with toughened stems to become tall, graceful trees. At least forty million years.

By the time we get to the Carboniferous Period (360 million years ago), trees were growing in huge numbers. The earliest, called lycophytes, were simple structures. They had roots and branches which divided into a 'Y' shape.[1] But they could also be very big, with some specimens, such as the lepidodendron tree, as wide as two metres and as high as a twelve-storey building.

Vascular tissues as seen through an electron microscope. Tiny tubes, called xylem, transport water and nutrients around the plant while thick lignin walls provide support.

Except for the wind, and maybe a scratching sound inside a hollow log or a faint buzzing in the branches, it was eerily quiet in this prehistoric world. There were few animals, and no birds – it was still far too soon for them. And the landscape looked pretty much the same in all directions – an endless thick, dark greenish-brown, a blur of identical-looking trees. Very few varieties existed at this time. There were also no flowers. The earth would have to wait at least another 150 million years before it could witness a first bloom. Compared to trees, flowers are a modern fad.

The lycophyte trees that dominated ancient forests lacked one ingredient that ultimately led to their graceful decline into extinction some 270 million years ago. They lacked true leaves. They mostly used scales on their trunks and thin green blades on their branches for photosynthesis instead. It was left to a relative of the vascular plants found at Rhynie to come up with the concept of creating little green solar panels attached to the tips of branches. These were the euphyllophytes – literally 'good leaf plants'. Most trees that are alive today descend from them. Euphyllophytes quickly grew into several varieties, including ferns and horsetails.

If it weren't for lycophytes, ferns and horsetails, our modern lives would be very different indeed. These early trees colonized the land in their millions. When they died, most of them sank into swampy marshes, where over millions of years they were compacted, hardened, chemically altered and metamorphosed by heat and pressure, ultimately becoming coal. This source of chemical energy eventually fuelled the Industrial Revolution (see page 336).

While lignin helped trees become strong and leaves trapped the energy of the sun to make food, trees still faced the difficulty of finding a steady and reliable supply of water to be channelled somehow all the way to the top of their canopies – water that was often dispersed many metres below the ground. Trees rose to

The lepidodendron tree is now extinct but it grew in massive numbers during the Carboniferous Period (360 million years ago). Some reached as high as forty metres, but they all lacked true leaves.

21:51:36

Much of the world's energy today comes from carboniferous trees like these that sank into swamps and gradually compacted into coal.

that challenge in two ways. The first relied on cultivating a good crop of friends to help. The second was down to clever design.

Tree roots grow downwards to find water. But to help them, they often enlist the support of another group of highly versatile living things. Neither a plant nor an animal (although for a long time they were grouped with plants), fungi form their own separate, almost invisible underground kingdom.

They came on to land from the sea because their tiny light spores were so easily blown about by the wind. They arrived at about the same time as the earliest plants started to grow on the shores. Since then fungi have developed into a huge variety of life forms, ranging from the smallest to the largest living things on earth. Small fungi are just one cell big. Yeast is an example, used in cooking all over the world to make bread and cakes. It grows by using a process called fermentation, which converts sugar into alcohol and carbon dioxide.

Most fungi live underground. They have elaborate networks of hairs, called hyphae, that gather together in clumps called mycelia. A mushroom or toadstool, which most people think of as a fungus, is simply the fruit of the mycelium, which occasionally pops up above the ground to spread spores so that it can reproduce.

Fungi can have massive mycelia. In fact, the largest living thing on earth today is a fungus. Found recently in the American state of Michigan, one hairy beast stretches underground for over five kilometres, and is estimated to weigh more than ten tonnes. It is one of the earth's longest survivors, too, having lived for well over 1,500 years.[2]

Fungi are the world's ecological dustmen. They process and digest dead and decaying matter, from the leaves on the ground to the dead skin in between your toes. When human dustmen take away rubbish it often just burned or thrown into a deep pit in the ground. Nature's dustmen, the fungi, not only rot away the dead rubbish of life, but turn it into materials rich in nutrients that fertilize plants and trees to help them grow. Fungi are vital links in the earth's ongoing cycle of renewal – of life and death.

As so often in nature, different groups of living things team up to mutual benefit. The fungus passes on some of the nutrients and water it gathers to the tree, and in return the tree feeds the fungus with sugars produced by its leaves. In this way the tree's capacity for gathering water and nutrients is dramatically increased, and the fungus gets fed. Sometimes a single fungus lives underground and attaches itself to many trees – so in this way the trees are actually connected together, linked up in a chain as if poised for a medieval dance. This relationship is called 'mychorriza'. It has been estimated that 80 per cent of all flowering plants today have some sort of mutually beneficial relationship with underground fungi.[3]

Even with the water in its roots, a tree still needs to transport it up its trunk to the all-important leaves which manufacture food – and that's a long way, in the case of some trees. For a long time no one could work out how trees did this. Of course, they have no moving parts like heart pumps to do the job for them. It was once thought that external air pressure caused water to flow upwards, as it does when you suck a straw, but air pressure simply isn't strong enough to pump water up a one-hundred-foot tree. The answer is down to the ingenious design of tree leaves, which contain millions of tiny perforations, or holes, called stomata. Trees open or close these pores depending on the weather and conditions of the day. When it is hot, water in the leaves evaporates through the stomata, making the sap in the tree trunk more concentrated. This drags more water upwards through the trunk and into the leaves at the top of the tree. The name of this process is transpiration.

The trees' final challenge is to find a way of spreading their offspring, even though trees cannot walk or move. The earliest species used the wind to spread spores in much the same way as fungi. The problem is that spores need exactly the right conditions to germinate – usually they must land in wet places, such as marshes or bogs. In dry climates, this is a big problem. Then, about 360 million years ago, trees came up with a much better solution. Seeds.

Unlike spores, seeds contain a partly formed tree embryo as well as a substantial food store of sugar, protein and fats. The embryo, with its larder stocked with food, is then encased in a coat (a 'testa'), ready for a sometimes epic journey, using one of a number of alternative transport systems (see page 63).

Seeds dramatically increase a tree's chances of successful reproduction. They are tougher than spores, they can survive droughts, they take their own food rations with them, and some can even float. The first seed-bearing trees were the cycads, which have been traced back to about 270 million years ago – just when the first dinosaurs appeared. About 130 species of cycads are still living today, although many are under threat of extinction owing to the destruction of their habitats. These trees also mastered the art of sexual reproduction, which usually requires the genes of two different trees to combine to make a seed (see page 28). But that leaves a final, apparently insuperable challenge: how could two different parents mix their genes when they are both, literally, rooted to the ground?

The solution came from other forms of life that had emerged on to the land. A few small, worm-like creatures probably emerged from the sea at about the same time as the earliest plants and mosses, about 420 million years ago. What tempted them ashore had something to do with the rising quantities of oxygen in the air. Oxygen levels had steadily increased over millions of years until they levelled off at the beginning of the Cambrian Period, about 500 million years ago. Then there was a 'blip' that occurred between four and two hundred million years ago.

This 'blip' was caused by the luscious green forests now covering the land. Plants and trees

A leaf stoma magnified 425 times. Evaporation through these tiny pores (a process called transpiration) drags new water up from the ground and also helps produce rain.

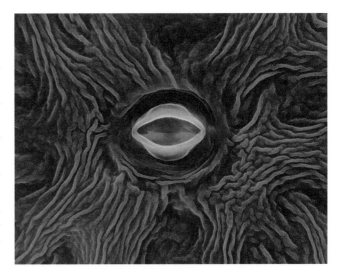

21:51:36

Rise of Oxygen

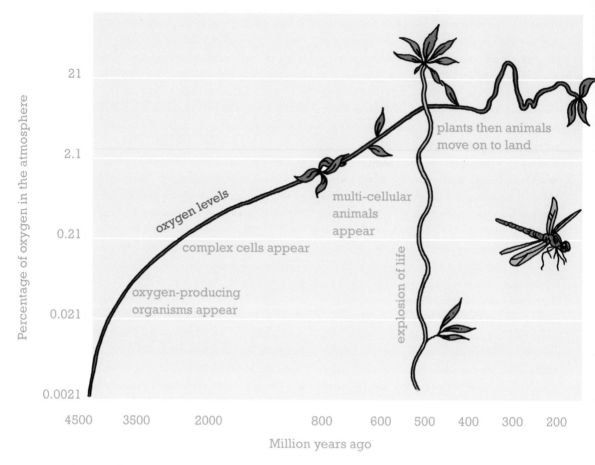

Percentage of oxygen in the atmosphere

21

2.1

0.21

0.021

0.0021

oxygen levels

complex cells appear

oxygen-producing
organisms appear

multi-cellular
animals
appear

explosion of life

plants then animals
move on to land

4500   3500   2000   800   600   500   400   300   200

Million years ago

dramatically increased the amount of oxygen in the atmosphere. The effect was a bit like lining the shores with sweets. Life in the seas just couldn't resist the temptation to come ashore for a taste. When the first sea creatures came crawling out they found that the adjustment wasn't too hard, with all that extra oxygen in the air to give them a boost. Today about 21 per cent of the air we breathe is oxygen. But 350 million years ago, with the arrival of the carboniferous forests, oxygen levels shot up perhaps to as much as 35 per cent.[4]

Extra oxygen explains why Stan Wood, a sharp-eyed commercial fossil-hunter from Scotland, did so well out of a dilapidated old limestone farm wall that he spotted next to a school football field in 1984. He thought the wall might contain some interesting fossils, so he bought it from some developers who were about to knock it down – for £25.

The fossils Wood found inside were so important that he ended up selling them for more than £50,000. He spent some of the money buying the disused old quarry in East Kirkton where the limestone in the wall came from. After bringing in some heavy digging machinery, he made some even more amazing discoveries. He found the fossil of a giant, air-breathing scorpion at least thirty centimetres long, with a vicious-looking barbed tail and a giant protective outer skeleton. This huge creature, an eurypterid, probably grew more than two metres long, bigger than most humans.

The higher concentrations of oxygen in the air meant many creatures could grow much bigger than their living descendants can today because energy-rich oxygen could diffuse further into an organism's breathing system. Stan Wood's scorpion, estimated at about 335 million years old, shows how this ancient beast had mastered two of the essential challenges for creatures that came ashore. It used primitive types of lungs for breathing air. These were adapted from its gills, and were protected by pockets of hard outer skin. The giant scorpion also had pairs of legs, so it could walk.

Some of the world's first insects also belong to this period. Dragonflies were the most spectacular of all. How they learned to fly is still a mystery. But it probably had something to do with the arrival of plants and trees. Wouldn't it make sense for an insect to just jump or glide from one tree to another, rather than climb all the way down and then up again? Something like this is what led the dragonfly to develop its wings. They grew out of the same kind of pockets of hard outer skin as those found in the giant scorpion.

Perhaps to begin with they used these small flaps just for jumping, maybe adding a little extra distance to make a big first leap. Gradually the flaps grew larger, until such acrobatics as gliding, diving and finally flapping became possible. Of course, flapping is an extremely energetic thing to do. Happily, in the oxygen-rich atmosphere of those days these first flyers were immersed in just the right stuff for trying something new and tiring.

The extra oxygen also made the air thicker, so it was easier for the dragonflies to achieve lift-off.

Extra oxygen also helped them grow big.[5] These colourful prehistoric flies were as large as today's seagulls. They leaped, jumped and flew from tree to tree totally unchallenged. They had complete command of the skies, feeding off other insects as and when they liked. They had no rival.

Smaller insects eventually developed an ingenious design to protect themselves. They evolved defences in the shape of sophisticated folding wings, just like those we see in house flies today. Folding wings allowed smaller insects to crawl into narrow spaces where the larger, fixed-winged predators, the dragonflies, could not go. Flying ('neopterous') insects are by far the largest group alive today, which means that the folding wing probably counts as one of nature's most successful ever inventions.

Another important requirement for land-based creatures is the ability to see. Dragonflies developed highly sophisticated compound eyes with 30,000 facets, each one a tiny eye, neatly arranged to give nearly 360-degree, or 'all-round', vision.

Dragonfly fossils have been found in many parts of the world, but the most spectacular came

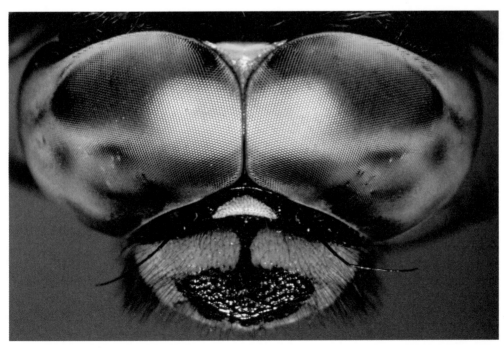

Compound eyes with up to 30,000 facets each gave dragonflies an all-round mosaic view of the prehistoric world.

21:51:36

from the small mining town of Bolsover, in Derbyshire, England, where a giant 300-million-year-old dragonfly fossil was discovered by two coalminers. With a twenty-centimetre wingspan, this is the oldest known dragonfly fossil, and far bigger than any dragonfly alive today. For a few days dragonfly fever gripped Britain, the newspapers had a field day and the legend of the 'Beast of Bolsover' was born.

The first ever land animal was probably a relation of the velvet worm. It wriggled out of the sea, feeding off the earliest plants and mosses that were clinging to the shore. Descendants of this creature, the common ancestor of the arthropod family, went on to develop legs, becoming the first millipedes and centipedes. Once on the land, these early arthropods gradually evolved into a wide variety of insects, combining the first few segments of their worm-like bodies to form a head, and adapting at least one pair of legs into feelers. Over time other segments merged to form the thorax (upper body) and abdomen (lower body and tail).

One of the most significant insects to emerge was the beetle. Today there are probably more species of beetle than of any other living creature. Over 350,000 different types have been discovered so far, which is about 40 per cent of all known insect species, but experts believe there may be between five and eight million types in all.

Beetles bring us back to the final engineering challenge faced by the world's first sexually active trees, the cycads. These were the insects that came to their assistance. As they rummaged in the undergrowth and up into the leaves of the trees, they transferred yellow pollen powder from the male parts of one cycad tree on to the female parts of another, so fertilizing the trees' genes to produce a new crop of seeds.

Beetles, other insects, worms and fungi are jointly responsible for attending to the land's most precious sustaining life force of all: the soil. Like constant gardeners, they recycle organic matter –

fallen leaves and rotting trees – into nutrients that fertilize the soil for tomorrow's plants and trees. Without living things, there would be no soil. The earth would be nothing more than dust and rock, like the surface of the moon, Mars or Venus. Some of the rock might weather and dissolve in the rain to be washed back into the sea in the form of mud and silt, but the crumbly black-brown stuff that makes vegetable gardens grow would never have formed were it not for life on earth. Over the course of millions of years all the soil on earth is renewed and regenerated. This is called the soil cycle.

There is nothing now left of the soil from the Carboniferous Period. The oldest soil today is just a few million years old. Wind, water, ice and the movement of the tectonic plates means that soil, like rock, is always being churned up or washed away. Soil, which is made up of weathered rock, minerals and organic matter, appeared first when plants and trees started to grow on the land in large numbers during the Carboniferous Period. Plants established themselves in cracks between rocks, pummelled by centuries of rain and weather. Their developing roots broke down the rock further.

Since plants were a rich source of food, they attracted fungi, worms and other tiny arthropods such as mites that live off organic matter. For the last 400 million years these creatures have been digging up the earth and turning it over, exposing it to the air and rain with their burrowing, allowing the weather and the elements to break up the soil so that it's always ready for new life to take seed.

With plentiful supplies of food in the form of plants and trees, more oxygen than ever before, a cooling climate and a landscape ideal for providing shelter (either in the branches of trees or in soil which could be burrowed into), the scene was now set for life's next major episode. What would the descendants of those backboned, four-finned creatures such as the lungfish make of this rich, earthy paradise now?

7

Great Egg Race

How the earth's restless crusts collided into a single, giant super-continent, provoking new forms of life to evolve and triggering the first mass extinction of land creatures.

LIFE ON LAND is about to take a big lurch forward. First, we will complete our tour of the Carboniferous Period (360–299 million years ago), and then head into the Permian when reptiles ruled the earth. Then, we must pick our way through a dramatic episode called the Permian Mass Extinction, when most of the life on earth was wiped out, before emerging into a bright new dawn. There we shall meet our first dinosaurs. At this time many of today's familiar insects will also arrive: bees, butterflies and moths. We will also see our first flowering plants and trees, and maybe even catch a glimpse of the world's first feathered bird.

Dinosaurs reigned supreme for 180 million years – through the Triassic, Jurassic and Cretaceous Periods. But, like all empires, theirs ultimately came to an end. Roughly sixty-five million years ago their world came crashing down, and in what seems like the blink of an eye these mighty rulers

were wiped off the face of the earth in a mysterious, sudden and absolute collapse. A quick check on our twenty-four-hour clock shows that in this part of our journey we will be travelling from about 10.24 p.m. to 11.39 p.m. – leaving just twenty-one minutes for the final evolution of all living things to unfold into human history.

Our tour begins with a rare and most important piece of evolutionary evidence provided by the fossils of a creature called tiktaalik which were discovered on Ellesmere Island, in Canada, in 2004. They represent what are known as 'transitional' fossils, and show features from two different species: one is the lobe-finned fish, the other the next-to-evolve tetrapod (meaning 'four-limbed creature'). Tiktaalik lived about 375 million years ago, towards the end of the Devonian Period (see the geological column on page 34), just as the early plants and mosses were adapting to life near the shores. This creature developed the first genuine arms, with shoulders, elbows and wrists – brilliantly suited for heaving its body on to the land or wading through shallow bogs.

Tiktaalik grew to about three metres in length. It had sharp teeth for hunting and, quite unlike any fish, it was able to turn its head from side to side to look for food or danger, using what was probably the world's first ever neck. Its skull was flat, like that of a modern crocodile, with big eyes bulging out on top of its head, suggesting that it spent much of its time lurking just beneath the water in small streams, lakes and shallow swamps.

Several million years later we come across the almost-impossible-to-pronounce ichthyostega, which lived between about 357 and 362 million years ago. This land-lover was about 1.5 metres long and had seven toes on each foot. It was one of the first genuine tetrapods, and certainly no longer a fish. Humans are also part of the tetrapod family, so this is definitely one of our ancestors. Young ichthyostegas were much better suited to life on land than their large, cumbersome parents. Being small, they could easily leap ashore and move around without having to drag an enormous heavy body with them. Being on land was also safer than the sea, because there was plenty to eat and fewer large predators.

As the generations passed by, these youngsters tended to spend longer and longer ashore before returning to the water. Ichthyostegas were important because they represented a true link between land-lumbering fish and the first highly successful family of animals that could live on land – the amphibians (e.g. frogs, toads, salamanders).

Amphibians were well adapted to life on land, although they nearly always needed to return to the water to lay their eggs and reproduce. Since their appearance this family has adapted successfully to life on every continent and in every type of climate, from arctic ice to sandy deserts, although many of the world's estimated 6,000 species are now on the edge of extinction due to global warming, pollution and the destruction of natural habitats such as wetlands, marshes and woodland (see page 375).

The transition from fish to amphibian was finally completed about 340 million years ago with the evolution of the first known family of amphibians, the temnospondyls. These could be as large as a fully-grown crocodile or as small as a newt. Eryops is a fine example, and first appeared in the fossil record about 270 million years ago. It had a stout body, wide ribs, grew up to 1.5 metres long, and had a strong backbone that prevented its body from sagging under its own weight.

For a long time amphibians ruled supreme on land. They were once the earth's biggest and fiercest creatures. But they had one significant disadvantage: they had to live near enough to the water to lay their eggs and reproduce. In severe droughts, or if the climate became too dry, this could prove a big problem. As time went by it became an even bigger problem, because the earth's continental plates were gradually converging into a single, giant super-continent.

An annual drift of five to ten centimetres sounds very slow, but over a million years that means a continent will move almost a hundred kilometres. Over a hundred million years, that's 10,000 kilometres – or a quarter of the way round the world. That explains why, between 250 million and 300 million years ago, life on land changed dramatically.

To begin with, it became a lot hotter and drier, as the land masses converged and distances to the sea became much greater. Amphibians, which had to migrate to the water's edge to breed, were confined to the shores, or to areas with large lakes. If some creature could develop a technology that meant its babies could be born tens, hundreds or even thousands of kilometres away from the water, it would have free rein over all other creatures. As the earth's crusts converged into a single super-continent, along came the next leap forward in the tale of life on earth. It is one that proves once and for all which came first, the chicken or the egg. It was definitely the egg.

Reptiles differ from amphibians because, if necessary, they can live their lives away from the water. They have waterproof skins, so that if the weather becomes very hot, the water inside their bodies doesn't escape, reducing the likelihood of dehydration. They were also the first creatures to be able to lay eggs on land. They evolved a way of surrounding their eggs with tough, waterproof shells that contain a liquid membrane to protect the embryo and give it all the nourishment it needs to grow until it is old enough to hatch out and survive on its own in the air.

The first known reptile dates back 315 million years ago. Hylonomus was about twenty centimetres long, and survived on a diet of millipedes and small insects, although it was often attacked by those super-huge dragonflies and other tetrapods on the lookout for a tasty meal. The earliest reptiles had skulls with no holes in them. Most of these are now extinct, although modern turtles, terrapins and tortoises still have this design. The next to emerge was a group that had two sets of holes in their skulls: one used for seeing (eyes), the others for eating (jaws). One such group were the mammal-like reptiles which would become the dominant force on land for many millions of years. This group, to which we humans are distantly related, emerged a very long time before the first dinosaurs.

One of the most successful species was the dimetrodon, which first appeared in the early Permian Period, between 260 and 280 million years ago. Growing up to three metres long, this lumbering giant walked on four side-sprawling legs and had a long, swaggering tail. It was the largest meat-eater of its time. But the success of this bizarre-looking creature was mostly due to the spectacular sail on its back, which it used as a radiator to heat itself up more quickly than other creatures at the beginning of each day.

Cold-blooded creatures such as reptiles and amphibians usually have to wait until their bodies are warm enough before they can move sufficiently fast to catch their prey. Not so the dimetrodon, which could venture out hunting while other creatures were still charging themselves up. So effective was this solar-panel-like sail that a dimetrodon could heat itself up three times faster with it than without it.[1]

This was the beginning of warm-bloodedness, a feature of all mammals, including humans. It allows living things to maintain the same internal body temperature regardless of the conditions outside, meaning such creatures could hunt at night or whenever it was safe to go out, and not just at the times when everyone else was outside, fully charged up. The dimetrodon also evolved other mammal-like features, such as different types of teeth. In fact, 'dimetrodon' means 'two-measure teeth'.

Bizarre creatures like these ruled the world on land for about sixty million years. Their success came to a sudden and dramatic end 252 million years ago, when life crashed into a deep abyss from which it

Spiny sails allowed these dimetrodon hunters to warm up faster than other creatures, ensuring they got the day's earliest catch.

almost never recovered. This was the single bleakest, most serious ever threat to prehistoric life on earth, a time known as the Permian Mass Extinction.

By now all the earth's land masses had collided, forming a massive single super-continent, Pangaea (the name comes from two Greek words: *'pan'* meaning 'all', and *'gea'*, meaning 'the earth'). The rest of the world consisted of a super-ocean, called Panthalassa. Weather systems and the circulation of the seas changed dramatically as a result of Pangaea, ushering in an era of huge seasonal monsoons and a much hotter, drier climate. Species that had evolved on isolated parts of the earth's surface were now connected with everything else. This meant that large, hungry predators like the dimetrodon could gobble them up, drastically reducing the number and variety of living things.

When enormous continental land masses crash and collide, one thing is certain – there will be a huge increase in the number and violence of volcanoes. Experts think this is one reason why such a horrific mass extinction occurred 252 million years ago, although recent evidence uncovered in June 2006 in the shape of a 480-kilometre-wide crater under the East Antarctic ice sheet (called the Wilkes-Land crater) suggests that a giant meteorite may also have had something to do with it. Perhaps both contributed to what is reckoned to have been prehistory's most destructive period ever for life on earth. An enormous super-volcano, located somewhere in what is now northern Russia, exploded with unprecedented fury. Initially it flooded about 200,000 square kilometres of land in Siberia with boiling-hot lava (forming the Siberian Traps – an area larger than Florida). Then, amazingly, it kept erupting for more than a million years.

This massive eruption devastated the earth's environment, triggering a catalogue of disasters. First, a thick, acrid blanket of poisonous dust blackened the land for miles around as highly toxic ash rose high into the air. After a few days, as the smoke reached up into the higher levels of the atmosphere, violent winds blew the deathly smog around the planet, plunging the rest of the world into a devastating dark age.

It was as if someone had just switched out the lights. A deadly cold winter clung to the world, day and night, all year round, for perhaps as long as fifty years. When the ash eventually cleared, temperatures see-sawed from intense cold to boiling heat. As the atmosphere was flooded by a massive dose of carbon dioxide blown out by the eruption, temperatures increased, heating up the seas and triggering yet another disaster. Frozen methane gas trapped in the ocean floor grew increasingly unstable, finally erupting to the surface. Gargantuan bubbles of gas blasted billions of tonnes of methane into the air. Methane, an even stronger greenhouse gas than carbon dioxide, chased temperatures even higher – high enough, scientists believe, to have extinguished as much as 96 per cent of all life on earth.

There have been at least five mass extinction events in the earth's history and 99 per cent of all species that have ever existed have since disappeared. The first two major extinctions mainly affected life in the oceans (444 million and 360 million years ago). The Permian Mass Extinction was the first to affect life on land, and the most devastating to date. Another major extinction occurred 200 million years ago and the most recent 65.5 million years ago (see page 59). According to a survey conducted by the American Museum of Natural History in 1998, 70 per cent of biologists today believe the world is experiencing a sixth mass extinction due to the activities of modern man (see page 375).

Living things naturally adapt to changes in the environment. But it takes generations of time for successful adaptations to emerge. Even though the Permian Mass Extinction took place over an estimated 80,000 years still only the very toughest, most well-suited creatures survived the darkness and the enormous swings of temperature. It was those which luck had blessed with specific genetic differences that improved their chances of survival in those extreme conditions lived on. Everything else perished. These were desperate times for the earth, her life-support systems ailing from a string of unmanageably chaotic events.

Pangaea: Land-Locked from Pole to Pole

Continental collisions c.250 million years ago led to the creation of a single global land mass, Pangaea, and its sister sea, Panthalassa.

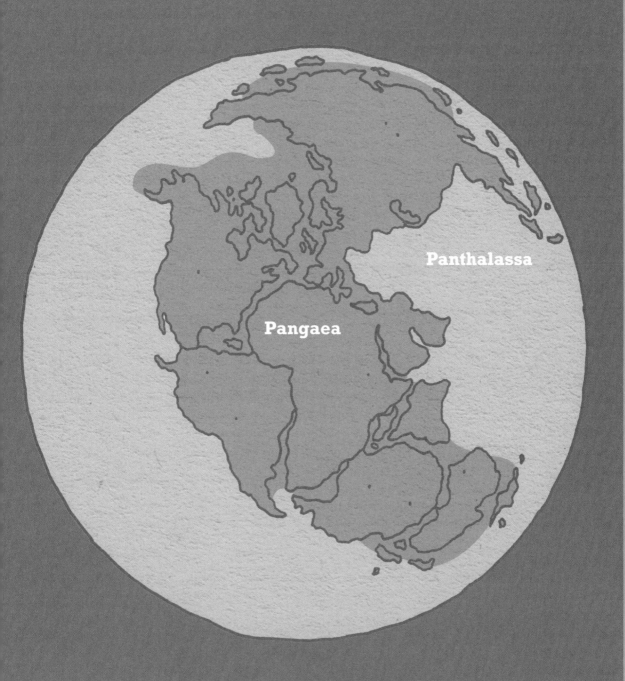

Panthalassa

Pangaea

22:24:00

Dino-Wars

*How a race of terrible lizards came to dominate life on land,
radiating from pole to pole until a freak extra-terrestrial
impact wiped them all out.*

NINE OUT OF TEN species that existed on the earth were killed off by the Permian Mass Extinction. After those catastrophic years, the most successful species were nature's loyal brigade of cleaners and recyclers – the fungi. When times get really tough, history shows it's best to be small – like insects, bacteria and fungi. For a while, the earth belonged to them.

They fed off the enormous piles of dead wood that were scattered across the hot, barren, mostly lifeless land. Records in the rocks show it clearly enough – almost all fossils that date from that time are fungi. Marine life was devastated. Only about one out of every twenty species of fish or arthropod survived. Everything else perished, including trilobites, sea scorpions, many types of coral and microscopic plankton. On land the situation was almost as bad. About 75 per cent of land life perished. That includes the large

amphibians and the massive dragonflies like the one found at Bolsover.

Reptiles fared little better – the dimetrodon with its ingenious sail was wiped out. In fact, only one species of mammal-like reptile survived at all: the lystrosaurus. If this creature had perished, evolution would almost certainly have shunted yet-to-come mammals and humans into might-have-beens, consigning our ancestors to the waste bucket of life, an experimental cul-de-sac. No matter, then, that our reptilian saviour wasn't exactly blessed with good looks, being something like a cross between a hippopotamus and a pig.

No one knows quite why or how the lystrosaurus survived the gigantic trauma of the Permian Mass Extinction, but for millions of years it thrived across all Pangaea. There is almost no record of any other vertebrate alive during this time, 230 million years ago, which is called the early Triassic Period. Indeed, the discovery in the 1970s of lystrosaurus fossils in Africa, India, China and Antarctica was what finally convinced scientists that Alfred Wegener's theory of plate tectonics was correct. Only if the earth's continents were once connected into a single giant land mass could the remains of these ancient creatures that could not swim have been spread out across the world (see page 27).

Over the course of many generations the lystrosaurus evolved into the even more mammal-like thrinaxodon. About the size of a cat, this creature ate insects and small animals, and seems to have had a primitive covering of fur. It was also warm-blooded, so it could feed and hunt effectively at any time of day or night. It may also have been one of the first creatures ever to have looked after its young.

Then, out of the barren sands of extinction, came a completely new generation of land-living reptiles – one that grew into the most fearsome and dominant force that had ever trodden the earth. Dinosaurs were reptiles that lived on land. They had rearward-pointing elbows and forward-pointing knees, with hips that allowed many of them to walk on two legs. Most were large – their average weight is estimated at 850 kilograms, whereas the average weight of a modern mammal today is only 863 grams.

The name 'dinosaur' was coined by the first curator of the National History Museum in London, Richard Owen, in 1842. It comes from two Greek words: *deinos*, meaning 'terrible', and *saurus*, meaning 'lizard'. More than 500 different species of dinosaur have been identified so far, although nearly 2,000 are thought to have existed. Some walked on two feet, some on four. Some ate plants, some ate other animals, some ate both. The first known dinosaurs were the prosauropods. These were plant-eaters that could grow up to ten metres long with small heads and enormously long necks. They usually walked on four legs, but sometimes climbed onto just two when reaching to nibble at the top of a tree.

Gideon Mantell was the first person ever to discover a dinosaur bone. He was a doctor and amateur fossil-hunter who lived in Lewes, in East Sussex. Near his home lay the remains of an ancient forest that was a goldmine for fossil-hunters of the period. In 1822 Mantell found several remarkably large teeth. He eagerly showed them to the leading experts of the day, but they all dismissed them as belonging to a known animal, perhaps a rhinoceros.

But Mantell was convinced they were wrong. He calculated from the size of the teeth that the animal must have been at least eighteen metres long – that's nearly as long as two back-to-back double-decker buses. After years of dispute Mantell's teeth were eventually recognized as belonging to a new type of creature, never before known. He called it the iguanodon, because it was reckoned to look like a much larger version of a modern iguana.

Mantell's discovery made people realize for the first time that the earth was once dominated by a family of huge, now-extinct monsters. In Mantell's day, as now, myths and legends the world

The iguanodon, named by Gideon Mantell, could grow up to ten metres long and had a large spike on its thumb that it used to ward off attackers.

22:43:12

over were full of dragons and dreadful beasts. All of a sudden these stories seemed to have a history based on fact, rather than fiction (for European beliefs in medieval monsters, see page 281).

The world went fossil mad. The place that went maddest of all was America. In 1858 a fossil-hunter called William Foulke discovered the first ever near-complete dinosaur skeleton in a quarry near his home of Haddonfield in New Jersey. This dinosaur was named after Foulke and the place where he found it – *Hadrosaurus foulki*. Soon after, two of America's leading palaeontologists, Edward Cope and Othniel Marsh, began scouring the land for all the dinosaur fossils they could find.

At first Cope and Marsh worked together, hiring a company of men to dig up the quarry where Foulke had made his discovery. They found several other almost complete dinosaur skeletons, but their friendship soon turned sour when it emerged that Marsh had secretly bribed the men digging out the quarry to tell him first whenever any new fossil discoveries came to light. A war – and not just of words – broke out between them. Both men were wealthy, and they spent their riches trying to out-smart each other with dinosaur discoveries.

From about 1870 attention shifted from New Jersey to Kansas, Nebraska and Colorado, where, because of the building of the first trans-American railroads (see page 339), fresh fossil discoveries were being made by the week. Armies of men were hired and equipped with mules, axes, spades and dynamite. They were sent deep into the caverns, quarries and hillsides of America's wastelands to blast away the rocks in an obsessive search for dinosaur remains. Spying, stealing and bribery were just a few of the

This twenty-six-metre-long diplodocus skeleton replica is on display at the Natural History Museum in London. These enormous sauropods were so large few predators could successfully attack them.

tricks each side got up to. At one stage Cope was so angry with Marsh for stealing his fossils that he had a trainload full of Marsh's fossils diverted to his own collections in Philadelphia.

When Cope published details of a new variety of dinosaur, but mistakenly attached its skull to the wrong end of its skeleton, he tried to cover up his error by buying every copy of the magazine in question and destroying them. Marsh, of course, did everything in his power to publicize Cope's mistake. Either despite or because of their extraordinary rivalry, by the time Cope died in 1897 the number of known dinosaur fossils in America had soared from nine to about 150. By the end of his life Marsh had clocked up eighty-six new species, including the now famous triceratops, diplodocus and stegosaurus. Cope discovered fifty-six, including the first ever dimetrodon (although that is not technically a dinosaur). The result is that we now know a good deal more about these mighty beasts that once ruled the world, and why they were so successful for so long.

Like the dimetrodon that thrived before the Permian Mass Extinction, dinosaurs benefited greatly from living in a era when it was literally possible to walk from pole to pole. Thanks to the arrangements of the earth's continents into one giant land mass, Pangaea, the strongest land-based creatures of this era had a unique opportunity to evolve into a position of unassailable dominance. Natural barriers, such as oceans, never got in their way. It was almost inevitable that their success would lead to a reduction in the variety of other land-living things.

The dinosaurs were the first creatures ever to have their legs directly under their bodies at all

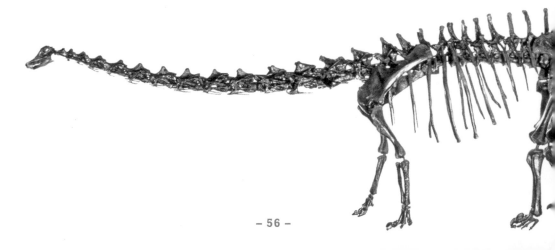

times, whether walking, running, galloping or hopping. This is called a 'fully improved stance', and meant that many of them could walk upright. This may have been the biggest single biological factor in the dinosaurs' success. Animals with a fully improved stance could grow bigger, move further and walk faster than other living creatures, allowing many body types and lifestyles to evolve.

Some of the biggest dinosaurs were the sauropods, such as the diplodocus. These beasts, which walked on all four legs, could grow up to 27.5 metres long and weighed up to eleven tonnes. Their survival strategy was simple: they grew so large that few other creatures were big enough or strong enough to kill them. So they did very well. In 1994 a 147-metre-long fossilized sauropod trackway was found along an ancient muddy estuary in Portugal. The huge footprints showed scientists how these creatures walked, confirming that they held their bodies above their legs and lifted their long tails to stop them dragging along the ground, allowing them to walk more easily.

Others were fast. Hypsilophodons grew only to about the height of an adult human's waist, but could run as fast as a modern deer. More than twenty of their skeletons have been found fossilized on the Isle of Wight, England, where they unwittingly wandered into a pool of lethal quicksand. With long, thin feet and short thighbones for rapid forward and backward movement, these two-legged creatures had self-sharpening teeth and lived off plants. They survived purely thanks to being able to run away from danger so fast.

Other dinosaurs could walk on either two or four legs, such as the iguanodon discovered by Mantell. This beast couldn't run out of harm's way. Instead its thumb was cleverly crafted into a terrifyingly sharp dagger which it used to defend itself, usually standing upright on its hind legs to fend off attackers.

But the strongest beast ever to have stalked the earth was the tyrannosaurus rex, or T. Rex for short. It belonged to the therapod family, which originated in what is now western North America.

T. Rex walked on two legs, and had a massive skull, balanced by a long, heavy tail. Its hands had just two fingers, and its forearms were quite short compared with its massive legs and tail. At twelve metres long and weighing the same as a modern elephant, this was one big dinosaur. It dined off either dead carcasses or live prey – possibly both.

Some experts think T. Rex could run fast despite its huge size – perhaps up to thirty miles per hour. Others think it lumbered around at a more modest ten miles per hour. No one knows for sure how powerful its muscles actually were, because all that remains are skeletons reconstructed from the more than thirty specimens found in rocks around the world. But with jaws more than eight times as powerful as a lion's, the T. Rex could literally pulverize bones to extract the nutritious marrow inside. Its teeth were like a shark's, continually being replaced throughout its life.

The most complete T. Rex skeleton to date was discovered on 12 August 1990 in a place called the Hell Creek Formation in South Dakota by Sue Hendrickson, an amateur fossil-hunter. This almost complete skeleton, called 'Sue' in honour of its finder, measures some four metres high and thirteen metres long. Finding fossils like this is a

Meet Sue. She is all that remains of a T. Rex found in 1990 in South Dakota and was later sold for a cool $7.6 million.

serious business. After a protracted legal fight over its ownership, Sue was ultimately declared the property of landowner Maurice Williams. He eventually sold her at auction for $7.6 million.

The majority of dinosaurs were peaceful, plant-eating creatures that developed the earth's first strong social communities. Herds, packs and families of dinosaurs were commonplace. Many fossils have been found where groups of dinosaurs died together, such as the ones which sank in the quicksand off the Isle of Wight. In Alberta, Canada, a mass grave has been discovered containing at least 300 grazing dinosaurs of all ages and sizes. This herd was swept to death by a flash flood while trying to cross a deep river. In Montana, in the USA, an even bigger herd of about 10,000 dinosaurs has been discovered. These creatures were poisoned by volcanic gases and buried in ash. Their fossilized bones stretch out in a straight line for more than a mile.

Many of these were hadrosaurs. Experts believe that these creatures used strange-looking crests on their heads, which varied slightly between different species, as visual cues for easy identification. There is also evidence to suggest that these crests were used for making sounds. By blowing air through their long, hollow tubes the animals were able to produce low, bellowing notes. Primitive forms of communication like this would have proved very useful to sound an alarm if one of the herd spotted an approaching attacker.

Modern understanding of how dinosaurs lived together was transformed by a fossil specialist called Dr Jack Horner, who made a most extraordinary discovery in the mid-1970s. Horner and his team were fossil-hunting in Montana when they came across America's first dinosaur nesting site. They found a number of dinosaur eggs as well as the fossilized remains of baby dinosaur embryos. They called these creatures maiasaura, which means 'good mother lizard'. Each season maiasaura returned to the same breeding grounds, where it refurbished its old nests. These dinosaurs lived in colonies and looked after their young in herds until they were old enough to start families of their own. Here, beyond all reasonable doubt, lie the origins of the family unit. More than 200 dinosaur-egg sites have now been found across the world. The eggs range in size from a small pebble to some as big as footballs.

Other dinosaur-like creatures, too numerous to mention, also lived through these times. In particular the pterosaurs (illustrated opposite), enormous flying reptiles that dominated the prehistoric skies. One of them holds the world record for biggest wingspan of any creature ever to have lived. At nearly twenty metres across, its wings were as large as those of a Second World War Spitfire.

Other reptiles headed back into the sea in search of a better life beneath the waves. No one knows which reptile first decided to quit the land, but about 290 million years ago, just before the first dinosaurs appeared, ichthyosaurs were living in the seas, coming up to the surface to breathe air but then diving deep to find food below. These are the first creatures known to have given birth to their young from inside their bodies, just as mammals do today – fossils have been found complete with their growing embryos inside. Their legs, which had previously evolved to cope with the challenges of a life on land, now evolved back into fins and flippers ideal for swimming fast and diving deep. In just the same way, but millions of years later, dolphins and whales also gave up life on land, re-adapting themselves for living back in the seas (see page 74). Evolution does not progress in straight lines.

What on earth happened to the dinosaurs? The rocks speak clearly for themselves: after dinosaurs dominated life on land for more than 160 million years, from about 65.5 million years ago another mass extinction occurred and all traces of dinosaur fossils disappeared completely. It wasn't just the dinosaurs that were wiped out. The last of the pterosaurs, the flying reptiles, also vanished, as did all the sea reptiles except for the turtles, which somehow survived. It was also the end of the road for the ammonites, those bizarre spiral-shaped fossils, relations of today's octopus and squid (see page 38). Many other species of mammals, trees and plants were also affected, although amphibians seem to have escaped lightly, as did the bony fish.

To survive this trauma, being small was an advantage, whereas anything over about a metre long was in big trouble. But then, why did crocodiles survive, but lizard-like dinosaurs die off? Why did bony fish live on, while the ammonites perished? Experts think that about 50 to 80 per cent of all plant and animal species were killed off this time.

This moment, 65.5 million years ago, marks the end of the Cretaceous Period and the beginning of the Tertiary. Life had been getting tougher for the dinosaurs for a long time. The climate was now warmer, thanks to increased levels of carbon dioxide in the atmosphere, and rainfall patterns were changing as the great super-continent Pangaea gradually broke apart. Traditional breeding grounds for the plant-eating dinosaurs were cut off from each other, and a new ocean, the Atlantic, began to divide the continents of Europe and Asia from America, Australia and Antarctica. As the continents moved, swamps and rivers created barriers between what were once huge feeding ranges.

Now separated from each other, big herds of dinosaurs had to compete over shrinking plots of land. As the population of plant-eating dinosaurs declined, the meat-eaters had less to feed on, jeopardising the whole dinosaur population.[1] But then, suddenly, the dinosaurs disappeared altogether. Whatever happened, it happened too quickly for nature to deal with. There wasn't enough time for sufficient generations to pass by, each one morphing into new forms, to cope with the rapidly changing conditions on earth.

Life on earth survives inside a thin, delicate layer of atmospheric skin on top of a lump of hard, lifeless rock. That's exactly how it would have looked to the six-mile-wide asteroid hurtling towards earth 65.5 million years ago at a speed of some 70,000 miles per hour. Far in the distance, a bright blue-green dot would have been gradually growing larger and larger by the day. As the dark chunk of deadly rock and ice made its final approach, planet earth would have resembled a sparkling bright jewel in a void, black sea of space.

Freakish misfortune powered it on its lethal journey, and when the end came it was as colossal as it was spectacular. Not that the earth isn't used to the occasional googly bowled by chance from outer space. Asteroids, meteorites, comets – they all fall from the skies in a constant rain. Usually they vaporize as they cut through the dense atmosphere, burning up with the shock of suddenly encountering the earth's thick, heavy air.

But not when they are this big. Down it came, possibly splitting up into several pieces before

Flying reptiles called pterosaurs dominated the skies from 228 to 65.5 million years ago, until they became extinct, along with the dinosaurs.

22:43:12

finally making its rendez-vous with an unsuspecting world. At the point of biggest impact, just at the edge of Yucatán Peninsula in Mexico, the asteroid would have blasted a crater more than a hundred miles wide, destroying in seconds everything within a 600-mile radius. The force would have been equivalent to thousands of nuclear bombs exploding all at once, vaporizing anything and everything into nothing more than a cloud of deathly, blisteringly hot, toxic gas.

Outside the immediate impact zone, where everything evaporated in an instant, the earth convulsed. Gigantic tidal waves hundreds of feet high spilled from shore to shore. The destruction these tsunamis caused on both land and sea would have been awesome. Weeks passed before the seas calmed once again. Worse luck came from the fact that the rocks directly under the impact zone were rich in sulphur, a highly toxic substance. The impact spread poisonous sulphur dust all around the world, adding yet another dimension to the devastation for everything that breathed the air.

The noise and sight of the impact would have deafened and blinded countless living creatures. Many of those not killed by the blast would have been drowned by the waves. For perhaps as long as a

year the earth, enveloped by heavy, acrid clouds, was without light from the sun. A blanket of darkness hung in the skies, and the seas spiked in temperature. Meanwhile, more misery spread to other parts of the world. An almighty series of volcanoes erupted in what is now India. Maybe the asteroid had split into several large pieces in its final, few moments, linking these events?[2] Devastating volcanic eruptions spewed boiling hot lava over more than a million square miles, triggering a series of colossal global conflagrations and creating a huge area of new land near Bombay, known today as the Deccan Traps.

Poisoned water and air, extreme heat, choking dust and total blackness prevented plants from growing and flowers from blooming. This is what caused the extinctions in the rest of the world beyond the impact zones. It was as if someone had grabbed the hands of our twenty-four-hour clock and wrenched them back from about 11.40 p.m. to about 10 p.m., returning life to what it had been at the beginning of the Carboniferous Period, almost 300 million years before.

After the Permian Mass Extinction 252 million years ago, it was the fungi which had briefly ruled the world. This time the honour went to the ferns, because their spores were so light and so

Big impact: the arrival of the deadly meteorite that led to a mass extinction 65.5 million years ago, which wiped out between 50 and 80 per cent of all species, including the dinosaurs.

indestructible that they were the first living things able to spread and thrive in the earth's barren wastelands. Relying on nothing more than wind power meant they could reach anywhere where it was possible for them to grow, regardless of death on earth, putting down their roots in the most inhospitable, hard, hot-black ground.

It's only quite recently that this whole saga has been properly pieced together. Until the late 1970s no one could agree on what happened to the dinosaurs. People had their own pet theories. Volcanoes? Disease? Someone even suggested that a nearby supernova explosion could have dealt the deathly blow. The puzzle was finally solved after scientists found a way of analysing minute traces of very rare substances in the rocks.

In the 1970s American geologist Walter Alvarez was conducting a rock study along the volcanic spine of Italy in a bid to find out what effect volcanoes had on ancient Roman settlements. He was looking for signs of magnetic changes that could provide clues as to the rocks' absolute age. His research took him to the small town of Gubbio in central Italy, which boasts the second-largest surviving Roman amphitheatre in the world. As you leave the town from the north, beds of limestone rocks line the winding road. These were once part of the ancient Cretaceous sea floor, thrust up as mountains millions of years ago by the African plate's slow but relentless march northwards into the Mediterranean (see page 85). Further along the road, the rocks become gradually younger, allowing experts to unravel the story of the late Cretaceous Period and the disaster that followed.

About a kilometre from Gubbio the rocks turn a pinkish colour. This is known as the Scaglia Rossa formation, and it represents the time exactly 65.5 million years ago when dinosaur fossils disappear from the record altogether.[3] These rocks are a feast for the fossil-hunter. Countless ammonites mingle with millions more microscopic plankton, or radiolarians, tiny sea-creatures famous for changing species rapidly, giving experts a clear indication of the age of the rocks they are studying. Then an unmistakable and profound change occurs precisely at the 65.5-million-year-old boundary. It is marked by a strip of clay about one centimetre thick. Above this point there is almost no sign of life. Look a few metres further up, and the bones of sea creatures reappear, suggesting a return to normality.

Alvarez was analysing the rocks for traces of the very rare metal iridium. When he got back to his laboratory he found that in this layer of clay the levels of iridium were hundreds of times greater than those found in the surrounding rocks. He called his father, a Nobel Prize-winning physicist, for help. While iridium is very rare on earth, it is extremely common in meteorites. The type of iridium found on earth differs from the type Alvarez found in the Italian clay layer, making it certain that this iridium originated in an object from outer space.

Alvarez and his father published their findings in a scientific paper in 1980. Since then there have been findings in several other parts of the world where the thin line of clay shows similarly high levels of the same kind of iridium – the tell-tale sign of extra-terrestrial impact. Alvarez's team weren't too concerned about finding any physical sign of impact, such as a crater. Chances are that by now all such evidence would have disappeared, churned up long ago by the movement of the earth's tectonic plates.

But in 1990 Alan Hildebrand, an American research student, struck lucky while rummaging through rocks in eastern Mexico. He came across the same strip of clay that Alvarez found in Italy, but this time it was laced with speckles of glass, called shocked quartz, which can only be formed by huge temperatures and pressures. It has been found at nuclear bomb testing sites and meteorite impact craters. But it cannot be produced naturally on the earth. After consulting with his university professor, Hildebrand realized that the meteor impact site must have been within a 600-mile range of his find. Now, thanks to the latest developments in satellite photography, the hundred-mile-wide Chicxulub crater can be seen in all its glory.

Most experts now accept that one or more meteorite impacts caused the sudden and final disappearance of the dinosaurs. The volcanic traumas that created the Deccan Traps just heaped on further misery. Whether or not these were also triggered by the impact, no one knows.

Flowers, Birds & Bees

How the earth's first flowers blossomed alongside feathered flight, and how new species of social insects constructed nature's first civilizations.

CHARLES DARWIN wrote in 1879 in a letter to a friend, the botanist Joseph Hooker, that he could not understand the sudden appearance of flowering plants in the fossil record. Where on earth did they spring from? 'The rapid development of all the higher plants in recent geological times is an abominable mystery … I should like to see this whole problem solved.'

To this day, no one has really come up with a decent explanation. Unlike some of those wilder theories about life's ingredients arriving on earth on a meteorite from outer space, there is no question of the same being true for flowers. Yet about 130 million years ago the world's first flower fossils suddenly start to show up. We are now approaching the start of the Cretaceous Period. (Cretaceous means 'chalky'. It lasted from 145 million to 65 million years ago.) The dinosaurs were still at their peak, but Pangaea, the massive super-continent, was in the process of splitting up into two large separate land masses, Laurasia (to the north) and Gondwana (to the south), with a big ocean, the Tethys, in between.

Some experts think flowers arose millions of years before, perhaps as long as 250 million years ago,[1] but fossils this old have never been found. Others think that several evolutionary phases occurred in quick succession, accounting for flowers' sudden appearance in the rocks.[2] The fact remains that flowers appeared for sure about 130 million years ago, and as yet there is no clear evidence that they lived much before then.

Flowering plants and trees made a massive impact on life on earth. Without them, life today would be very different indeed. More than 75 per cent of all the food humans eat (directly or indirectly) comes from flowering trees and plants. No longer was the earth dominated by endless streaks of browns, greens and blues. For the first time there were blooms of red, yellow, orange, purple and pink.

A flower is a powerful technology used by many plants and trees to reproduce so that they can spread all over the world. Evolution must have been at its magical best when the first flowers evolved, because the designs it came up with to aid fertilization and spread seeds are among the most spectacular of all.

They did it by using that tried and tested strategy that the older trees knew best: they put all their effort into making friends. Flower power helped plants and trees recruit armies of other creatures to help them spread to all corners of the earth. It maybe no accident that flying insects such as bees, moths and butterflies first appeared alongside this first spring of flowers.

The question of which came first, the chicken or the egg, was easy. But when it comes to the flower or the bumble bee, that's a lot harder. The best guess is probably that they evolved together – a process called co-evolution. Flowers needed bees as much as bees needed flowers. Each developed ways of helping the other survive better, because they both stood to make gains from mutual cooperation – one providing food, the other a means of transport. With the help of a pollinator such as a beetle or bee, genes from male and female flowers could mix to produce new seeds with their own unique genetic code. Flowers now developed a huge range of incentives to get animals on land and in the air to carry their pollen and seeds to other places.

Fruit is the female part of a flower, the ovary, which once it has been fertilized changes its shape and form to help disperse seeds. Sometimes these seeds travel on the wind, sometimes by water, or sometimes by sticking to an animal's fur. So the downy white parachute of a dandelion seed is a fruit. And so is the acorn from an oak tree, or a prickly burr. The most ingenious transportation method of all is by burying seeds in a ready-made meal. A passing animal might help itself to the morsel, digest it for a day or two, and then obligingly spread the indigestible seeds as it moves along by scattering them in its dung, giving them an additional growth shot in the shape of a godparent's gift of manure. Seeds are built to be tough. They can survive the most upset, unpleasant of stomachs.

An evolutionary bargain: flowers feed bees, while bees transport pollen, thereby spreading a flower's genes from place to place.

Not all fruits rely on being eaten. Some trees, like the coconut palm, developed other strategies, such as floating their large seeds over thousands of miles of water, from one coast to another. Or there's the curious sandbox tree, whose fruit explodes like a firework, scattering its seeds up to a hundred metres away. Cotton fruits produce fibres that stick to animal skins as a way of spreading their seeds. Nuts are edible seeds designed to be carried off by animals which hoard them for the winter. Almost always they leave some uneaten, allowing these seeds to grow into new plants in a new place.

Before fruit there was the wind – nature's most traditional method of spreading pollen and dispersing seeds and spores. The wind is still a favourite among many flowers, especially grasses (e.g. wheat and barley). Dandelions have tiny parachutes to catch the breeze, and the helicopter-like wings of the sycamore seed work well too. Flowers that use the wind for pollination as well as seed dispersal have no need to attract animals or insects, so they don't bother with big, showy flowers. They save energy by keeping their petals small and discreet.

An important new group of plants evolved during the Cretaceous Period, called monocots. Unlike most other plants (called dicots), they came up with the ingenious trick of growing back to front. Instead of new growth being added to the tips of the leaves – a conifer's greenest, youngest shoots are always at the ends of each branch – a monocot's leaves grow upwards from a central, often submerged, bud. The new design was an instant success, because it meant that if a plant's leaves got nibbled by a passing dinosaur it didn't lose its most recent growth, because this was safely tucked away at the bottom. Grasses are monocots that use this design to recover quickly after being grazed by animals. In fact, many grasses *like* to be grazed. It strengthens their stems, but doesn't damage their potential for new growth, since the growth bud (called the apical bud) is always kept beneath the ground, out of harm's way.

So successful was this design that grasslands have come to cover as much of the earth's surface as all the other plant and tree species combined. What's more, by the end of the Cretaceous Period a completely new type of tree had evolved. Unlike the ancient cycads and conifers, monocot palm trees grow from a bud at the top of a thick, scaled trunk. There are more than 2,600 types of palm trees alive today. The most ancient palm tree fossils – from the nipa palm – date from around 112 million years ago.[3] These are rather special, because their trunks and roots are sunk in marshy swamps or riverbanks. Their way of spreading themselves

takes some beating. These are trees that swim. They tie themselves together by the roots and, using the force of the tides and water, break loose into floating islands that can carry cargo in the form of small groups of animals, which use them as rafts to float from one place to another.

Although today an enormous amount of research has gone into studying the fossil record and the genetic ancestry of modern plants and trees, many pieces are missing in a puzzle that was, and is still, Darwin's 'abominable mystery'.

Here's another puzzle: where did birds come from? Hermann von Meyer, a German fossil-hunter, thought he'd got the answer in 1861 when he announced the discovery of what he claimed was the first ever bird. He called it archaeopteryx. It was about the same size as a modern magpie, and was definitely bird-like. Its feathers were arranged in just the same way as modern birds', with an aerodynamic configuration to make flight possible. It even had bird-like claws on its legs, and a wishbone.

Von Meyer's fossil dated back some 140 million years, at about the same time that the first flowers were beginning to bloom. Since then a further ten archaeopteryx specimens have been found in a region of Germany called Solnhofen. But for years a mystery remained. From what did these creatures descend, and how did they learn to fly? So perplexing was this puzzle that the birds were shunted into their own separate group, neither reptiles like the dinosaurs, or mammals like man. They were just, well, birds.

Then, about thirty years ago, a solution to the puzzle began to unfold. American palaeontologist John Ostrom was professor in charge of the impressive fossil collection started by the competitive American collector Othniel Marsh (see page 56) and now kept at the Peabody Museum of Natural History in Cambridge, Massachusetts. Ostrom was convinced that dinosaurs were more like birds than reptiles, and in southern Montana in 1964 he discovered several hundred examples of a very special fossil indeed – the deinonychus. This was a truly terrifying beast. It was an active, fast-moving killer, called a raptor. It caught its prey by stabbing it to death with

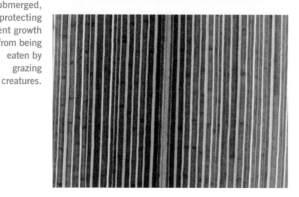

Dicot plants (above) grow from their tips and have leaves with veins that branch out sideways. Monocots (below) have veins that grow out from a single central bud, often submerged, protecting recent growth from being eaten by grazing creatures.

terrible claws on its hind feet. Its strong tail provided balance, better turning ability and the strength to jump, run and mercilessly kick its prey to oblivion.

Previously it was thought that dinosaurs were slow plodders because of their enormous size, but this fossil showed that at least some were fast movers – and that probably meant that they were warm-blooded, which they would need to be to have enough energy. But were they fast enough to take to the skies? Flight had already evolved several times, first with the dragonflies, then reptiles such as the pterosaurs mastered the art. But neither had used the ingenious invention of feathers, the hallmark of birds today.

In 1970 Ostrom returned to the idea that the archaeopteryx was the first known fossil of a bird. He discovered that the flexible wrists on the forelegs of the deinonychus, which could move sideways as well as up and down, were of exactly the same design as those he found on re-examining the archaeopteryx. In 1976 he made a most astonishing claim: that birds are the living descendants of dinosaurs. For years Ostrom's ideas were dismissed as interesting but eccentric – after all, the dinosaurs became extinct more than 65 million years ago. Until someone could physically prove that, say, dinosaurs had feathers, then birds would have to stay out on a limb, their origins mysterious, separate and unknown.

In the early 1990s fossil experts found what are possibly the world's most important ever fossil discoveries, in the Liaoning Province in far north-eastern China. They dug up what has been called a 'dinosaurian Pompeii' – literally millions of plants, insects, molluscs, fish, frogs, turtles, lizards, mammals, birds and dinosaurs. All these creatures had suffocated and were rapidly entombed in a torrent of hot volcanic ash and dust following the massive eruption, about 130 million years ago, of a volcano nearby. Quietly grazing by the water's edge, they never had a chance. They were buried so quickly that even the oxygen from the air was sealed out, preserving them and their soft tissues until their discovery only a few years ago.

In 1995 Chinese scientists announced that they had dug out a fossil which proves conclusively that some dinosaurs did indeed have down-like feathers. Sinosauropteryx caused a sensation. It was a small, 1.5-metre-long, two-legged dinosaur with the jaws and flattened teeth typical of meat-eating dinosaurs. It had clawed fingers, and its hind limbs show that it must have been a fast runner. At last, the puzzle of where birds descended from seemed to have been solved. They are indeed the last surviving branch of the dinosaurs. But this still left several questions unanswered. Sinosauropteryx's feathers show no signs of the aerodynamics needed for lifting a body into the air. It couldn't fly.

So what were its feathers for? Experts began to think that perhaps feathers were not originally intended as equipment for flight at all. Perhaps it was only later, by some freak or accident, that they became adapted for the purpose. Small dinosaurs may have developed feathers as a way of keeping warm – a form of insulation.

Feathers are now thought to have been common on all kinds of dinosaurs. The reason this has only recently come to light is that feathers are not normally preserved in fossils.

Lots of dino-birds have now been discovered in China's treasure trove of fossils and bones. One of the most famous is Dave, the fuzzy raptor, unearthed in 2000. He was covered from head to tail in a coat of fine feathers, although they weren't designed for flight.

Archaeopteryx lived c.140 million years ago and was the ancestor of modern birds. It is the oldest creature known to have adapted feathers for flying.

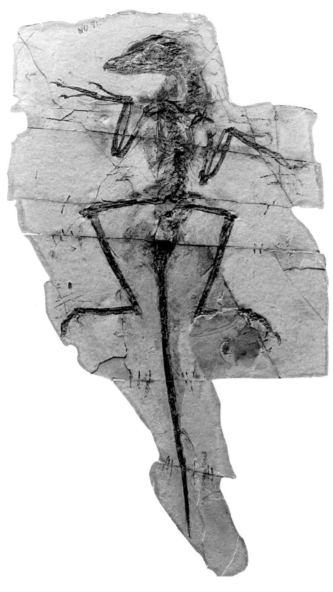

This 130-million-year-old fossil, discovered by Chinese farmers, shows a dinosaur with its feathery body covering intact, linking the evolution of birds to the dinosaurs.

sort of claws for climbing trees. But just how and when its feathers of flight evolved is still a mystery, whose answers probably lie somewhere in rocks still waiting to be found.

We are now roughly a hundred million years from the present day, that's about 11.20 p.m. on our twenty-four-hour clock, and at last we see the first signs of civilization. Not human civilizations these, but insects in the form of wasps and their descendants, bees and ants. There were also termites, a different branch of life altogether, related to beetles. From these creatures new, more advanced social ways of living developed that became the prototypes for all subsequent civilizations to evolve on earth.

Wasps arrived with the first dinosaurs, back in the Jurassic Period (200 to 145 million years ago), although they evolved fastest after the appearance of flowers in the early Cretaceous. Many wasps lived alone, but some developed simple social habits, which became more pronounced in bees and ants.

Paper was first manufactured by social wasps tens, if not hundreds, of millions of years ago. A queen wasp hibernates over winter, and emerges in early spring to find a suitable nesting site. She starts by constructing a paper nest, using wood fibres which she softens by chewing and mixing them with the saliva in her mouth. The paper she makes is used to construct cells in a comb, each one housing an egg which grows into a female worker wasp. After hatching, the workers finish off the job of building the rest of the nest on the queen's behalf.

Other than starting off the colony, the queen of wasps has no special status in the group, so these colonies represent only the most basic forms of ordered insect society. Bees, the descendants of the wasps that emerged alongside the first flowers, switched from dining on other insects to a diet of pollen and nectar instead. There are about 20,000 different species of bee alive today. Some of these – especially honeybees, bumblebees and stingless bees – formed highly social groups, offering a deep insight into how nature's civilizations work.

Being 'eusocial' means that a group of living things divide jobs up between themselves. They pass knowledge and learning on from one

The final part of the bird riddle remains unsolved. How and when did they adapt their insulating feathers into wings for flight? Did they jump from the trees after spying a tasty meal from on high, swooping down to snatch their prey? Or did they run along the ground, perhaps trying to escape some other hungry beast, and flap up to safety? No one is certain, although another recent fossil from China's 'Pompeii', a microraptor, suggests the 'top-down' theory, because it has both the right kind of feathers for flight and the right generation to another, care for their youngsters and even, in certain circumstances, sacrifice their lives for the benefit of the group. Such characteristics were for a long time thought to be unique to mankind when it first organized itself into tribes and eventually cities and states. But, as any beekeeper will tell you, that is not so.

Queen bees are in overall command of a group of male 'drones' and a third set of sterile female workers. Bees communicate with each other through the language of dance. When they return to their hives they dance to inform the others as to the whereabouts of good sources of food. The 'round dance' means that food is within fifty metres of the hive. The 'waggle dance', which may be vertical or horizontal, provides more detail about both the distance and the direction of the located food source. Then there is the 'jerky dance', used by the bees to decide whether to increase or decrease the amount of food gathering they need to do, depending on the hive's overall requirements.

Bees are so socially advanced that they even vote on critical issues such as relocating the hive. Each spring about half the bees depart with the queen to begin a new colony elsewhere, leaving a number of other potential queens to vie for control of the existing nest amongst themselves. But where should the emigrants locate their new nest? Getting this right is crucial for their future well-being. It must be not too exposed to predators such as birds, not too far from good sources of food, nor liable to be flooded or blown away by the wind.

Usually bees will identify as many as twenty possible locations within a one-hundred-square-kilometre area, and about 90 per cent of the time they appear to choose what seems to be the best place. The system for their decision-making is remarkably effective. Each bee has the chance to vote, giving the colony the best chance of making the right decision for its future survival. About 5 per cent of the bees scout out to find the prospective sites, reporting the location of each back to the rest of the community using their various dances. Other bees then check out these sites, returning to the nest and dancing longer and harder in the direction of the site they think is best.

After about two weeks the site with the strongest, most vigorous dance is the winner.[4] Having voted with their feet, the bees then swarm.

Bees weren't the only remarkable civilizations that began to evolve back then. Ants belong to the same family as honeybees: the oldest ant fossil yet discovered is a specimen trapped in amber, estimated at more than eighty million years old. Their civilizations include the earth's first schools, what look like the beginnings of slavery, and even a bizarre attempt at behaving like an early type of computer. There are many similarities between ants' nests and honeybees' hives, but ants do not dance. Instead, they communicate through chemicals called pheromones that other ants can smell. When an ant finds food it will leave a trail of scent along the ground all the way home, to lead others to its source. It then finds the way back by remembering certain landmarks often using the position of the sun as its guide. As other ants follow the first trail each one leaves more scent, until the food source is completely exhausted. Once the ants no longer leave a scent the smell evaporates, and the trail is lost for ever.

Ants' smells say other things, too. For example, if an ant gets squashed, its dying gift is a smell that triggers an alarm to all the other ants nearby, sending them into a mad panic as they run around in a rush trying to avoid the same fate. Ants mix pheromones with their food so they can pass on messages to each other about health and nutrition. They can even identify themselves through smells as belonging to different groups or castes within the nest, each of which is responsible for certain activities in the civilization. The queen produces a certain pheromone without which the workers would start to raise a new queen. It acts as a safety device in case the queen dies.

Ants were also the first creatures we know of that learned to teach each other. When a young ant makes its first journey out of the nest an older ant will tutor it in the art of finding and fetching food. These tutors slow down to allow the pupil to catch up, and speed up when it draws close (this behaviour has been observed in the species

Temnothorax albipennis). They have even been known to link together over gaps to form chains, allowing other ants to scale streams of water like an army with its own portable pontoon bridge.

The ingenuity of ants' teamwork knows almost no bounds. Weaver ants build nests in trees by attaching leaves together. Leafcutter ants are gardeners. They collect leaves which they feed to a special fungus that grows in gardens in their nests. The ants then dine off the fungus when it is ready to eat. Sahara desert ants have an incredible system for finding their way back home in a landscape without landmarks. They seem to be able to keep track of the number of steps they have taken, so they can turn round and, recounting each step carefully, find their way back home.

But in the world of ants, just as in the world of humans, teamwork has its limits. Some species live off attacking other colonies, leading to epic battles. Huge ant wars can break out between different colonies, sometimes leaving thousands dead on the battlefield. The victors take booty in the form of eggs and larvae, which they then raise as their own workers and slaves. Amazon ants are incapable of feeding themselves, relying completely on captured ants to provide for and serve them.

Leafcutter ants busily gathering food to feed their fancy for fungi.

Termites evolved alongside the dinosaurs during the Jurassic Period – fossilized nests are thought to date from as far back as 200 million years ago – and grew more numerous from the Cretaceous Period onwards. These creatures can create the biggest micro-cities of all. Termites often live in colonies that number several million individuals. A termites' nest is a monarchy, but this time the king rules with one or more queens by his side. A pregnant queen can lay several thousand eggs a day. She gets so large (sometimes up to ten centimetres long) that she is often unable to move. If she needs more space, thousands of worker termites heave her up and push her to a newly built chamber. In return she rewards them with a special form of milk.

Worker termites have a number of important jobs. They look for food, create storage areas, maintain the nest and defend the kingdom against attack. They also serve as mobile canteens. They are the only termites in the kingdom able to digest plant food, which they then pass round to all the others, either from their mouths or their bottoms. It is a hallmark of any advanced civilization to care for and feed one another.

Workers build their nests using a combination of soil, chewed wood, saliva and dung – a mixture

not that different from the daub walls early man used to build his first huts and houses more than 8,000 years ago. These walls are extremely tough, and can sometimes be as high as a double-decker bus. This world incorporates many of the mod cons you would expect of an advanced society, including air conditioning (provided by special tunnels that control the temperature), water-collection systems and fungus gardens. The fungi provide food for the termites; in return the termites spread spores in their dung.

Soldier termites repel attacks from their worst enemies – the ants. Some have jaws so large they cannot feed themselves, so they are cared for by other workers. They defend the nest by squirting poison from their heads. Others use their large heads to block narrow tunnels to prevent ant attack, providing the equivalent service of a heavy locked door. When under attack, reserve battalions of soldier termites line up behind the front one. When a soldier is killed, another immediately takes its place. If the ants do manage to breach the nest, the soldiers will form a line shooting toxic liquid at them while others repair the breach from behind. Once the hole is filled the brave defenders in front have no escape route: their fate is to die for the sake

of the others, but their efforts will most likely have saved the nest from annihilation by ants (for a parallel in human history, see page 178).

Termite civilizations show very high levels of collective intelligence. Compass termites build tall nests that point north–south, to help drive hot-air currents through their elaborate network of tunnels, because temperature control is vital for the good establishment of their growing gardens of fruiting fungi.

The story of civilized insects is largely untold in histories of the earth, but their worlds go on around us, with important consequences. Sometimes they are under the control of mankind – bees pollinate our orchards and make us honey – and sometimes they cause humans annoyance and distress. Termites are seen by scientists as a source of future energy supplies because of their efficient fermentation process that can produce up to two litres of hydrogen per sheet of paper. However, they can also burrow so deeply into a wood structure that they can cause buildings to collapse. Whatever their reputation today, these creatures' powers of organization, intelligence and self-sacrifice for the greater good of their kind are the very hallmarks of what we humans have come to call civilized.

23:14:48

Prime Time

How a minor family of nocturnal forest dwellers became the next masters of life on land, spilling out on to the earth's drifting continents and morphing into a new cornucopia of species.

HAD THAT ASTEROID missed the earth 65.5 million years ago, would many of the animals we know and love today, including mankind, ever have evolved? No one can possibly know the answer, because life on earth cannot be rerun through a computer program to see what would have happened if the dinosaurs had survived.

What *is* clear is that evolution took a dramatically different path following their extinction. The disappearance of the dinosaurs provided huge new opportunities for another group of animals to take centre stage and become the next masters of the earth. When the dinosaurs were in charge, mammals clung on at the edge of the world's families of living creatures. Although the mammal-like dimetrodon dominated the land for millions of years before the Permian Mass Extinction finished it off, its successors, the mammals, were almost completely eclipsed by the rise of the great lizards.

Mammals back then were mostly small and squirrel-like. They emerged from their underground burrows only when it was safe to do so, often at night. When the coast was clear, they'd scuttle off to feed on insects – easy food for ground dwellers or for those who chose to hide out of harm's way safely tucked up in a tall tree.

Many of them developed evolutionary tricks to help survive the terrifying dinosaurs. Most gave birth to live young, so they didn't need to leave eggs lying around as food for someone else's breakfast. They were so scared of leaving their burrows that they developed their own personal soup kitchens, in the form of breast milk, so they could feed their young at any time, day or night, without leaving the nest. They also grew fur, which helped them keep warm. Finally, warm-bloodedness let them hunt at night when it was cold outside, and safer from the predatory dinosaurs. So when

disaster struck 65.5 million years ago, these small, furry mammals were well placed to survive the impact and the dreadful blackout that followed.

Once the dust had settled, mammals could enjoy the sunshine without the threat of being eaten by reptilian monsters. It was as if the gates of a dark, terrifying prison had suddenly been flung open and the mother of all curfews had come to an abrupt end. Birds were now the only living branch of the dinosaurs left, filling the world with song as the earth moved into a fresh new dawn.

The last 65.5 million years of our journey represent just over twenty minutes on the twenty-four-hour clock. What's to come is the saga of how humble, herbivorous mice were transformed into mighty, meat-eating men.

Things got off to a quick start. Within three million years shrew-sized mammals had evolved into creatures as large as dogs. Within five million years, mammals of all shapes and sizes roamed the land.[1] With the dinosaurs gone the land was a place

of relative peace and tranquility, somewhere nature could adapt itself to fill every available niche for new life.

This period, from fifty-six to thirty-four million years ago, is called the Eocene (see the geological column on page 34). The name comes from the ancient Greek word meaning 'new dawn'. Extraordinary mammals filled every available corner. In the forests there roamed deer, boars, bears, koalas, pandas, monkeys and apes. On the grasslands grazed cattle, bison, oxen, sheep, zebra, pigs, horses, giraffes, kangaroos and donkeys. In burrows dwelt rabbits, badgers, hedgehogs, mice and foxes. In rivers and swamps splashed rhinos, elephants, beavers, otters and hippos. In the deserts trekked camels, llamas and rats. In the oceans swam whales, dolphins, seals, walruses. And in the air flew bats.

How did these new rulers develop into such an extraordinary and bizarre range of different creatures? Could an elephant, a hippo and a giraffe really be descended from the same common

Horses evolved in North America from this dog-like creature, the hyracotherium. They grew into larger animals from c.35 million years ago as grasslands replaced forests.

ancestor as a mouse, a rat and a bat? Are whales and dolphins really distant cousins of the koala and the kangaroo? How all this happened has a lot to do with the earth's restless crusts.

By this time Pangaea had split up into two large continents, Laurasia and Gondwana. These were now in the process of splitting up themselves to form the large continents we know today. Africa, Asia, Europe, North and South America, Australia and Antarctica were slowly drifting into isolation, each one carrying its own precious cargo of plants and animals. An underground rift in the earth's crust started spewing out massive upwellings of lava that were (and still are) pushing the giant continental land masses of Europe and America further away from each other.

It was this same heaving that pushed Britain – which had been under the sea for sixty million years – back above the waves, finally surfacing about forty million years ago, like a giant submarine bobbing up for air. Britain's long underwater history explains why so much of its land and shores are covered in chalk. Layer upon layer of calcium carbonate was deposited by microscopic sea creatures – a bit like the inside of a kettle being caked in a little more limescale with each boil.

As the continents drifted apart, islands were born and cargoes of plants and animals adapted to their new environments. This is why, as the generations went by, they evolved into huge numbers of new and varied species. Biodiversity increases through geographical isolation.[2]

Mammals evolved into three main groups. Two of them, including nearly all of today's mammals, originated in Laurasia, the north part of Pangaea, and one from the south, Gondwana. It is only the monotremes that survive from Gondwana. They are a very strange lot. And the most famous of them all is very strange indeed.

Monotremes

Is it a duck? Or perhaps a weird fish? Or maybe a bird? Or even a beaver? The question baffled experts when the creature was first discovered by Europeans in Australia in the late 1700s. To begin with the duckbilled platypus was dismissed as a hoax, a kind of half-bird, half-mammal. Surely someone was having a laugh: naturalists suspected that perhaps it was a duck's beak sewed on to the body of a beaver – they even tried to find the stitches with a pair of scissors. But by 1800 people accepted that this was something completely new, unknown but most definitely natural.

Duckbilled platypuses are now very rare and found only in eastern Australia and Tasmania. They have thick brown fur, webbed feet and a large rubbery snout. Their legs stick out slightly, causing them to waddle sideways when they walk. They are also one of the only types of mammal still to lay eggs. Monotremes like the duckbilled platypus branched off early in mammalian evolution, and do not seem to have changed a great deal since. But this creature is also famous because it is the only mammal to have an electric bill. In its beak is a series of about 40,000 electric sensors that 'feel' for food, as if its nose were a hand.

Duckbills are definitely mammals, because their babies are fed by milk produced by their mother's mammary glands. Platypus fossils dating back 110 million years show that these early mammals originally waddled over to Australia from South America before the two land masses fully broke up into separate island continents.

Until about forty million years ago, Antarctica, now 98 per cent covered in ice more than a mile thick, was a warm, balmy land of forests and fruit trees. Animals regularly wandered across the connected land masses, journeying from what is now Argentina to the eastern side of Australia via Antarctica. Until about eighty-five million years

The duckbilled platypus, for a while thought to be a hoax, is now one of the rarest creatures on earth and belongs to the only mammal family that lays eggs.

ago it was even possible to walk across land to New Zealand, because it was still attached to Antarctica. The biggest wanderers of all were the second main group of mammals which developed their own radically different way of life. These were the marsupials. The kangaroo and the koala are their most famous survivors today.

Marsupials

The name means 'pouch' in Latin. These mammals rear their young in a special pouch on their outer skin. They give birth to extremely tiny babies which crawl up the mother's belly and tuck themselves inside her pouch, latching onto her nipple for food and drink. There they stay for a number of weeks, until they are big enough to leave the pouch on their own, returning for warmth and food as and when they want.

Marsupial species developed in similar ways to other non-marsupial mammals that we are familiar with today. There is still a marsupial version of the mole, and a marsupial mouse. Once there were marsupial cats, dogs and lions on the Australian continent too. But their luck began to change with the arrival, about fifteen million years ago, of bats and rats which came over from Asia as it drifted close enough for them to fly or raft across the sea. Then, from about 40,000 years ago, men arrived in

canoes. In time they would introduce dingoes, rabbits, camels, horses and foxes (for hunting), all of which have undermined the native ecosystem for indigenous Australian and Tasmanian marsupials.

The saddest story of all is probably that of the Tasmanian tiger, one of the world's largest ever meat-eating marsupials, that looked like a cross between a wolf and a large, predatory cat. When it was first discovered by Europeans in 1792 there were only about 3,000 left. It survived until the 1930s, when it became extinct after human bounty-hunters killed the last remaining few for their hides. Extinct, that is, save for one last specimen, a female (bizarrely called Benjamin) who died in Hobart Zoo on 7 September 1936 – probably of neglect and exposure, because she didn't have any shelter. Shortly before she died a black-and-white film was made of her pacing up and down in her enclosure. As a final farewell, she bit the cameraman on the backside.

Placentals

The third, and by far the biggest, group of mammals is the placentals. These creatures (to which humans belong) keep their babies inside their bodies until they are well developed. The baby's blood and the mother's blood are brought into close contact by a food-exchange organ called the placenta, through

Benjamin, the earth's last Tasmanian tiger, who died in 1936.

23:20:55

which nutrients pass from the mother to the baby, and waste passes from the baby to the mother's blood. The first placentals to evolve, the afrotheres, were probably relatives of today's elephants and sea cows – not that they looked like elephants, since all mammals in the early days were small, furry, and fed at night for fear of being eaten by dinosaurs.

One or two small members of the elephant family still survive – the elephant shrew, with its elongated snout, looks a bit like its much larger relative. Recent molecular analysis seems to confirm the link.[3] Today only two elephant species are left – the African and the Indian – although until about 12,000 years ago others such as the woolly mammoth and the mastodon roamed the ice caps and parts of northern America.

About fifty million years ago, several members of the same family as the elephants returned to the sea. These are the sea cows. They use their front feet for swimming underwater, and their tail as a rudder for steering. All that remains of their back feet are two vestigial bones that float deep inside their bodies. Like all mammals, sea cows breathe air, surfacing when necessary. But they never leave warm water, not even to give birth. Today's sea cows are in extreme danger of extinction because humans keep destroying their natural river and estuary habitats.

Placentals took a completely different evolutionary direction on the isolated land of South America, where the marsupials also lived. Ancestors of today's armadillos, sloths and anteaters established themselves as plant- and insect-eaters; this family of mammals is called the xenarthrans. South America also saw the evolution of the world's first ungulates. These are animals that learned to walk on their nails, more commonly known as hoofs. There was one creature called a toxodon, which looked like an African rhino. Charles Darwin identified bones from this beast on his voyage to South America in the *Beagle*: 'Perhaps one of the strangest animals ever discovered … In size it equalled an elephant.' There were also horses and camel-like animals (litopterns). All these South American ungulates are now extinct.

But ungulates weren't found just in South America. They also evolved independently from about fifty-four million years ago in Africa, where they fared a great deal better. They have survived to become today's deer, sheep, antelopes, pigs, goats, cattle, giraffes and hippos. Some of them also returned to the water. Today's whales and dolphins are actually related to ancient seafaring hippos that lost contact with the land, shedding their legs in a process of evolutionary 'tidying up'.

Meanwhile, in North America (not yet connected to its southern counterpart), ungulates evolved for a third time. Their descendants are today's horses, camels and rhinos. These 'odd-toed' ungulates – so-called because a horse walks on one toe, while rhinos walk on three – include the hyracotherium, a tiny horse only about the size of a small dog (see the picture on page 71). Originally these creatures lived in forests, but as the climate dried out, from about thirty-five million years ago, many forests were replaced by open grasslands. Open spaces meant that small horses were more exposed. Only those with longer legs, that could gallop more successfully out of harm's way, survived. So, over millions of years, the dog-sized hyracotherium evolved into today's much larger and faster steeds.

Camels also originated in North America. They eventually spread to other parts of the world, along with horses, crossing a land bridge from Alaska into Asia and then into Mongolia, finally reaching the Asian steppes and the deserts of the Middle East about two and a half million years ago.

Fat-rat: the giant capybara lived in South America from c.five million years ago. This rodent grew as large as a donkey.

Next came paws. Carnivora is a mammal family whose living descendants include cats (e.g. lions and cheetahs), dogs (e.g. wolves and jackals), weasels, bears (including the panda), hyenas, seals, sea lions and walruses. All these creatures are descended from a common ancestor that evolved rapidly from about fifty million years ago into these different forms we know today.

Fossils from bats, the only flying mammal, also date to this time.[4] Bats developed an ingenious way of 'seeing' in the dark by producing high-pitched surround-sound signals by clicking their tongues. Echoes of these sounds allow bats to work out exactly where everything is, even at night. Bats 'see' in the dark by calculating how long each sound echo takes to bounce back, building a picture in their brains of where objects are located. The less time the echo takes, the nearer is the object. A similar system called sonar is now used by humans for navigating underwater in submarines. But bats developed the world's first sonar system a good fifty million years earlier.

Shortly after the paws came claws. These are the rodents – today's descendants include mice, rats, rabbits, hares, gerbils, voles, hamsters, beavers, squirrels, marmots and guinea pigs. They are notorious for spreading disease such as plague, including the dreaded Black Death (see page 265). In the last 1,000 years rats are reckoned to have carried diseases that have killed more people than all the human wars and revolutions put together.[5] In South America some rodents grew to an enormous size, such as the giant capybara (called a proto-hydrochoerus), which grew as large as a donkey (see opposite).

Primates

These were the world's experts in swinging, leaping and jumping from one tree to the next. Life was safest in the trees, far from carnivorous reptiles like crocodiles, which survived the fate of the dinosaurs and still oozed through the swamps below. Fruit and nuts were within easy reach. And up in the trees the view was much better. Getting around was also easier than on the ground, provided you had the skills of a leaping Tarzan.

Primates appeared on the island of Madagascar, which broke off the tip of Africa 165 million years ago. It now contains some of the most protected and endangered species of plants and animals in the world. Descendants of these primitive primates include lemurs. Their name comes from the Latin word for 'spirits of the night', and like many early primates they are mostly nocturnal creatures, with excellent night-time vision. Lemurs have survived thanks to Madagascar's isolation, protecting them from their descendants, monkeys and apes, which later overran them on the mainland of Africa. Lemurs are, however, endangered species today because of illegal logging, destroying their natural woodland habitat. Bush babies are other nocturnal tree-dwellers. Big eyes, excellent hearing and long tails adapted them perfectly for a life in the shadows of the forest. But the biggest group of all was the monkeys. They evolved into three main varieties: 'Old World' monkeys, 'New World' monkeys, and apes.

Modern science has found that the genes in monkeys and apes are very similar indeed. Although monkeys have tails and apes do not, biologically they are all part of one family.

Monkeys first evolved in Africa, but they now live in both Asia and South America. How did they get there? There was no overland route, they can't fly and they can't swim – at least, not across an ocean thousands of miles wide. That means there is only one possibility. They must have hitched a lift. About twenty-five million years ago one or more groups of African monkeys somehow found

Bats are mammals that fly. They 'see' in the dark by clicking their tongues and listening out for echoes.

Mammal Migrations

Without the dominance of dinosaurs, the range and diversity of mammals grew rapidly. They migrated throughout the world as the continents continued to shift and sea levels fell.

Camels migrated from North America reaching Asia and South America three m.y.a.

Ancestors of today's horses evolved in North America fifty m.y.a. and eventually arrived in Asia via Alaska three m.y.a.

Primates migrated through Africa fifty m.y.a. and then rafted to South America twenty-five m.y.a., roamed to Asia eighteen m.y.a. and migrated back to Africa evolving into gorillas seven m.y.a.

Marsupials hopped across South America and Antarctica to Australia fifty-five m.y.a. when the continents were still attached

m.y.a. = million years ago

Sea cows evolved from the same ancestors as elephants and returned to the water fifty m.y.a. Whales and dolphins returned to the seas thirty-five m.y.a.

A technique called brachiation is what lets these apes 'walk' from branch to branch using only their arms.

themselves on a raft bobbing across the Atlantic Ocean, eventually to be washed ashore on the coast of modern-day Brazil. This is when the first South American monkey fossils appear, providing evidence that by then they were thriving up in the trees of a newly adopted home.

These New World monkeys are distinctive for their flat noses, and for using their tails to help them swing and balance in the trees. They can happily hang from a branch by their tailbone alone, using it like an extra hand.

As if one perilous transatlantic crossing wasn't risky enough, other African monkeys set off in the opposite direction trekking though Asia, across the land bridges of the Middle East and Arabia. Gibbons and orang-utans are found today in places such as India, Malaysia, China, Borneo, Indonesia and Java. Whether they lost their tails before they left, on the way, or later, isn't entirely clear.

Gibbons are the world's best acrobats. Their incredible skill comes from a technique called brachiation – the word comes from the Latin *brachium*, meaning arm. These creatures specialize in being able to 'walk' from branch to branch using nothing more than their arms, and being able to glide upright across dazzlingly high wooden tightropes. The secret of their acrobatic success lies in their wrists, which have a specially adapted ball-and-socket joint, allowing complete freedom of movement sideways as well as forwards and back.

Asian orang-utans and African gorillas are two of the four members of the great ape family. Orang-utans live in the forests of Indonesia and Malaysia. They are highly intelligent and very like humans – in fact their name comes from the

Malaysian words *orang hutan*, meaning 'person of the forest'. These apes, which evolved about fourteen million years ago, live only in the trees. Every evening they build a new nest for themselves and their families.

Like humans, these playful animals have a sophisticated language and culture.[6] Just before bedtime they have been seen blowing raspberries at each other, the orang-utan equivalent of saying 'night-night'. They also play games like surfing down fallen dead trees, grabbing as many leaves as they can on the way. Wild orang-utans have even been known to visit their orphaned relatives in rescue centres. They seem to communicate with them, and when the orphans are released they help them re-adapt to a life back in the trees.

Recent genetic evidence and fossil finds suggest that African great apes evolved from species such as gibbons and orang-utans, which once lived only in Asia. This means that about ten million years ago one of these groups of apes, either the orang-utans or the gibbons, made their way back across the

Asian land mass into Africa.[7] Here they evolved into today's gorillas and chimpanzees.

Gorillas are mild-mannered vegetarians that live on grasslands, and not up in the trees. Only two species survive today. Both are endangered. Several hundred died in 2004 of ebola virus, for which there is currently no known vaccine or cure. Gorillas are highly intelligent. Koko, born in 1971, is a captive female gorilla living in California. She has been taught sign language from the age of one. Her trainer, Dr Penny Patterson, claims she can communicate using a vocabulary of up to a thousand words. Something of a scientific debate has been raging for years since Koko first showed off her language skills. Does she really understand what she is saying? Or is she just prompted by the prospect of a reward if she says the right thing? In August 2004 Koko indicated that she had a toothache. According to her handlers she communicated that she was in pain. She could even indicate its level on a scale of one to ten.

Koko shows other human-like traits, too. She is one of the few other animals ever to have been known to keep and care for a pet. In 1984 Koko asked for a cat. She selected a grey male kitten from an abandoned litter. She named him All Ball. Koko cared for the kitten as if he were a baby gorilla, until All Ball escaped from Koko's cage and was hit by a car. Koko cried for the next two days. Since then she has adopted a number of other pets, including two more kittens, Lipstick and Smoky.

Humans are apes. Until the 1960s it was thought that mankind split from apes about twenty million years ago – mostly because so few fossils had been discovered to prove, one way or another, what happened and when. There was also a strong feeling that the split had to be at least that far back, or there could never have been enough time for us

Koko with her pet cat, All Ball. She cared for him as if he were a baby gorilla.

humans to have evolved into such apparently superior beings. We talk, we build things, we invent amazing machines, we are clean (generally), ingenious, and we appear to have mastered nature, tailoring it to our own ends.

But in the early 1990s, molecular biologists discovered that we humans share at least 96 per cent of our genetic code (DNA) with the other great apes (chimpanzees, gorillas and orang-utans). Their analysis showed that humans are descended from an ape which probably lived sometime between four and seven million years ago – just ninety seconds before midnight on our twenty-four-hour scale of all earth history.

This ape's offspring divided into chimpanzees and their cousins the bonobos on the one hand, and into early human beings on the other. Exactly who this ancestor was, and where he or she lived, is one of the greatest mysteries in all human knowledge, one that is still waiting to be solved.

23:20:55

Part 2
Homo Sapiens
(7 million–5000 BC)

HUGE CLIMATE CHANGES caused by massive collisions of the earth's continental crusts created ice caps that plunged much of the planet into brilliant white winters lasting thousands of years. As the temperatures fell and rains reduced, grasslands replaced trees and forests, birds evolved new patterns for migration, and animals either adapted or died.

About four million years ago apes, a branch of primates living in the trees, began to experiment with a new way of life, venturing into the now extensive grasslands, where some of them mastered the trick of walking on two feet rather than four. Their brains swelled with the experience of using their now freely available hands for making tools to help them survive the harshest conditions that nature's climate changes could throw. Two-legged chimps evolved rapidly into creatures like you and me – apes that learned how to talk, sing and light campfires. They even learned how to draw.

For at least two million years several different species of humans roamed the earth's continents, travelling from place to place, carrying with them their small children and sharing their possessions – the fewer the better, to lighten the load. There were no laws, no private property, no places where it was forbidden to go. Living within nature meant giving the land time to recover by moving on. Human populations were stable, reaching a global peak of about five million individuals.

Disaster struck in Australia, and then in the Americas, where between 40,000 and 10,000 years ago most large mammals suddenly disappeared into extinction, leaving these ecosystems without the benefit of some of nature's most powerful creatures, such as cattle, horses and camels. Then, about 12,700 years ago, a sudden period of dramatic climate change following the last Ice Age melt caused people living in the Mediterranean and the Middle East to attempt new, experimental ways of living in a desperate bid for survival. They discovered how to manipulate nature through artificial selection, learning to cultivate easy-to-farm crops and to tame animals for keeping in captivity.

Once the climate stabilized, some people stuck to the wandering way of life, but now they could take their newly domesticated flocks of cattle, sheep, pigs and goats with them for food. Others built more permanent settlements in towns and cities, their new-fangled farming providing surplus food for feeding people who were no longer tied to the land. Enter the world's first priests, kings and administrators, artisans, merchants and domestic slaves. With this restless march into the brave new world of human civilizations, the traditional relationship between mankind and the rest of the natural world began to change beyond all recognition.

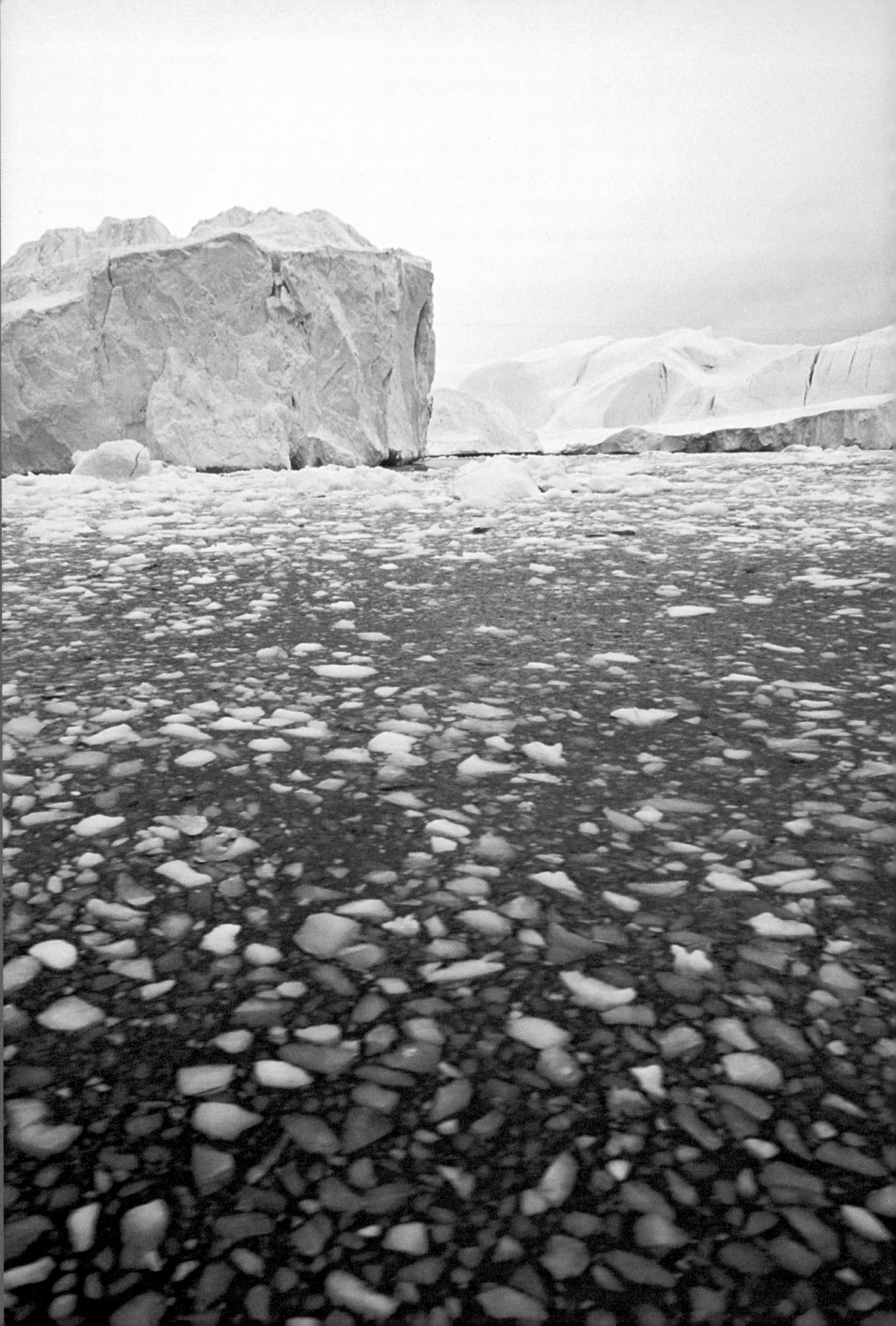

11

Ice Box

How climate changes caused by cyclical variations in the earth's rotation and random movements of the continental crusts created expansive grasslands and caps of polar ice.

REMEMBER THOSE old-fashioned rides at traditional funfairs? Carousels, the helter-skelter, and, best of all, the dodgems or 'bumper cars'? Bizarre as it sounds, we are all riding on top of the earth's very own natural set of dodgems. They aren't quite like those in man-made funfairs, because the earth's bumper-car ride travels extremely slowly. Also, whenever its dodgems crash into each other they seem to have a rather dramatic impact on the world's weather.

Welcome to nature's fairground. Gradually, over the course of the last 200 million years, the earth's restless crusts of land have broken up to form today's giant continents. Actually, this was quite good for life on earth, because it made it much harder for any single type of living thing, like the dinosaurs, to dominate the land. Seas and oceans forged natural barriers as the crusts split apart. New types of living creature evolved, many learning to thrive in different habitats with less competition. Scattered continents also increased the number of beaches, sea shores, wetlands and salt pans – all serving to bolster life on earth.

Since the extinction of the dinosaurs 65.5 million years ago, the amount of carbon dioxide gas in the earth's atmosphere has plummeted from nearly 3,000 parts per million (ppm) to just 284 ppm by 1832 (to see how they have risen in the last 175 years as a result of human activities, see page 377). Less carbon dioxide (CO_2) reduced global temperatures significantly during the same period, despite the sun warming by as much as 30 per cent since the beginning of our twenty-four-hour journey. Such changes in the atmosphere and the temperature have, in large part, been caused by the constant collisions of the earth's crusts, which, although they move slowly in human timescales, have had dramatic consequences for life on earth.

Polar ice floes, such as this glacier off the coast of Greenland, have been a big feature of the earth's climate for the last forty million years.

Continental Collision

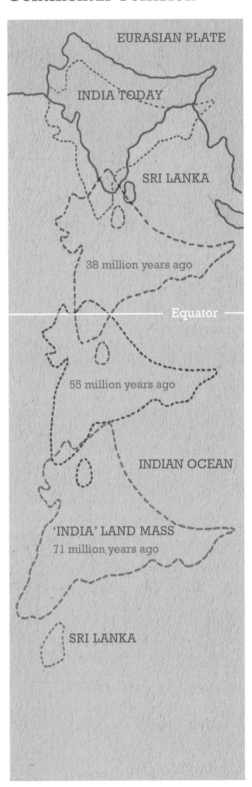

After a record-breaking ride across the oceans, India smashed into Asia, forming the Himalayas.

EURASIAN PLATE

INDIA TODAY

SRI LANKA

38 million years ago

Equator

55 million years ago

INDIAN OCEAN

'INDIA' LAND MASS

71 million years ago

SRI LANKA

About ninety million years ago, the Indian continent sheared away from Africa, becoming, in effect, a giant dodgem on the loose. It raced north at the unprecedented speed of fifteen centimetres a year, swivelling upwards and crashing into Asia about forty million years ago, having travelled 3,000 kilometres. Scientists believe it moved at such a speed because the plate is so thin compared with most others. Having crashed into Asia, India is now moving north-east at the more leisurely pace of five centimetres per year.

When it collided with the larger, slower-moving Asian continent, it created the biggest mountain range on earth. The Himalayas, and the enormously high plateau of Tibet, are thought to have been responsible for reducing global temperatures in a dramatic way, by removing large amounts of CO_2 from the atmosphere.[1]

Air cooled by the high Himalayan peaks condensed huge volumes of water vapour from the warm Indian Ocean into massive monsoons, which tipped their loads on to India and the southern part of the Tibetan plateau. As the rains fell, carbon dioxide in the air dissolved into the water, eventually to wash off via rivers and streams and to settle as sediment on the sea floor. With the CO_2 now safely removed from the atmosphere, the earth got cooler still.[2]

Meanwhile other land masses were slowly but chaotically making their way across the globe, crashing together with no less dramatic results. At about the same time that India rammed into Asia, Africa pushed up against the seas that separated it from the same continent. The ocean floor buckled upwards, forming a series of land bridges across the ancient Tethys Sea, a stretch of water that once connected today's Middle East to the Indian Ocean. These bridges are what probably gave the African monkeys their land route into Asia, where they then evolved into the first members of our family – the apes (see page 78). The first horses and camels came across from North America into the grasslands of Asia, via another land bridge that connected Alaska to the eastern tip of Russia, allowing them to settle finally in the deserts of the Middle East.

Africa bounced off from this collision, pushing northwards into Europe – a process that threw up the European Alps, which stretch from France through Switzerland, Italy and Austria. Then, in the middle of this multiple vehicle pile-up, which began about twenty million years ago, Africa crunched up further into the Middle East, closing up the Tethys once and for all (it was eventually reconnected by the building of the Suez Canal, see pages 177, 356).

By six million years ago Africa had come so close to what is now southern Spain that its enormous weight threw up a new range of mountains that encircled the Mediterranean Sea by land on all sides. With no connection to the Atlantic Ocean, the water in this enormous lake gradually dried up, leaving layer upon layer of dirty white sea-salt on its bed. In some places today, these salt deposits are more than a mile thick, leading experts to believe that this basin may have dried up and refilled as many as forty times over the course of more than a million years.[3] Here the earth's natural process for removing massive quantities of salt from the sea was working overtime, powered by the movement of the earth's crusts in a fairground ride for giants (see pages 25–26).

Finally, about 5.3 million years ago, water burst for the last time through the chain of mountains which once joined Spain and North Africa. Colossal cliffs plunged downwards perhaps as much as 3,000 metres to the valley floor – that's more than fifty times higher than the Niagara Falls. Just a hundred years later the whole Mediterranean basin was filled with water once again. With 170 cubic kilometres of water cascading down the huge waterfall every day, this must have been one of the most extraordinary and dramatic episodes in the natural history of the world.

The earth had already slipped into an Ice Age thanks to another land mass on the loose, on the other side of the world. When Antarctica split away from South America, about forty million years ago, it slid south towards the South Pole. As it broke away a new sea channel opened up, known today as Drake's Passage after its discovery by the famous English explorer, pirate and circumnavigator Sir Francis Drake (c.1540–96).

Cold water from the Southern Ocean was now forced to circulate around Antarctica instead of moving north to mix with the warmer waters of the Pacific and Indian Oceans. As the region cooled, a massive ice sheet developed over this once tropical land. Today's Antarctic ice sheets are more than a mile thick, and stretch across a land mass fifty times larger than the United Kingdom. Such an enormous ice desert caused sea temperatures to plunge by as much as ten degrees, its bright white ice reflecting the sun's rays back into space, causing temperatures to cool even more. With a gleaming ice cap on its South Pole, the earth now entered a new Ice Age – its first in at least 250 million years – all due to the chaotic journeys of its out-of-control land masses.[4]

Cooler temperatures tend to mean that less water evaporates from the seas, resulting in reduced rainfall in many inland areas. Such conditions gave

The Antarctic ice sheet as seen from space at its winter maximum, covering Antarctica and parts of the surrounding oceans.

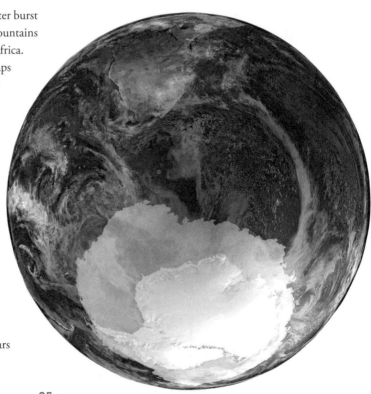

23:46:48

The pampas of Patagonia, South America. Grasslands replaced forests as the earth grew colder and drier.

birth to the great grasslands of today, which replaced forested areas that need larger quantities of rainwater to sustain them.[5] The prairies of North America, the pampas of Argentina and the grassland steppes of Europe and Central Asia all emerged during this Ice Age.

Animals adapted to these new conditions, and evolved into larger species and more numerous herds as competition for survival increased in the wide-open spaces. The spacious grasslands offered feeding grounds for birds, which rose to the challenge of the conditions of an Ice Age by migrating in vast flocks to find the best (and avoid the worst) places to live. Grasslands rich with seeds provided convenient stopping-off points for feeding. Blackbirds, meadowlarks, sparrows, quails and hawks all thrived thanks to these new steppes, which eventually came to cover more than 25 per cent of the world's land surface.

The last continental dodgem of significance in the earth's giant fairground ride was South America. About three million years ago it collided with its

North American neighbour, joining up via a thin sliver of land – today's Panama. The effect of this collision was as cataclysmic as the rest. For the first time ever, mammals that had evolved in complete isolation could now walk between these two continents.

Opportunities for some spelled disaster for others. Llamas, dogs, cats, lions, bears, horses and rats roamed down to the south, while armadillos, opossums and sloths moved up to the north. New predators were introduced, competition for feeding grounds increased. This event, called the Great American Interchange, reached its peak three million years ago. The biggest casualties were the great marsupials of South America, the pouch-bearing lion, and the rhino and elephant look-alikes whose fossilized bones still stud the earth. They were no match for the violent invaders from the north – the giant cats, dogs and bears. Extinctions were the inevitable result.

The collision of these two continents had its own dramatic effect on the earth's weather. Atlantic sea currents were forced northwards, and were prevented from mixing with the waters of the Pacific because their path was now blocked by land. As a result, a new weather system spluttered into life. The Gulf Stream began pumping warm air northwards, heating up north-west Europe by at least ten degrees. It also brought more water vapour, freshly evaporated from the Atlantic sea. As its clouds blew northwards towards the cold Arctic, rain turned to snow, which settled over time in layers on the cold seas, where it turned into thick packs of ice. The effect was that by three million years ago the earth had donned a second ice cap, this time in the Arctic.

With ice caps now on both poles, the earth grew colder still. Much of it was plunged, literally, into a deep freeze. Colossal ice sheets groped down from the poles, sometimes as far as modern-day London, Paris, Berlin and Moscow. They stretched past the plains of Canada, over the Great Lakes and as far as New York. Most of Russia and all of Greenland were iced over. For thousands of years the seas all around were thick packs of ice, many soaring skywards more than a mile high.

Sea levels plummeted because so much water was trapped as ice. It was possible to walk from where Dover, England, is today across to Calais, France. There was no English Channel, just flat, desiccated tundra. Even beyond where these enormous walls of ice ended, it was still too cold for anything much to grow.

In northern Europe temperatures often fell as low as –80°C, while winds whipped up to speeds of 200 miles per hour. Massive frozen rivers of ice were also on the move. They advanced like gigantic bulldozers, slowly but relentlessly scratching their way down the sides of the earth and then gradually retreating as temperatures rose again, melting as they went and leaving freshwater lakes in their wake. These glaciers carved out many of the world's most spectacular lakes and valleys – from the Lake District in England to the Swiss mountain valleys, and from the Great Lakes of North America to the fjords of Norway.

Conquests of ice like these have occurred as many as thirty times in the last two million years (see page 107). On each occasion huge boulders of rock have been carried miles away from their mountains of origin by the ice, rockfaces have been squeezed and scratched into the most contorted shapes, and the ground has been pushed down – only slowly to bob upwards again after the massive packs of ice fell into retreat. The United Kingdom, for example, is still rising after the retreat of the last massive ice sheet 10,000 years ago – but only by about one millimetre a year.

Life thrived despite these enormous and dramatic events, because nature's changes were slow enough for succeeding generations of species successfully to adapt. Not all life was affected. It was still warm in the tropics, and some rainforests remained, although they were far fewer than before the Ice Age began approximately forty million years ago.

It was now a cooler, grassier age. One where the seas rose and fell, opening and closing causeways to other lands, and where relentless thick packs of ice came and went like clockwork. It was into this dramatic world that mankind was born.

Food for Thought

How some creatures, called apes, came down from the trees, learned to walk on two feet, began to fashion tools for hunting and evolved into species with larger than average brains.

CHAD IS KNOWN as the 'dead heart of Africa'. Today the country is mostly barren, with broad, arid plains in the centre, a desert in the north, dry mountains in the north-west and tropical lowlands in the south. It is in the hottest, dustiest and most inhospitable part of this country that some of the most intriguing historical objects of all time have been found.

In July 2001 the skull of a creature that could be the ancestor of all humans was discovered by a team of four scientists led by a Frenchman called Michel Brunet. These bones are from a human-like creature – nicknamed 'Toumai', which means 'hope of life' in the local language – that lived approximately seven million years ago. Sometime

shortly before then, some experts believe, the human branch of the apes, *Homo*, split off from the branch that became today's chimpanzees.

The bones show a creature – its scientific name is *Sahelanthropus tchadensis* – with thickened brows, short teeth and a face that strongly resembles that of modern humans. Its head is small, measuring just 350 cubic centimetres (ccs), about the same size as a chimpanzee's. By comparison the skull of a modern human measures 1,350 ccs. Unfortunately it has been impossible to tell from just its skull if this human ancestor walked on two feet or four. Some scientists think this is the missing link between apes and humans, a transitory species, a kind of Eve from which all humankind descended.

Others think the bones are nothing more than the skull of an early species of female gorilla.

This find in Chad came only months after another spectacular discovery, this time in Kenya. Leg and arm bones dating back about six million years were discovered alongside teeth and skull fragments. This was a creature (*Orrorin tugenensis*) strong enough to climb trees, but did it walk on two legs or on all fours?

A protracted scientific controversy has raged since the discovery of these two sets of bones, because they seem to fly in the face of recent genetic analysis. According to what is called the 'molecular clock', the split between man and chimp cannot have occurred much more than about five or six million years ago. Any longer than this, and our genes – which are at least 96 per cent the same as chimpanzees' – could not be so similar.[1]

A recent re-examination of some bones that were found back in 1974 has provided genetic scientists with just the kind of fossil evidence they need to support their theory that the split between humans and chimpanzees took place more recently than seven million years ago.

Lucy lived in Ethiopia about 3.2 million years ago. She was discovered by an international team of scientists headed by Donald Johanson, an American fossil expert. On 30 November 1974, next to the Awash River, Johanson and one of his students, Tom Gray, were searching for human fossils when they found the fragments of an arm bone poking out from an upper slope. After scrabbling through the earth they found further relics – a jawbone, more arm fragments, a thighbone and ribs. Gradually, piece by piece, more than 40 per cent of a complete skeleton was unearthed.

They called her 'Lucy', because she was clearly a woman, and at the time of the find Johanson was listening to the Beatles' song 'Lucy in the Sky with Diamonds'. She was about 1.1 metres (three feet eight inches) tall, and weighed about twenty-nine kilograms (sixty-five pounds). When her discovery was announced to the world, Lucy caused a sensation, because scientists could tell from the shape of her pelvis that she was the earliest known ape to have definitely walked on two feet.

Four years later another team of fossil-hunters made a second extraordinary discovery in nearby Tanzania. At a place called Laetoli they found a series of footprints perfectly preserved in powdery volcanic ash that further backed up the idea that, like us, Lucy-like creatures walked upright.

Were these footprints made by a family of early humans walking towards a nearby watering hole? Soon after they passed, a volcano erupted, its mountains of ash preserving their tracks as trace fossils in the rocks. These footprints have been dated to 3.7 million years. There is no mistaking: they were made by creatures that walked on two feet.

Until the discovery of Lucy, scientists had assumed that it was because of the superior intelligence of early humans that they decided that walking on two feet was a good idea. Walking upright meant they could use their free hands to make technology, in the form of tools and weapons, that would help them prosper and survive. What was so surprising about Lucy was not simply that she walked on two feet, making her species a candidate for the earliest human ancestor, but that her head was not much larger than that of a chimpanzee.

Generally speaking, a smaller head means a smaller brain and less intelligence. What Lucy tells us is that walking upright came long before humans got big brains and big heads – probably long before they had the nous to work out

Toumai is thought by some experts to be the oldest known relic of the human race. Others think it's simply the skull of a female gorilla.

that walking upright was a smart idea. So what was it that made her want to walk upright, when a life on all fours must have made perfectly good sense? What's the big advantage of two feet over four? Escaping from danger or chasing after prey are just as fast on four feet as on two (witness the fleeing of a deer or the hunting of a cheetah), and it's just as easy to climb trees with four legs as with two and a pair of hands – look at a squirrel or a monkey. There are also some serious drawbacks to living a life on two feet. To walk comfortably, two-footed females have to have a much narrower pelvis, which makes childbirth extremely painful and much more dangerous to both baby and mother.

No one is sure what it was that made Lucy-like creatures stand upright. Perhaps it was as simple as the fact of squatting on the forest floor searching for food, which over time made their feet became flatter so they could balance better. Then, generations later, they got into the habit of walking upright. If this was the case, the long-term consequences of such an apparently simple evolutionary adaptation were probably the most monumental in all history.

A recent study has compared humans walking on treadmills to chimpanzees. It showed that walking on two feet uses only 25 per cent of the energy used by walking on all fours, suggesting that two-legged creatures may have enjoyed a significant survival advantage when times got tough.[2] Walking upright meant that these creatures could feed on the move – just as people do today with take-away hamburgers and hotdogs. With their hands free, these creatures could now carry food more easily to makeshift stores, helping them survive harsher environments. This gave them the confidence to spread out beyond the trees and hunt in the vast grasslands that were still displacing many forests as the world's weather turned worse.

Since Lucy's discovery the bones of many other similar creatures have been found. A recent discovery was of a three-year-old child called 'Salem', meaning 'Peace', unearthed in Ethiopia in the year 2000. Salem has since been painstakingly excavated to reveal a complete skull, collar bones, ribs and kneecaps.[3] These are the ancient remains of creatures called *Australopithecus*, the oldest of

which, 'Little Foot', dates back as far as 3.9 million years, after being accidentally discovered in 1994 by Ronald Clarke, a fossil-hunter who was rummaging through a bag of old cow bones. For molecular scientists Lucy and her ilk provide excellent support for their theory that chimps and humans descended from a common ancestor some four to five million years ago. Lucy and her newly found trick of walking on two feet also provide definitive evidence of the difference that marked the first major distinction between humans and modern chimpanzees.

But was Lucy human? If being human is just about being ape-like and walking on two feet, then Lucy's our woman. But if it's about having larger brains and much greater intelligence than other creatures, then we must wait another 800,000 years, until about 2.4 million years ago, when the bones of the first *Homo habilis* appeared. He was smaller than us (about 1.3 metres tall), but a lot taller than Lucy, and a more upright walker too. More importantly, perhaps, his brain was nearly twice the size of Lucy's (although at 650 ccs still only half that of us *Homo sapiens*). *Homo habilis* makes Lucy look and seem much more like an upright-walking chimpanzee than the first human being.

Homo habilis were the first human creatures for which there is evidence of the intelligent use of tools – sharpened flint stones for cleaving meat from the bone. This marks the beginning of what is known as the Old Stone Age, and *Homo habilis* is the point where we can safely say humans join the earth's nature park. He was the first of our kind, the world's first *Homo*.

Walking upright allowed *habilis* to start developing craftwork skills. Carpentry or stone sculpting requires very precise hand-to-eye coordination. Skills like these need well-developed motor skills and sophisticated hand and finger controls – processes that probably provoked the evolution of bigger brains. Recent studies have shown that relative to other mammals, *Homo habilis* had a brain at least four times larger than it should have been for his size and weight.[4]

Bigger brains use lots of energy. To power our brains we need about twenty watts of energy, or

400 calories a day – that's about 20 per cent of our total energy consumption – just to be able to think. There's a good deal of truth in that old expression, 'food for thought'. So began a most important evolutionary spiral. Big brains need lots of energy, which is best supplied by eating meat. The most successful means of getting meat is by hunting for it, using tools and weapons. The creatures that were most well adapted to making such tools were those with the biggest brains. An avalanche of evolutionary changes kicked in, all thanks to Lucy's fortuitous swivel up onto her two feet. These were adaptations that led, almost inevitably, to hunting, weapons, tools and intelligence, to the genus *Homo*, the species *habilis* and beyond.[5]

Some of Lucy's ancestors stayed in the trees, so they didn't need to bother with walking upright. They evolved into today's chimpanzees and bonobos. Without free hands, the evolutionary spiral that led to bigger brains and modern human intelligence never took off and their brains stayed small. After all, if you don't need a bigger brain to survive, don't have one. It's far more energy-efficient.

Chimps and humans are genetically so close because these evolutionary changes happened so recently in history – probably no more than about four million years ago. Despite this short space of time, a big difference has emerged in terms of intelligence and brain size. An apparently simple change in circumstances, such as having freely available hands, seems to have triggered an evolutionary revolution.

But just how big is the difference between humans and their closest animal relatives? Chimps, like gorillas, have been known to communicate. Kanzi is a bonobo ape, born in 1980, who now lives in Georgia, USA. He can understand more than 3,000 spoken English words, many more than Koko the gorilla. When Kanzi wants to 'talk' back, he points to a series of pictures so he can be understood by humans. It was reported in November 2006 that Kanzi was taken for a walk in the woods after having touched the symbols for marshmallows and fire. Once in the woods, he then proceeded to snap twigs and make them into a pile, light a fire and toast his own marshmallows on a stick.[6]

The original handyman: Homo habilis evolved in East Africa more than two million years ago and was probably the first human to have created and used tools.

23:58:43

Humanity

How several species of early humans adapted to Ice Age conditions, learned to tame fire and cooked freshly hunted meat as they spread across Africa, Europe and Asia.

MOST AFRICAN people today are desperately poor and suffer terribly from disease, poverty, famine and war. You'd think there was good reason to flee – and they do. Thousands of Africans try to escape the continent each year, many of them across the Straits of Gibraltar into Europe (for some reasons why, see page 374). History is repeating itself.

About two million years ago *Homo habilis* had evolved into a new species of human, *Homo erectus*, who looked much more like us *Homo sapiens* today. For a long time experts thought that the ancestors of modern humans originated in China, or maybe Java, because that's where the bones of *Homo erectus* were first discovered, which dated back at least 500,000 years. Now, thanks to the discovery of a ten-year-old boy mysteriously swept to his death in an African swamp near the shores of Lake Turkana in Kenya some 1.8 million years ago, we know that

he first appeared in Africa at just about the same time as *Homo habilis* died out.

Turkana Boy was found in 1984 by a team of fossil-hunters led by Richard Leakey, a British palaeontologist who lives in Africa. The implications were far-reaching, as Leakey himself explained soon after the find: 'In 1984 his bones were painstakingly excavated to reveal a species on the brink of becoming human. All people on earth have one thing in common. We share a single African ancestor; the same as this young boy.'

When alive, Turkana Boy was black-skinned and sweaty. His kind, *Homo erectus*, had lost their body hair because of the blistering African heat, which meant there was no need for fur. Dark skin and sweat glands helped these early people survive the harsh, arid heat of the African grasslands. Like us, Turkana Boy would have had hair on his head as a natural sunhat, because a decent crop on top protects upright walkers from the sun's burning ultraviolet rays.

The boy had a long, protruding nose unlike his habiline ancestors, again useful for helping cool the blood. His pelvis shows that he walked more upright, and his skull shows a marked increase in size: now as big as 1,100 ccs, nearly twice the size of *habilis*. Bigger brains need more food. Unlike his evolutionary ancestors, who often fell victim to a hungry cheetah or a pouncing lion, *erectus* was the first human to make spears. In a contest with wild beasts, he almost always won.

Homo erectus had several key advantages over anything else alive in the wild: his hands, his brains and, perhaps most important of all, his control of fire. This would have helped scare away the large animals that so bothered his ancestors, and also meant that *Homo erectus* eventually became the world's first cook. Well before they became extinct 70,000 years ago, these early people had figured out that cooked food releases energy more quickly than raw meat. It also takes less time to digest. The remains of human campfires nearly 1.5 million years old have been found in Africa and Asia (man-made fires magnetize the soil leaving tell-tale signs of human kindling). Who taught man how to light a fire? How did he learn to control it?

The ancient Greeks had a myth that tells the story of a Titan called Prometheus, who stole the fire of the gods and smuggled it down to earth in the stalk of a fennel plant. He paid dearly for his crime. When the king of the gods, Zeus, found out, he had Prometheus tied to a rock. Each day an eagle was sent to pick out his liver. Each night his liver would grow back so that it could be picked out again when the eagle returned. Zeus also took his revenge on man for gaining the knowledge of how to use fire. He sent a box down to earth along with a pretty girl called Pandora, who was told never to open it. Of course temptation eventually got the better of her, and when she lifted the lid suffering and despair were unleashed upon mankind for ever.

Fossil evidence shows that *Homo erectus* mastered the art of making fire by using stones. Pieces of scorched flint dating back 500,000 years

A cast of the fossilized skull of Turkana Boy, unearthed in 1984. Its discovery means Homo erectus must have first evolved in Africa about 1.8 million years ago and then migrated across Europe and Asia.

have been found at several of their campsites located in northern Israel.

Homo erectus lived in groups of about a hundred, hunting together using sharpened flints, following wherever the scent of blood led them, chasing and trapping wild animals for food. Their tools were more sophisticated than those first made by the *habilines*. The biggest difference was that axes were worked on both sides of the blade into a sharp point. These 'biface' tools have cutting edges up to four times bigger than those of the older technology, ideal for hacking wood, digging out roots, butchering animal carcasses and skinning hides.

Could *Homo erectus* talk? Scientists think Turkana Boy's bones suggest that he could not, because the nerve openings in his vertebrae were not large enough to contain the complex nerve systems needed to control breathing required for speech. Maybe he developed a sign language of some kind, or perhaps something similar to today's teenage grunts, a verbal equivalent of modern-day txts? F we cn gt by ths wy, I hve no dbt thy cld 2. With their portable toolkits, the protection of their communities and the magic of fire, these were people ready to walk wherever needs must to make sure they were fed. *Homo erectus* was the first human

23:59:21

These hand-axes, dating back about 400,000 years, show how man had become highly skilled at sharpening tools for better hacking, butchering and skinning.

species to explore life outside Africa – humanity's first migrants, Africa's ancient Marco Polos.[1]

With the earth's continents arranged more or less as they are today, it was possible then, as now, to travel overland from Africa across the Middle East into south Asia, India and China. Could Stone Age man really have made it all that way without roads and tracks, let alone cars, boats or planes? Unlike most of us, these people had one huge advantage: they were not in a hurry.

Dragon Bone Hill is found on the east coast of China. Fossils from as many as forty *Homo erectus* were found there in 1927. They were estimated to date back about 400,000 years. Unfortunately, these bones are now lost. Just before the Second World War they were hidden away for safe-keeping. But in 1941 they were packed up to be sent to the United States until the end of the war. Despite a lot of searching, they have never been seen since. One theory is that they sank on the Japanese hospital ship *Awa Maru* in 1945.

Homo erectus lived on average for about thirty years. Even at a supremely leisurely pace of travel,

say ten miles a year, it still would have taken them only 600 years to wander across the 6,000 miles of land from Africa to China. That's about thirty generations. The earliest *Homo erectus* fossils found in Africa date back 1.8 million years, and since the bones from Dragon Bone Hill are no more than 400,000 years old, these people had more than enough time to make the journey. In fact, they could have walked there and back dozens of times.

Early humans walked their way around the habitable continents of Africa, Europe and Asia from the time of *Homo erectus*. The first evidence of humans in Britain dates from 700,000 years ago. Boxgrove Man, a human skull found in Sussex, is a *Homo erectus* descendant, known as *Homo heidelbergensis*, dating back about 500,000 years. Did he build a raft or wade across when the sea was shallow, or simply walk across to Britain when there was no English Channel? Either is possible.

While *Homo erectus* was populating Asia the climate turned much worse, as the beginning of a deep Ice Age chill caused glaciers to scoop down over the continents. For *Homo erectus*, who had wandered so far and wide, such cold conditions were a big problem. Even the miracle of fire wasn't always enough to ensure survival in the bitterly cold temperatures that sometimes descended on much of the European and Asian continents for thousands of years at a time. Nature had its own answer, of course, in the form of Neanderthals.

In 1857 quarry workers near Düsseldorf, north Germany, found what looked like human bones in the Neander Valley. Their discoveries first alerted naturalists to the possibility that there may have been several species of humans that lived before our own *Homo sapiens*, and that man was almost certainly descended from apes (what they didn't know then was that, genetically, humans were actually a branch of the ape family itself).

Since then, bones from many Neanderthal sites have been found. The oldest date back about 350,000 years, which means that several species of humans must have lived simultaneously for a very long time – at least until about 70,000 years ago,

when the *Homo erectus* line disappeared, probably as a result of climate changes and the arrival of a yet more powerful species, namely us, *Homo sapiens* (see page 97). Experts believe that during this period at least five different species of humans were living on the planet, including: *Homo erectus*, *Homo ergaster* (the African form of *erectus*), *Homo neanderthalensis*, *Homo heidelbergensis* and *Homo rhodesiensis*. There may have been more, and scientists are still unsure if some of these are subspecies rather than separate species. Did they fight? Did they live together or in their own separate communities? Did they interbreed? Could they talk?

There is still a lot of disagreement and confusion about all of this. What does seem clear is that once *Homo erectus* migrated out of Africa from about 1.7 million years ago, several different species of humans evolved in separate parts of the world, and that geographical and climatic differences caused small but significant evolutionary changes. Chances

A Neanderthal burial site dating back c.60,000 years located near Mount Carmel, Israel. Similar sites have been found mainly across Europe and the Middle East.

23:59:21

are that these species didn't mix much, because back then there were very few humans around – maybe a million or so spread out widely across the continents of Europe and Asia, which hold more than four billion people today.

Neanderthals first appeared in Asia about 350,000 years ago, and then spread northwards and westwards into Europe as the weather permitted, even crossing to Britain where the remains of a jawbone in Kents Cavern, Torquay, have been found dating back as recently as 35,000 years ago.

Neanderthals have had a bad press. To call someone a 'Neanderthal' is usually meant as an insult, implying that they are thick, old-fashioned or brutish. Pictures of Neanderthal people were, until recently, more like apes than men. They were shown walking with a stoop and bent knees.

This is wrong. Neanderthal brains were at least the same size as those of modern humans, if not a touch bigger. They also walked as upright as we do, although they were more hairy and usually shorter. They were stronger than us, with broad noses and foreheads that jutted out above the eyebrows. Almost all these adaptations helped to reduce the surface area of these early humans, helping them conserve heat in the bitterly cold Ice Ages.

Neanderthals were highly skilled at using tools. Recent archaeological evidence shows that their hands were at least as nimble as ours.[2] The most famous site where tools have been found is at Le Moustier in the Dordogne, France. In 1909 archaeologists found an almost complete Neanderthal skull there, less than 45,000 years old. Hundreds of sharp, skilfully crafted stone tools were found alongside the bones.

Neanderthals used some of these as weapons. Their spears were not designed to be thrown, but were used for stabbing and clubbing. Stone tools helped them build impressive shelters – the first human houses – and they are the first people known to have buried their dead, often leaving ornaments in the graves of those they loved to help them on to the next life.[3] This means that these people almost certainly had beliefs, perhaps religions, and developed societies in which some

individuals were thought to be more important than others.

Perhaps the most significant discovery of all was dug up in 1995 by Dr Ivan Turk, a fossil-hunter from Ljubljana, Slovenia, next to a fireplace in a Neanderthal house. He found a hollowed-out bear bone with several holes bored out in a straight line. This could be a fragment from the world's oldest known musical instrument – a Neanderthal flute.

What tunes were played on this prehistoric pipe? It's hard to say, because no one knows for sure how long the original flute was, or how many holes it had, as only a small section remains. Some researchers believe it would have played what we would recognize as a minor or 'blues' scale today, with a flattened third note.[4]

In 1983 a Neanderthal bone was found in a cave in Israel which is almost identical to a bone in modern humans called the hyoid, which connects our tongues to our throats. This means that Neanderthals almost certainly had the capacity to speak. Also, the size of openings in Neanderthal vertebrae for the nerves that control the tongue for speech are about the same as in modern humans – unlike Lucy's people – meaning that they could produce a wide range of sounds.

Music, ceremonies, weapons, tools and conversation are the stuff of people possessing intelligence, brains, culture and a love of beauty. The Neanderthals were certainly big and burly enough to cope with the hardship of living in caves through an Ice Age, but there's nothing to suggest that they were any more brutish than us. What happened to these strong, intelligent, well adapted folk? To understand this, we must look at ourselves properly, face-to-face in the mirror.

The Great Leap Forward

How a single species of humans, Sapiens, became the last to survive on earth, colonizing previously uninhabited lands and learning to talk and hunt with new types of flighted weapons.

ANY IDEA HOW CLOSE we are to midnight on our twenty-four-hour clock of earth history? A few minutes left to go, perhaps? After all, we've still got all human 'recorded' history to tackle, starting from when the first large-scale human civilizations emerged in the Middle East through to today, with approaching seven billion of us living off the planet.

How about three seconds? That's it. Imagine how short sitting doing nothing for just three seconds would seem compared to doing nothing for the length of a full day. Anyone attempting that rather daft experiment would get a good impression of how much earth history actually took place before the first *Homo sapiens* could be heard calling across the hot, dusty African plains.

We now know it was from Africa that *Homo sapiens*, or modern humans, came. We did not descend from the Neanderthals. The two species lived at the same time for thousands of years, and while there was probably a little interbreeding, recent genetic evidence suggests that there was not much. Red hair, freckles and pale skin are features that may have been handed down from the Neanderthals.[1]

Instead we must look back to Africa – to the descendants of *Homo erectus*, Turkana Boy's people. Bones from twenty-one sites across Africa have been found, stretching back almost 500,000 years, and they paint a picture that shows how small evolutionary steps along the *Homo* line led to who we are today.

The oldest *Homo sapiens* fossils yet found came from southern Ethiopia. In 1967 two human skulls were discovered buried deep in mud at the bottom of the Omo River. They have recently been re-dated, and are now thought to originate from about 195,000 years ago. These skulls, called 'Omo I' and 'Omo II', certainly look as if they belong to the immediate predecessors of modern humans. They are slightly larger than modern human skulls, but otherwise strikingly similar. They have been classified as a subspecies of *Homo sapiens* called *idaltu*. They have flat faces and prominent cheekbones, but not the protruding brow ridge of the earlier *Homo erectus* and the Neanderthals.

In 1997 three more *Homo sapiens idaltu* skulls were found near the Ethiopian village of Herto – two adults and one child, dating back 160,000 years. More than 640 stone tools were also found at this site, showing sophisticated workmanship. They were probably used for butchering hippos, crocodiles and catfish, which then lived in the shallow lakes next to the nearby Awash River.

Genetic research supports the idea that all humans alive today emerged from a single evolutionary line in Africa from about this time. But one strange feature thrown up by genetic analysis remains unsolved. There is surprisingly little genetic variation between people alive today – much less than in most species of mammals. Even our closest cousins, the chimpanzees, show genetic differences ten times greater across their species than we do.

Such a small genetic variation can only mean one thing: at some point our species, *Homo sapiens*, must have shrunk to a very few individuals – perhaps between 1,000 and 10,000 people – all of whom shared a very similar genetic code. This idea has set experts off on another hunt, to find some 'event' that fits in with the idea of *Homo sapiens* suffering an almost fatal collapse in population early on in its history.

One candidate for such an event is a massive Category 8 volcano – a super-eruption – that occurred about 75,000 years ago at Toba, on the Indonesian island of Sumatra.[2] A giant hot-spot of molten lava is thought to have burst through the earth's crust, releasing energy 3,000 times greater than that of the huge eruption of Mount St Helens in Washington state, USA, in 1981. One site in India today still has ash deposits from the eruption of Toba that are six metres thick. Such an enormous explosion would have created a blanket of dust in the atmosphere, blocking out the sunlight for months, or even years, triggering a sudden drop in global temperatures and possibly even starting an Ice Age all by itself.

Maybe there are other reasons. A disease, perhaps, that wiped out large parts of the human population? At the moment, no one knows. But from a very small stock of African ancestors, *Homo sapiens* walked its way around the world copying the migrations of his ancestors, eventually to supplant all other species of humans. The last of these, the Neanderthals, held off from extinction until about 24,000 years ago, which is the date of their very last known traces, in Gorham's Cave, Gibraltar.

Did we kill them? Did we eat them? Or did they die because of climate change, or through lack of food during an especially nasty Ice Age snap? That's another thing no one knows for sure, although it seems that changes in the climate about 30,000 years ago were especially severe in central Asia and northern Europe, which is where the Neanderthals lived.

Various scientific studies have now traced the genes of thousands of ethnically diverse modern humans to work out where their ancestors came from, and when.[3] This new science, called phylogeography, is revealing more and more about the origins of living things. Small genetic variations can often be read like a transcript of what happened where and when.

It was probably a warm interglacial interlude within the Ice Age, between about 130,000 and 90,000 years ago, that initially triggered large-scale *Homo sapiens* migrations across Africa. Then, from about 70,000 years ago, the climate cooled, causing glaciers to form on the tops of mountain ranges so that parts of north-west and north-east Africa were cut off from each other, as well as from the south. As Darwin discovered, whenever a species is physically separated, small variations begin to creep into its respective gene pools, creating diversity. So it was with modern man, giving us our four main ethnic groups: Khosian (African), Caucasian (European), Mongolian (Chinese and American Indian) and Aboriginal (Australian).

From about 60,000 years ago, these four groups of humans emigrated from Africa separately and in their own time across the world, exporting their small genetic differences with them. Some *Homo sapiens* swept across Asia, displacing the last of the Neanderthals either by depriving them of food, hunting them, or maybe occasionally absorbing them into their own species through limited interbreeding. Some turned south and reached India and China. They learned to build rafts. So from about 40,000 years ago, Australia, for many millions of years the preserve of marsupial mammals, became another human hunting ground as the first people paddled ashore.

Aboriginals made the journey to Australia when sea levels were at a historic low, although it is estimated that they would still have had to cross about eighty kilometres of open sea to reach the Australian coast from Indonesia. They lived off the bush in tribes of between 500 and 800 people (tribes needed to be large enough to prevent inbreeding, which can occur if populations fall below about 475 individuals), developing their own distinctive lifestyles as well as art, ceramics and tools. Amazingly, they lived undisturbed until the first European explorers arrived, guns in hand, to claim the land as their own less than 250 years ago (see page 351).

The first *Homo sapiens* to arrive in Europe walked eastwards out of Africa about 50,000 years ago, and then came north via the Middle East. These people are sometimes called Cro-Magnons – the name comes from an early discovery of *Homo sapiens* remains in the Cro-Magnon rock shelter in the Dordogne, France, in 1868. They brought with them enormous changes in lifestyles, technology and culture, including the world's first spears that were specially designed for flight. The bow and arrow was probably invented somewhere on their travels, perhaps in North Africa or the Middle East, which were lush places at the time, full of game and large wild animals that could be more easily

Carvings on these mammoth bones, found at a Cro-Magnon settlement in the Ukraine, show signs of early human art in the form of markings and drawings – perhaps for tracking the cycles of the moon.

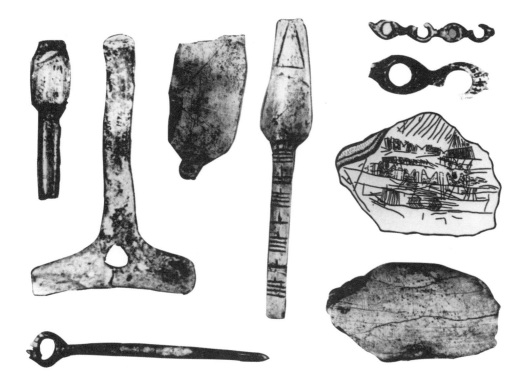

Migrations of Man

Humanity's appetite for globe-trotting has ancient origins going back more than one million years.

Homo erectus: These early humans migrated out of Africa c.1.7 million years ago, travelling across Europe and Asia and reaching as far south as Indonesia, where bones have been found dating back 500,000 years.

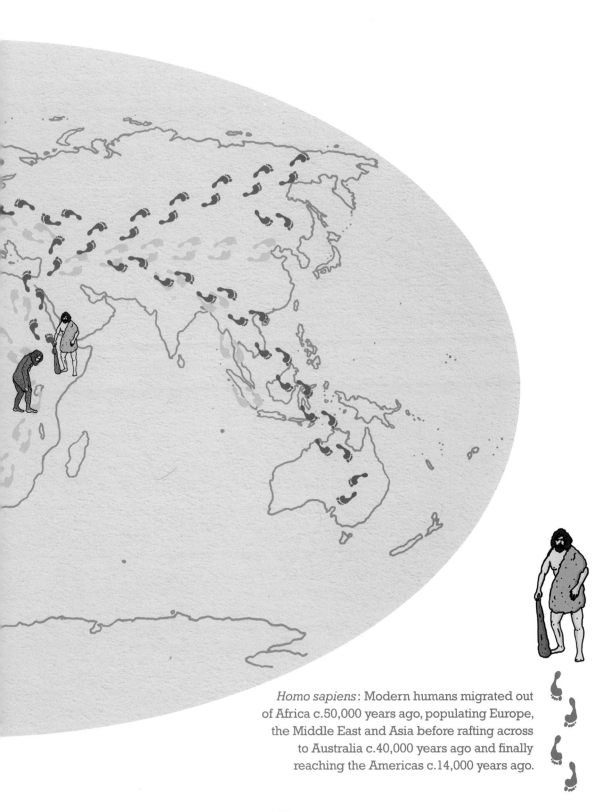

Homo sapiens: Modern humans migrated out of Africa c.50,000 years ago, populating Europe, the Middle East and Asia before rafting across to Australia c.40,000 years ago and finally reaching the Americas c.14,000 years ago.

23:59:57

and more safely killed from a distance, rather than at close range with Neanderthal-style clubs.

The time from about 50,000 years ago marks the beginning of the final one second to midnight on the twenty-four-hour clock of earth history. It has been described as 'the Great Leap Forward', because the complexity of human tools increased dramatically.[4] Bones, tusks and antlers were used for the first time to carve out ornaments as well as to craft useful household items such as needles for sewing, and spoon-like oil lamps that burned animal fats. Jewellery, in the form of necklaces and pendants, has been found buried inside Cro-Magnon graves.

The 24,000-year-old Venus of Willendorf is one of the earliest known human sculptures and suggests that femininity and fertility were at the heart of prehistoric human spiritualism.

The first ceramic pots date from this period, as do the world's first known sculptures such as the Venus of Willendorf, a female fertility figure found in Austria in 1908 which is thought to date from about 24,000 years ago.

Whether or not these people actually learned music or language from the Neanderthals, no one knows, but the flute also seems to have found its way into Cro-Magnon caves. Their taste for music and sculpture was equalled by their appetite for art: some of the first known cave paintings are Cro-Magnon in origin and date back as far as 25,000 years ago. They can be seen to this day in the caves of Lascaux, in the Dordogne region of France.

The earth was far cooler back then. The last of the great ice sheets swept down from the North Pole about 22,000 years ago, to disappear rather quickly 12,000 years later. During this time Cro-Magnon people adapted to the changes in climate by developing paler skin that helped produce sufficient quantities of vitamin D for bone formation despite the weaker sunlight of the Ice Age.

Cro-Magnon people arrived in Britain about 20,000 years ago. They walked across the Channel from France, since it didn't flood until the end of the last big Ice Age melt, about 10,000 years ago. But they weren't the first to arrive. We now know that up to seven previous attempts were made by earlier people to populate the British Isles, starting with *Homo erectus* some 700,000 years ago.[5] Each time, the populations of humans died out, probably because of the horrendously icy conditions that periodically swept over the islands as far south as present-day London. Even in the very south the cold would sometimes have been too much for any type of human to bear.

About 15,000 years ago giant glaciers still locked up much of the earth's waters, sinking her sea levels so that a massive land bridge the size of Poland, called Beringia, connected the eastern tip of Russia to Alaska across what is now a ninety-five-kilometre-wide stretch of sea called the Bering Strait. In those days people could cross by foot from Asia to North America – a land that was until then probably free from human habitation (although some scientists think people may have rafted there a few thousand years before from south Asia, via the Pacific islands). North and South America were the last of the great habitable continents to be populated by man, and are still appropriately called 'the New World' even today. It was an opportunistic walk all the way across Asia, following big animals, hunting on the move, making the most of nature's twisting and turning climate changes.

With another land bridge via Panama linking the two great Americas, it wasn't long before the first people from North America wandered down to the southern American continent, where the climate was warmer and the land rich in vegetation and game.

The arrival of Stone Age humans in this part of the world – as in Australia – came with dramatic consequences for much of the world's wildlife. Although a few of nature's ecosystems lingered on without any human representation – New Zealand and Iceland were untouched by humans until about 800 AD and 1000 AD respectively – many of the world's living creatures were by now beginning to succumb to mankind's growing influence as he spread out to envelop the whole of planet earth.

Hunter-Gatherer

How humans lived in a state of nature for 99 per cent of their history on earth without permanent homes, full-time jobs or private possessions.

IMAGINE A WORLD where you can have anything and everything you want. There is plenty of fresh food every day, and of immense variety. There is some work to be done, maybe three or fours hours a day on average, but no more. In this world you can sleep and rest as much as you wish, or spend time with friends and relatives, cooking, talking, dancing, or just having fun.

And don't worry about money, mortgages or debt. No need for those. No exams, qualifications or career to grapple with, no reviews, promotions or demotions. In this place there's no such thing as losing your job. You can't get into trouble with the law or the police here, because there are none. There is no need. If you want something, nearby friends and neighbours will help you find it – or let you borrow it, if they have it.

This world has very little risk of disease. Most illnesses we have today don't exist. War and violence are also rare, because there is plenty of food and little competition for natural resources. Sounds good? Fancy moving in? I'm afraid today that's not really an option. But, amazingly, this is the kind of lifestyle that we human beings have lived for 99 per cent of our history. Much of the evidence we have suggests that Stone Age man lived well, happily and mostly in peace.

It is called the hunter-gatherer way of life. Until about 10,000 years ago there were few, if any, permanent homes or villages. People moved around

23:59:59

all the time, from place to place. Men would hunt animals, and women gathered wild fruit and nuts. Sometimes the women helped out hunting, too, especially when trying to catch an animal like a deer which needed to be surrounded on all sides to prevent it from escaping.

Living the life of a traveller, a nomadic life, meant people had few if any actual possessions. All they had was what they could carry. In cooler climates they wore animal skins and furs; in hotter areas many went around almost naked. Why carry what you don't need? They took essential provisions with them, such as water inside gourds, vegetables belonging to the pumpkin family which can easily be hollowed out to make bottles.[1] They would also carry spears or

What Maria saw: 20,000-year-old paintings of bison still adorn the ceilings of caves near Altamira, Spain, although for a time experts wrongly believed they were an elaborate hoax.

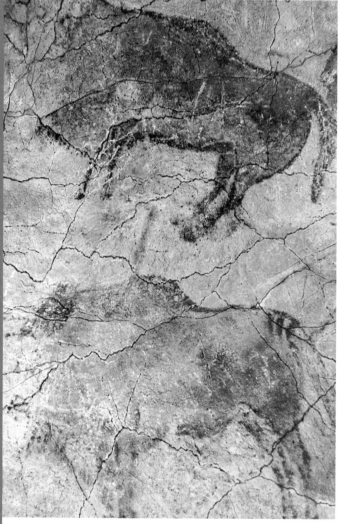

bows and arrows for hunting, and flint implements for skinning dead animals and lighting fires.

These people needed little else. The whole idea of owning anything at all was completely alien to them. Their habit was to share things with each other, because it meant there was less to carry. There was no need for money, because they hunted and gathered whatever and whenever they needed. They had no use for storage areas or farm buildings. They had no property, and no one could tell them 'PRIVATE – No Hunting Here', because no one owned any land. It was, like the air we breathe today, something common to us all, a resource to be shared between all living things: plants, animals and people.

People lived in this state of nature from the time of their first appearance as *Homo habilis*, or even as far back as the *Australopithecus*, Lucy's people, dating back at least three million years. They existed in harmony with each other, and with nature. They hunted when they were hungry, slept when they were tired, and when the land was void of fruit and meat they moved on elsewhere, giving the earth a chance to restore, recover and renew.

They were also highly artistic and cultured. Hunter-gathering people are the first whose culture we know of, because their art survives to this day, deep in the prehistoric caves of south-west France and northern Spain. These first etchings of hunter-gathering humans are where the history of human art begins.

Probably the finest surviving cave paintings of all were discovered by an eight-year-old Spanish girl called Maria Sautuola, quite by chance, one autumn morning in 1875. She was exploring some curious-looking caves with her father Marcelino close to their family's home in Altamira, near Santander. As they began walking down a gloomy, 270-metre-long cave, she looked up and saw that the ceiling was covered in pictures of what looked like cows. Her father, a keen amateur archaeologist, immediately recognized these as bison, and set to work trying to find out more. He enlisted the help of an expert friend, and soon declared they were the oldest paintings in the world.

Unfortunately, the world was not ready to believe that such skilful paintings could possibly be so old, or have been the work of such 'primitive' people. Experts rounded on Sautuola and accused him of having paid someone to paint the caves so he could deceive the world and become famous. It wasn't until 1902, fourteen years after Sautuola had died in disgrace, that experts started to find other caves with similar paintings, and then agreed they were genuine after all. Modern dating technology suggests they are nearly 20,000 years old.

But what exactly did these Stone Age people paint, and why? What materials did they use? What did their works of art mean?

It seems almost certain that they were not meant for display, since so many of them, such as the pictures found by Marcelino Sautuola in Spain and the famous Lascaux paintings in southern France, were painted deep inside caves, where there is very little light. Historians now think that ancient holy men, called shamans, would retreat into the caves to perform magic rituals. By painting on the rocks of the caves, the shamans hoped to summon animals, food and good fortune from their mother, the earth. They used various types of clay, sometimes mixed with iron oxide, to create different colours or pigments which they blended with animal fat to make sticky paints.

Were the people living this wandering, possession-free lifestyle happy with their lot? Before people started living in towns and making weapons out of metals like copper, bronze and iron (see page 128), there is little evidence of extensive warfare or violence. Perhaps the best guide is to look at tribes that still live the hunter-gathering way, even in the modern world.

There are very few true hunter-gatherers left, because modern farming and industry have almost completely overrun these ancient human societies. But in a very few almost forgotten corners of the world, in the Australian and African bush, they cling on – just – where the land is no good for farming, or too remote for towns and cities to prosper. These people survive, perhaps a few thousand of them, by the most slender of threads. But are they happy?

The Hadzabe tribe lives in the depths of central Africa, in what is now Tanzania. Until recently these people have thrived in the forests, hunting wild game, gathering fruit and berries, moving from one place to the next. By 2006 there were just 2,000 of them left, squeezed into a tight corridor of bushland surrounded by encroaching farmers and creeping urbanization. They have no history of aggression, so rather than fight against the invading farmers and townsfolk, they have instinctively withdrawn into the thorny forests below their traditional hunting grounds, where they hide from the modern world as best they can.

The Hadzabe tribe own nothing, and share everything. It was reported recently that on a rare occasion when a member of the tribe was given employment by an outsider, as a guide showing curious tourists round the bush, the entire clan pitched up on his payday to share out the money between them. While modern societies depend on order, regimentation, law, police, administrators, rules and rulers, the Hadzabe rely on little more than intelligence and cooperation, organizing themselves into small, manageable groups. Being flexible and willing to help each other is key to their survival. Owning nothing and having no fixed homes gives them the light-footedness to make this possible.

Some experts believe that their language, or at least a form of it, may be similar to the original language used by Stone Age man. It is like no other existing language, and relies mostly on clicking sounds, quite unlike the vowels and consonants we use today. These clicks are particularly effective when the tribe is out hunting, because they allow them to pass information to each other across large distances, without having to shout and give away their positions.

More evidence of these people's ancient heritage comes from recent genetic studies which show that their DNA is the most diverse of any human population yet studied. Greater diversity means their bloodlines are old, because genes differ from one generation to another at a predictable rate. Experts believe the Hadzabe lineage split off from the rest of the human family tree early on in the evolution of *Homo sapiens*, meaning that they are among the oldest survivors of our species. Soon these ancient people are likely to disappear and their lineage will be lost, becoming merged into the modern world.

What is so remarkable about the Hadzabe way of life, and what it tells us about Stone Age living in general, is that it is so efficient. Since everyone is involved in the process of food production, and the whole tribe lives on the move, there is no ruler or group that sits around waiting to be fed. There is no need for money, banks, loans or wages. No need for accountants, lawyers, merchants or the taxman. There is no need for writing, electricity or modes of transport other than just two feet each.

Because these people know the forest so well, every one of them is an expert in what is good, or not good, to eat. Knowledge of how to treat sicknesses and other health problems using the plants of the forest has been passed down orally through hundreds of generations. Most modern medicine is ultimately derived from ingredients found in nature. The unschooled Hadzabe, as with Stone Age man, have a prodigious knowledge of herb and plant remedies that even the best-informed modern pharmaceutical scientist would find tough to match.

A deep regard for all things natural was the basis of hunter-gatherers' mythology, or religion. For them the woods were full of magic and wonder.

They contained the spirits of their dead ancestors, who returned in the afterlife to protect, guide and comfort the living – or so they thought. The woods were their ultimate source of food, warmth, habitation, medicine and shelter. To them, nothing was more important than looking after nature's forests. They trusted completely in her abundance and her resources.

Perhaps the biggest long-term strength of the hunter-gatherers' lifestyle was that it provided an inbuilt control on the overall level of human population. This is because hunter-gatherers relied on travelling by foot so it was necessary for them to have their children well spaced apart – one every four or five years at most – because it is impossible to carry too many children at once. A stable population of about five million hunter-gathering humans lived on earth for tens of thousands of years, without the population increasing significantly overall. It was a natural limit, a sustainable level, founded on a nomadic way of life.[2]

So what happened? Why did five million humans who had lived for tens of thousands of years as happy hunter-gatherers change the habits of generations and turn to a radical, new, and much more demanding way of life?

16

Deadly Game

How a coincidence of modern humans and climate changes upset the ecological balance first in Australia and then in the Americas, leading to a mass extinction of many large animals.

DRAMATIC CLIFFS of ice have repeatedly bulldozed their way down from the ice caps over the last three million years. Sometimes they have covered as much as 30 per cent of the earth's entire land mass in glaciers and ice sheets more than a mile thick. But because such climate changes happened reasonably slowly, life generally adapted well.

Large animals became furrier, like the woolly mammoth, so they could live in the cold. Humans became smaller and hairier, like the Neanderthals, and some of them even turned white, preserving heat and helping to reflect away dangerous ultraviolet rays of light. Such natural evolutionary adaptations saw them through the worst of the cold. The last cold glacial period ended rather suddenly. About 14,000 years ago average temperatures on the earth rose by six degrees, easily enough to flip the climate from severe Ice

Age to the much balmier interglacial conditions we have today.

Swings in temperature on earth normally happen because the planet veers slightly nearer to or further from the sun during its annual orbits, making the climate hotter or colder. Another factor is the way in which the earth spins on its axis. It actually behaves like a slightly out-of-control spinning top, so that over long periods of time its tilt can vary from twenty-one to twenty-seven degrees, making big differences to the temperatures of the poles. Scientists are fairly sure that it was thanks to these orbital variations and spinning-top cycles that, between about 14,000 and 11,000 years ago, temperatures warmed up enough to give a hot spell of a few thousand years, in between the bitter conditions of the Ice Age.

Warm episodes within an Ice Age are called interglacials. We are experiencing one now, which began about 14,000 years ago. Thanks to human-induced global warming, this one may last much longer than others; or it may even spin the world's climate out of its forty-million-year-old Ice Age altogether if the polar ice caps melt completely.

About 18,000 years ago, when the last ice glacial was at its height, the North American landscape south of the enormous ice sheet that covered the Great Lakes looked like parkland – a mixture of trees and grasses, a paradise for roaming wild mammals. Large carnivorous mammals such as lions and sabre-toothed cats fed on mastodons and massive woolly mammoths. Horses and camels were other tasty meals for these king carnivores, which first evolved in these American woodlands (see page 74).

Bison, like those Maria Sautuola saw painted in that Spanish cave, roamed across the rich pastures. But these bison were no cows – they were almost as big as elephants. Beavers, which we know as smallish river creatures, grew as large as today's biggest grizzly bear, and the bears back then were almost twice the size they are now. The earth was full of enormous mammals, known today as 'megafauna'. Big beasts fared better in the bitter climate because their large bodies helped protect their vital organs from the extreme cold.

There have been at least thirty ice-overs and melts in the last two million years – some more severe than others. Each time, nature and her living things bounced back as individual species adapted to different climates, hot or cold. The last big melt, which reached its height about 14,000 years ago, should have been just like any other. But for some reason this time something went catastrophically wrong. Scientists and historians are still trying to figure out exactly why. The consequences of this disaster continue to shape human and earth history to this day.

At least eighty species of mega-large mammals were alive in North America when the ice sheets melted. Some of them had survived tens of millions of years. Yet, suddenly and mysteriously, they died out. Horses, big cats, elephants, mammoths and mastodons, camels, giant beavers, peccaries (American pigs), sloths and the glyptodont, an armadillo the size of a pick-up truck – all of them disappeared. In all, thirty-three out of forty-five species became extinct, leaving most of the animals in North America no bigger than a turkey. Even the beavers and bears that survived became dwarves in comparison to their ancestors. Today's North American bison are the smallest that have ever lived. In all, experts reckon that more than 80 per cent of America's large animal population disappeared within about a thousand years.

Pretty much the same thing happened in Australia, which lost thirteen species of large mammals, although the extinctions there started

Woolly mammoths like these only became extinct about 13,000 years ago, along with other large American animals, an event that coincided with the arrival of Homo sapiens.

earlier. Victims included the giant kangaroo, the giant horned tortoise, the rhino-sized wombat and its relatives, the diptorodonts, as well as the fierce marsupial lion. In the end, nothing larger than today's kangaroo survived. Yet in North Africa, Europe and Asia, even as the glaciers retreated and the seas rose, most of the large mammals survived: elephants, horses, camels, wolves, big cats – they all made it through.

What on earth was going on? Why did Europe and Asia escape when the New World and Australia suffered dramatic and sudden extinctions?

Some experts think that the climate was to blame. As the temperatures rose, large animals were at a disadvantage, because their big bodies meant it was harder for them to keep cool. These creatures may have died of heat exhaustion. Yet large mammal species had survived many previous interglacial periods where the temperature was warm; and what about the African elephants, lions and tigers? How did they survive?

Another theory is that some mysterious disease swept over the New World, devastating its animal populations. But how could such a bug target only the big animals, leaving the smaller ones and humans as survivors?

The most popular theory was first put forward by American scientist Paul Martin, nearly forty years ago.[1] He put it down to the arrival of *Homo sapiens*. In both America and Australia these mass mammal extinctions followed shortly after the arrival of the first humans. In Australia they began about 40,000 years ago, in the Americas about 13,000 years ago. According to Martin, because animals in these continents had never come across humans, they were vulnerable. Read the diary of any explorer who encounters a natural habitat where no man has ever been, for example Charles

Glaciers smothered the roof of the world about 22,000 years ago as the last Ice Age reached its coldest grip.

23:59:59

Darwin on the Galapagos Islands, and you will find that they always comment on the lack of timidity of the wildlife. It's still like this today in the few parts of the world which have no humans living near them.

So when the first human wanderers arrived, flint weapons, bows, arrows and spears in hand, the animals they came across were fearless. They may have looked on with curiosity at these half-hairy, two-legged apes clambering ashore, but chances are that the horses would have just munched on. Even the lions, provided they weren't too hungry, would probably have just fallen back to sleep. Thus they were easy prey for hunter-gathering man with his sharp spears – so much so that in less than a thousand years most of the big game had been slaughtered, and many species were on the verge of extinction.

The theory also explains why in North Africa, Europe and Asia many similar animals survived the presence of mankind. Animals here had evolved alongside human species for over two million years, and had grown used to their appetite for meat and hunting. The experience of their ancestors had evolved into a well-honed instinct that allowed them to survive in sufficient numbers, avoiding contact with humans by running away and hiding. This meant that the mass extinctions seen in Australia and the Americas simply never occurred.

So, the theory goes, in just a few years *Homo sapiens* single-handedly deprived nearly half the world's land masses of all their large creatures by hunting them to oblivion.

Recently this theory, called the Pleistocene Overkill, has itself come under attack. For example, it doesn't explain why some species not generally eaten by humans (e.g. sloths) became extinct, while others that were hunted (e.g. bison) survived. Mass slaughter by humans also doesn't explain why beavers, bears and bison all became so small.

The best theory seems to be one that blends the arrival of humans with the effects of natural, cyclical climate change. It goes like this: when humans first arrived on the virgin continents of Australia and the Americas, they indeed found big game were easy prey. Many of the key predator species, such as lions, tigers and wolves, were killed off in massive numbers by the two-legged hunter-gatherers. At the same time, temperatures rose rapidly, causing the glaciers to melt and the seas to rise. What was once a rich American landscape of parkland trees and pastures gave way to huge stretches of arid inland savannahs with dried-up waterholes, that turned into thick conifer forests near the much wetter coasts.

Because humans killed off so many of the larger, carnivorous predator species, the populations of these animals' prey – herbivores such as bison, deer, sloths, horses and camels – grew hugely, because there was nothing left to eat them. They became so numerous that there simply wasn't enough food to go around. Combined with the changes in vegetation caused by rapid climate change, the effect was catastrophic. Herbivores were wiped out in their millions through starvation because the landscape couldn't support them any more, and only small species which could endure long periods consuming little food and water survived.

The intensive over-grazing of these huge overpopulations also contributed to the effect of climate change, accelerating the transition from parkland to grassland, making the landscape even less suitable for supporting future generations of large animals.

How fragile are nature's ecosystems. Add a new bit of something over here (humans), and see them remove something else over there (lions and sabre-toothed tigers). Now throw in a bit of random climate change, and devastation sets in on a massive scale. The role of humans in the annihilation of the large herbivorous marsupials and placental mammals of Australia and the Americas between 40,000 and 12,000 years ago – at the beginning of the last second to midnight on the twenty-four-hour clock of earth history – was humanity's first big impact on the earth's fragile, changing natural environment. It was not the last.

Food Crops Up

How men and women experimented with survival techniques following the last Ice Age melt, leading to humanity's first attempts to manipulate evolution for its own self-interest.

MANKIND HAD BEGUN to make its mark. But the reckless slaughter of carnivores, its full consequences yet to play out on the stage of human and earth history, was matched by the start of another revolution that wove together the fate of humans and the earth's living systems as never before.

Starting from about 12,000 years ago, human history begins to reveal the first attempts by people to control and adapt natural evolution to suit their own needs. It starts with the beginning of farming – the artificial breeding of animals and the intensive growing of particular plants, or crops, for food.

Natural selection changed life on earth over billions of years, from simple single-celled microbes into everything from fruiting fungi to jumping jerboas, and slimy slugs to venomous vipers. These changes were caused by small genetic differences between generations that increased a species' chances of survival in the earth's many constantly changing environments.

But about 12,000 years ago, when humans first started to cultivate the land and tame wild animals, they hijacked the process. They began what is now called 'artificial selection'. Instead of nature choosing and breeding the most successful specimens in the wild, humans started to choose, breed and grow those that suited them best.

Artificial selection allowed people to settle – to live permanently in one place – because all the food they needed could be cultivated in one spot. They started to live in villages all year round, to build the first houses, which then grew into towns, which then grew into cities, which grew into states, and which, ultimately, turned into civilizations. With the advent of farming came the first sedentary lifestyles; and with them, massive increases in human population, the re-sculpting of

the earth's landscape to suit food production, and the beginnings of modern diseases, almost all of which originate from humans living in close proximity with domesticated animals (see page 116).

The change to farming also marks the beginnings of all those jobs that aren't associated with food production, because, for the first time ever, there was usually enough food to support people who weren't directly involved in its provision. Over time farming became at least ten times more productive than the hunter-gathering lifestyle that people all over the world had always practised before.

Farming meant that people could have more offspring, as they no longer needed to carry their children with them. They could store their food in granaries, and still afford to give birth every two years, if not more often. There was the added benefit that living in a village or small town meant that there were more people around to help look after young children, encouraging families to grow.

As the populations of villages and towns increased, those not involved in farming could become artisans – skilled workers – who made artefacts like pottery, jewellery and clothes for settled people. They could also explore new technologies such as wheels, chariots and armour made from pliable raw materials, which they learned to extract from the ground in the form of copper, bronze and iron.

Then there were merchants, who began to trade the products that artisans manufactured, along with any surplus food left over from the farms. Trade meant travel, ships, writing, accounting and money. Other jobs for non-food-producers included making sure that the village or town stayed on the right side of the gods, giving the best possible chance of good harvests and fending off evil events. These early priests or holy men eventually helped give birth to many of the major religions of the world.

Growing numbers of settled populations required new forms of organization and control. The world's first kings and emperors emerged, with their aristocracies and bureaucrats whose jobs were to collect taxes, issue laws and administer justice for all to see. Kings could afford to protect their power

with armies because, thanks to farming crops and animals, it was now possible to feed thousands of troops using stored grain made into bread, or tamed animals that could be milked, eaten or used for pulling carts or carrying soldiers into battle. Farming permitted campaigns of attack to glorify and expand these new urban cultures which soon began to spawn all over the ancient world.

How, why and where did this radical shift into farming crops and animals take place? One thing is for sure: there was no bright Stone Age individual who one day came up with the idea of breeding crops and animals because they might serve humans better. Nor did the idea of farming the land begin at one time and in one place.

Some history books make what happened read like a revolution, because it had such far-reaching consequences for both humankind and the earth. They call it the Neolithic Revolution. But to begin with it was a slow process, even though today about 99.9 per cent of humans live in societies that rely on agriculture and animal farming (modern towns and cities are dependent on agriculture from overseas imports or the surrounding countryside to make living in their urban environments viable). The change in human lifestyle from mobile hunter-gatherers into settlements where farmers could support and feed big towns and cities was as remarkable as it was profound.

Common sense suggests that humans didn't take up farming because they wanted to. After all, who would have wanted to be a farmer? These days it's comparatively easy thanks to modern technology, with tractors, ploughs, machines for milking, threshing, bailing and harvesting. But that was not the case 12,000 years ago, when people began to sow wild seeds hoping for a half-decent crop from which to make their first loaves of bread. Compared to the easy life of the hunter, with plenty of game around, and where one decent kill could feed a family for a week, the lot of a crop farmer was painful and arduous. For a start, crops could be harvested only at certain times of the year, so arable farming was certainly no substitute for the traditional fast-food culture of meat on demand.

Planting, weeding, digging and harvesting were just a few of the miserable tasks that had to be endured long before a single loaf of bread could be baked. Seeds from barley, wheat and rye, which were the first crops cultivated by man, had to be collected by hand from the stalks of the grasses and ground up into flour using the most primitive of food processors, a pestle and mortar.

And these weren't the seeds we are used to today. They were natural and wild, not the product of generations of artificial selection. For good reason, nature had designed them to be as light as possible, loosely attached to the stalk so that wind power had the best chance of blowing them far and wide, spreading them to other areas where they could germinate and reproduce in the wild.

But, as any arable farmer will tell you, small seeds that easily fall from the stalk are a bread-winner's nightmare. A large quantity of this type of wheat is needed to make bread, not to mention the backbreaking task of picking up all the loose seeds that tumble irritatingly to the ground.

Unpredictable, unpleasant and just plain hard work – that's what farming crops was like 12,000 years ago. Skeletons of early farmers tell the story: twisted toes, buckled, arthritic knees, and in some cases lower backs that are completely deformed due

to the exhausting task of grinding grain into flour between slabs of stone. Persuading wild animals to do what you wanted them to wasn't much easier. Thousands of years of genetic modification have led to sweet, rich, easy-to-harvest crops and obedient, compliant domestic animals; but back then it was an uphill struggle all the way.

A single word can explain the reasons for the rise of agriculture: stress. The cause 12,000 years ago was exactly the same as it is in many parts of the world today, and probably will be for many generations to come: climate change.

The most recent Ice Age was at its coldest 22,000 years ago. So much water was then locked up in the ice sheets that covered much of northern Europe that sea levels were approximately 130 metres lower than they are today. That's equivalent to dropping the seas by about thirty-three metres more than the height of Big Ben.

Back then there was no English Channel, no Black Sea, and the Mediterranean was low flatland rather than deep blue sea. The land of Beringia, now the Bering Strait, stretched out as icy tundra between Alaska and Russia, providing that vital bridge across which humans and animals passed from Asia to the Americas. And the Red Sea, which now carries ships from the Mediterranean through

Neolithic food processing: the pestle and mortar were vital implements for early farmers, whose livelihoods depended on grinding domesticated wheat into flour.

23:59:59

to the Indian Ocean, via the Suez Canal, was just a flat tract of dry land. As the glaciers melted over the course of thousands of years, rising sea levels caused massive flooding all over the world.

The big melt increased global temperatures by more than 7°C, and was probably caused by cyclical changes in the earth's rotation (see page 107). It reached its peak about 14,000 years ago. At that point the oceans rose by a massive twenty-five metres in just 500 years after a huge ice shelf collapsed into the rising seas.[1] By about 8,000 years ago the major melting was all but over, and the seas were at nearly the same level as they are today. One of the last areas to flood was the English Channel, cutting off Britain from the rest of Europe for the first time in more than 100,000 years.

Such a dramatic and rapid change in the natural environment was bound to have profound effects on living things. For mankind it meant that many traditional hunting grounds simply sank beneath the oceans. Regions of the world that were once rich forests, ideal for hunting and gathering, were reduced to barren deserts as patterns of rainfall and weather systems rapidly changed.

In many parts of the world people were forced to move upwards into the hills, or closer to freshwater lakes and rivers. In some areas the traditional style of life of moving on from place to place became just too risky. There was either too little good hunting ground, or the land was too dry to sustain sufficient vegetation.

One example of how climate changes forced people into a new lifestyle can been seen in what is called the Fertile Crescent, the area that extends from upper Egypt, then down the Nile to lower Egypt, Israel and Syria as far north as central Turkey, and then down towards the Gulf along the Euphrates Valley, through the ancient land of Mesopotamia (modern Iraq and Iran). Fourteen thousand years ago this was a rich land, with forests of oak and pistachio trees, plentiful rainfall and nutritious vegetation. It wasn't at all like the dry, barren land we are familiar with today.

At about that time, people called the Natufians had settled near the water's edge around modern-day Lebanon, because the sea provided them with a good source of fish for food. Others went higher up into the hills, where the soil was richer and where wild grasses grew. In fact, they found that this land was so rich in resources that it wasn't necessary to be on the move all the time. In some seasons they would hunt for wild animals, such as gazelles; in others they would settle in small villages, where they lived in round mud and clay huts for some or all of the year. Several Natufian sites have recently been discovered and excavated in Lebanon, Syria and northern Israel.

What happened next was a freak of nature. It's one that scientists predict could happen again, perhaps soon. Instead of temperatures continuing to rise as they had for the previous 8,000 years, the climate suddenly plunged into another Ice Age. In less than fifty years, most of the world reverted to a state of deep freeze. It was as if the land was suddenly gripped by the evil power of a Snow Queen. This time her spell lasted for about 1,300 years.

This episode, called the 'Younger Dryas', occurred approximately 12,700 years ago. We know it happened because of the evidence in the Greenland ice-cores, which scientists use to gauge global temperatures going back more than 800,000 years.[2] Experts now think they know *why* it happened, too. The Gulf Stream, that conveyor of warm water that brings mild temperatures to Europe, relies on the additional density of salty water to force air and sea currents up and down the north-east coast of America and across the Atlantic. As fresh water locked up in the North American glaciers flooded into the Atlantic Ocean, it massively reduced the saltiness of the sea water, weakening the effectiveness of the Gulf Stream, or even shutting it down altogether.[3]

This would explain how temperatures could see-saw so hugely in just a few years. With much of the fresh water now locked up as ice once again, about 1,300 years later, global temperatures rose by as much as five degrees in just ten years as increased salinity in the North Seas helped the Gulf Stream splutter back into life. Such dramatic and rapid climate changes had probably never been experienced in all human history.

Prehistoric Climate Change

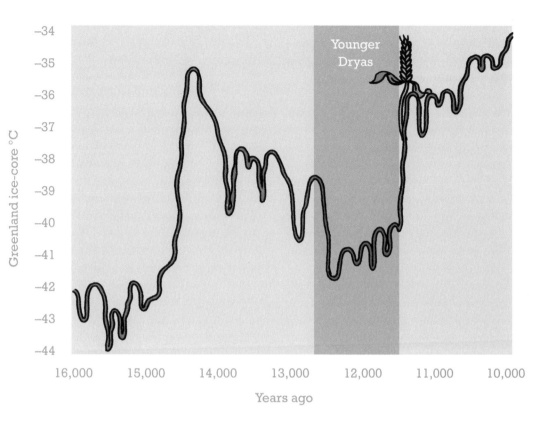

The effect, particularly on people living in Europe and the Mediterranean, was catastrophic. For the people living in the so-called Fertile Crescent, not only had their hunting grounds been drowned by rising sea levels following the Ice Age melt, but now a severe drought set in and much of their remaining rich and fertile woodland was transformed into barren scrub.

Wild grasses such as wheat were an important part of the staple Natufian diet, but in the now sweltering scrubland they simply withered away. Some experts think this is what may have led Natufian women to experiment with sowing seeds themselves, and deliberately clearing the land to make it suitable for cultivating grasses such as wheat, barley and rye.[4] In the face of starvation, these women stored the best seeds they could find, the biggest, sweetest and most easy to harvest, to sow on specially prepared land as a crop for the following year.

Was it their handiwork – an agricultural insurance policy – that triggered a chain of events that eventually led to the spread of crop farming all over the Middle East, Europe, and northern Africa? Seeds are easy to store and transport, and the Natufian women's crop cultivation seems to be the earliest known to history. Evidence of the Natufian people's inventiveness comes from the discovery by modern archaeologists of farming tools, in the form of picks and sickle blades used for harvesting cereal crops. Alongside these ancient farming implements are pestles, mortars and bowls, all essential instruments for gathering and grinding up seeds.

Archaeologists have painstakingly sifted material excavated from one Natufian site called Abu Hureyra in modern-day Syria. What they have found suggests that here was a culture that had learned how to domesticate wild crops by selectively sowing the best-looking seeds. As the wild grasses that people relied on for food died out, they were forced to start

Abrupt cooling and warming after the last Ice Age melt is thought to have provoked some people to experiment with the domestication of crops and animals.

23:59:59

cultivating the most easily grown seeds in order to survive. From the location of seed finds, it seems they planted them on slopes where moisture collected naturally. They then actively managed these hillside terraces and slopes by keeping the weeds and scrub at bay, so giving their crops the best possible chance of producing a good yield.

Natufians were also among the first people known to have started domesticating animals – in their case wolves. By choosing the tamest grey wolves, they eventually bred them into domestic dogs which could help them hunt other animals that lived in the regions nearby – in particular wild sheep, boar, goats and horses. With the help of dogs, it was a relatively small step to tame these other wild animals and breed them in one place for their meat and milk.

Why do some animals become pets (e.g. dogs), or live peacefully in a farmyard environment (e.g. cows), while others (e.g. zebras) do not? Recently scientists who have been studying the histories of these animals have discovered that for a wild animal to be successfully domesticated it must have at least these three characteristics: it must live in a herd; it must live in a community with an organized social structure (so that in its mind a human can become the leader of its pack); and finally, it must not mind sharing its grazing territories with animals of other species.[5]

Out of approximately 148 large mammal species, only about twelve meet these three criteria (i.e. dogs, sheep, goats, cows, pigs, horses, camels, alpacas, donkeys, reindeer, water buffalo and yaks – some other animals have successfully been tamed, but not bred into more tame varieties). All these species have been selectively bred in a way that suits human beings. In the Fertile Crescent, at least three of these twelve species lived in the wild – wolves, wild sheep and wild goats. By 8,000 years ago these early farm animals were joined by the first domesticated pigs and cows, and then, about 6,000 years ago, by donkeys and horses which roamed in the wild grasslands and forests of the Eurasian steppes connecting Europe to the Far East.

Natufian people loved their dogs. Graves have been found in which they and their dogs are buried side by side.[6] Their graves also reveal another tell-tale sign of early animal domestication: a high infant mortality rate. One third of all Natufian graves unearthed so far contain the skeletons of children under the age of eight. Were these victims of the first diseases to mutate from animals and jump across to humans? If so, it points to the beginnings of a new type of human selection – people naturally more vulnerable to these new diseases died more often than those less susceptible. As generations passed, people who lived in close proximity to domesticated animals gained a greater immunity from the diseases they spread (for some of the effects of selective immunity on human history, see page 287).

Once the Younger Dryas period ended, about 11,400 years ago, the climate recovered its previous balminess, and within the space of just a few years people in the Fertile Crescent were once again living in a land of plenty, with enough rainfall to support rich, diverse vegetation. But now there was a big difference. These people were equipped with a raft of potent new technologies, in the form of breeds and seeds that gave them the opportunity of living a radically different way of life.

From about 9000 BC permanent new human settlements began to appear throughout the Middle East. These 'Neolithic' farming people were now able to live in larger communities thanks to an abundance of stored food, gained from a knowledge of farming and the benefits of domesticating animals for their meat, milk and pulling power. Hunting and gathering was for some now becoming a tradition of the past.

Jericho is one of the oldest Neolithic towns. It is up to eight times larger than earlier Natufian sites, and is thought to be one of the first to have city walls. Excavations have revealed rounded houses, many with more than one room, and open spaces for domestic activities such as cooking and washing. These early buildings have stone foundations, cobbled floors and walls made of mud/clay bricks. Every site has its own stone or clay silo for storing food and grain, a sure sign that, for these people at least, the days of living on the move were now long gone. Necessity had

forced them to adapt nature to their own needs, resulting in a new way of life.

Walls were erected on Jericho's western side – not, as once believed, to defend the city from attack by jealous neighbours, but as a means of protecting it from mud flows and flash floods that frequently swept in from the still rising seas. It was another sign that here the human spirit was newly focused on trying to control and tame nature.

That these people were in touch with other emerging cultures is beyond doubt. Obsidian is a form of natural glass which forms when volcanic lava cools quickly. It was a highly sought after material because it made the sharpest, most effective arrowheads. Obsidian occurs naturally in the rocky hills of central Turkey, but has been found hundreds of miles away in Neolithic Jericho, showing that long-distance trade routes were already well established. Why not exchange glass for precious seeds, already modified to be excellent

to eat and easy to harvest, the fruits of more than a thousand years of special selection and hard graft? It is not difficult to imagine how, once under way in one place, agricultural know-how, seed supplies and domesticated animals quickly diffused throughout Europe, the Middle East and beyond.

Obsidian is a naturally occurring form of volcanic glass, highly prized in ancient times because it was sharp and tough – ideal for making tools, blades and arrowheads.

Part 3
Settling In
(5000 BC – c.570 AD)

MAN BECAME SKILLED at harnessing the power of evolution to make his civilizations more powerful. Technology for protecting against nature's tantrums grew more sophisticated, and man learned how to pass on knowledge from one generation to the next by expressing himself clearly in languages that could be written down.

Populations increased. Close to the great river valleys, the first human civilizations emerged. Pyramids were built for burying men's god-like kings, ziggurats let holy priests stargaze into the future, and circular henges hosted festivals for worshipping the mother goddess, the sun and the moon.

Wheels, chariots and horses, reinforced by the discovery of bronze, iron and steel, led to fierce contests fought between people who travelled from place to place with their flocks, and those who chose to root themselves to one particular spot in newly built towns and cities. Inequalities between militarized nomads and defenceless settlers led to a widespread dash to make arms and build defences. Forests were hacked down, metals were smelted in smoky fires, and new roads cut scars into the earth's supple crust. Ships, horses, camels and feet carried goods made by townsfolk from one settlement to the next. Peasants, priests, kings and armies jockeyed for position as the world's natural riches surrendered themselves to humanity's changing forms.

Clever religions explained why sometimes nature's violent power bit back. Greek philosophers discovered a set of universal laws that they could use to predict the passages of the planets, the sun and the moon, leading them on further quests to find out how to build ever more powerful artificial worlds. By 2,000 years ago, human numbers had swelled to more than 200 million individuals, directly challenging the balance of nature's traditional ecosystems. Former beasts of the forest became humanity's main source of food, clothing and power, as well as real-life characters in deadly dramas that titillated the masses in theatres of entertainment.

Some people still lived as tribes, in a state of nature. They were best-placed to weather whatever storms or droughts nature threw at them, their lifestyles staying simple, nimble, mobile, sustainable and usually at peace. Taboos regulated what natural resources they used, and helped them to live together within collaborations of small, semi-independent clans. Farming was independently discovered in parts of the Americas a few thousand years after its first appearance in Europe and Asia. Civilizations there learned about the seasons and the stars, mathematics, writing, building and crafts. But theirs were 'wheel-less' worlds, whose lack of large animals limited their power over nature and made them vulnerable not just to changes in the climate, but to bloodthirsty conquistadors from the other side of the world, who finally hunted them down in a ruthless search for gold.

Written Evidence

How the art of writing ushered in an era known as 'history' in which merchants, rulers, artisans, farmers and priests established the first human civilizations.

THE DIFFERENCE between history and prehistory can be summarized in a word. Writing allowed people to keep records of what happened and when. It meant they could pass on stories to other generations. With writing came the beginning of what is called 'recorded' history. Everything before is 'pre-history', or prehistoric.

Of course, before writing there were already plenty of stories, but they were told by word of mouth. Sometimes stories that are passed down from one generation to the next survive well enough without being written down, but they easily become confused, and often turn into fantasies or myths. The only way they can be properly analysed and interpreted accurately is if they are written down, so they do not change over time. Recorded human history really begins at the same time as writing first appears – and that happened in the earliest human civilizations of the Middle East, about 5,000 years ago, which takes us to within a tenth of a second to midnight on the twenty-four-hour clock of earth history.

Try to imagine a world without the written word. No books, no newspapers, no letters, no laws. Without writing, human civilizations and empires on a large scale would probably never have emerged, let alone survived. Written words allowed rulers to control people from far away through indelibly inscribed royal decrees, civil laws and military orders.

The Rosetta Stone, found in Egypt by Napoleon's troops in 1799, unlocked the key to deciphering Egyptian hieroglyphics thanks to the same text being chiselled in three different languages.

23:59:59

No one knows who actually invented writing. It is highly unlikely that any one person did. People may have begun to experiment with the first scratches and scribbles as long as 10,000 years ago, as a way of keeping track of the cycles of the moon and stars.[1] However, it was only about 5,000 years ago that the first clear use of written symbols by a settled civilization appears as a way of keeping commercial records and accounts. They may have been merchants who traded valuable goods like obsidian, crops like barley and wine or precious raw materials such as rare stones, gold, silver, copper, tin and iron.

Merchants of the Middle East drew simple pictures on clay tablets to identify particular goods, and next to them they scraped counting marks to show a quantity. These tablets were baked in ovens to make their marks permanent, creating an unchangeable set of records showing exactly who had received what goods. Writing helped people manage their accounts of trade and exchange.

But making drawings on clay was a time-consuming and laborious business. It made more sense to come up with a shorthand code to speed the process up. Over time, wedge-shaped strokes replaced the pictures, because they were easier and quicker to mark onto the tablets. These strokes were made using a kind of pen made out of reed,

This 4,350-year-old Sumerian merchant's tablet, inscribed in cuneiform, details accounts of commercial transactions in goats and sheep.

in the shape of a modern-day cutting knife. This style of writing is called cuneiform, and it forms the basis for three of the oldest written languages in the world: Sumerian, Assyrian and Babylonian.

Sumeria cuts right through the heart of modern-day Iraq, as far south as the Persian Gulf, where it connects to the Indian Ocean. It was one of the first regions where mankind's new itch to control nature extended into the business of building artificial worlds, in the form of cities and states. It is also where experts believe writing emerged.

Sumeria was a perfect dwelling place for early settled communities of humans. By 10,000 years ago sea levels had risen by nearly 130 metres from their low point, and in this part of the world the climate was wetter, and therefore better for growing crops, than it is now. It is only in the last 5,000 years or so that temperatures have increased and rainfall reduced to make the Middle East the sandy, barren land we know today.

A wetter climate was ideal for growing crops such as wheat, barley and grapes, that need winter rainfall. And just the right kind of wild animals – those perfect for domestication, such as goats, sheep and oxen – lived on the slopes and hillsides of the region. Such animals could be used as a source of food, as power for pulling ploughs and carts, and to provide raw materials for making clothes, bottles and leather goods.

The ancient region in which the first Sumerian cities emerged is called Mesopotamia, and its name gives a big clue as to why it was here that humans were able to build their first states. In Greek it means 'between the rivers'. The Euphrates and the Tigris proved ideal for supplying water to nearby land through systems of man-made irrigation channels, dykes, reservoirs and dams. These meant people could purposely flood their fields to provide just the right conditions for their artificially chosen crops to thrive. The river valley also provided a large, long, flowing superhighway to carry people and their possessions from one riverside city to the next.

Two other great ancient river valley civilizations also arose in the Middle East. One was along the banks of the Nile, in Egypt, the other along

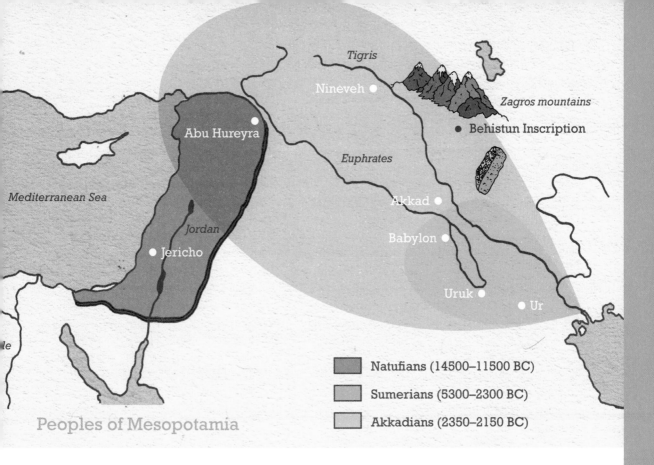

Natufians (14500–11500 BC)

Sumerians (5300–2300 BC)

Akkadians (2350–2150 BC)

Peoples of Mesopotamia

the Jordan in Israel. Exactly which society emerged first is a matter of some debate, because archaeologists are constantly finding new evidence of complex early civilizations going further and further back in time.

Writing is one of the best signs of an advanced early human civilization. Its emergence meant that at least some of the people living there were traders who didn't have to bother with farming, and that others were manufacturers who made things worth trading. Writing is a sure indication that a society has in it people with time to think (e.g. priests), to invent (e.g. artisans and craftsmen), to organize (e.g. bureaucrats) and, of course, to rule (e.g. kings). Written words are hallmarks of civilizations that have mastered the means of mass food production and have successfully divided up tasks between their populations. While some still specialize in gathering food, others keep order, or develop skills in craftwork or trade (note the similarities with nature's oldest civilizations, such as bees and termites – see pages 66, 68).

We know that a writing system developed in ancient Sumeria about 5,000 years ago thanks to one of the most remarkable archaeological discoveries of all time, which was made in the 1840s by a young amateur British archaeologist named Austen Layard. Rather than stick around in London to practise law, as was his training, Layard decided to head off for a life on the island of Ceylon (now Sri Lanka), off the southern tip of India, which was then under British rule.

Layard never got to Ceylon. He stalled in the Middle East, in Persia, where he became fascinated by the history of the region, and in particular by a strange large mound near the town of Mosul, on the banks of the Tigris. He was so curious about this odd-looking man-made hill covered in dust and sand that he persuaded the British Ambassador in Turkey to pay for an archaeological dig to see what lay beneath it. On 9 November 1845, Layard, with a team of local tribesmen, started excavations. Within hours their brushes and spades revealed the walls of an ancient palace covered with stone slabs,

Following early attempts at plant and animal domestication by the Natufians, large-scale human civilizations emerged in nearby river valleys.

each one tightly inscribed with a curiously shaped form of unknown ancient writing.

This wasn't just a fancy royal palace with a few old graffiti marks that Layard and his team had stumbled across. After a series of excavations, two palaces and a huge royal library had been unearthed on the site, which turned out to be what remains of the ancient biblical city of Nineveh.

The library was built by Ashurbanipal, the last great king of Assyria, who died in 627 BC. He was a scholar and an avid collector. Unlike most kings of his day, Ashurbanipal could read and write, and he was well known for being able to solve complicated mathematical problems. Layard and his team uncovered a staggering 20,000 clay tablets from Ashurbanipal's library, including king lists, histories, religious texts, mathematical and astronomical treatises, contracts, legal documents, decrees and royal letters. They provide a fascinating insight into ancient times that has transformed our understanding of when and where the first civilizations emerged, and what they were like.

Of course, before anything useful could be learned from these precious tablets of clay, someone had to work out how to read the texts on them. As luck would have it, at about the same time as Layard and his team were uncovering the tablets from Nineveh, a British army officer serving in Persia made another extraordinary discovery – one that unlocked the key to understanding how to read the cuneiform writing.

The Rosetta Stone (see page 120), currently in the British Museum, is a famous rock found in Egypt by Napoleon's troops in 1799 that contains the same passage of writing in three different languages, two of them in ancient Egyptian hieroglyphic writing and the third in classical Greek. In 1822 the French scholar Jean-François Champollion deciphered the stone, which became the key to translating hieroglyphics. What Henry Rawlinson discovered high in the foothills of the Zagros mountains on the Iran–Iraq border was no less spectacular. A hundred metres up a limestone cliff alongside an ancient road leading out of Babylon in Iraq, he found a series of statues cut into the rock, each with a passage of writing beneath.

These carvings, called the Behistun Inscription, tell the story of the conquests of King Darius, a Persian king who ruled the area from 522 to 485 BC. As on the Rosetta Stone, the stories are told in three languages, but because they were so high up, no one before Rawlinson had tried to read what was inscribed. In 1835 Rawlinson scaled the cliff and copied down the inscriptions. He found that the first part of the text was a list of Persian kings, which exactly matched one found in the histories of the Greek writer Herodotus. This gave Rawlinson the clue he needed to work out how to read the letters of ancient cuneiform writing, and thanks to him experts have since been able to decipher most of what is written on the precious tablets discovered by Layard in Ashurbanipal's library.

Probably the most famous tablets from the Nineveh hoard are those that tell of the adventures of an early king of Sumeria called Gilgamesh. He ruled over one of the first Sumerian cities, called Uruk, situated on the east bank of the Euphrates now in southern Iraq. At its height as many as 80,000 people lived in Uruk, making it then the largest city in the world. Gilgamesh built a series of thick, high walls to defend the city against attack. Recent excavations by a German team of archaeologists show that the people living there also constructed an intricate network of canals inside the city walls. It was like an ancient Venice, with a series of elaborate temples and towers dedicated to the gods.

Gilgamesh was the fifth king of the city, and ruled in about 2650 BC. He became a highly revered figure for all Mesopotamian people, and a series of famous myths and legends tells of his dangerous and daring deeds. Gilgamesh was not a good king at first, so the gods made a wild, hairy man, called Enkidu, to fight him. But just as men had learned to domesticate wild animals, Gilgamesh was able to tame Enkidu, and the two soon became great friends, going on many adventures together. When Enkidu died Gilgamesh was heartbroken, and realized that although he wanted to live for ever, he too would die some day. By the end of the tale Gilgamesh had decided that being remembered for creating a beautiful city with impressive walls

and fine temples to the gods was his best chance of immortality.

Parts of this legend appear in stories found in later civilizations, showing how ideas and tales, as well as language and writing, travelled far and wide. In one of Gilgamesh's adventures an old king called Uta-napishti tells him of a terrible flood sent by Enlil, the god of the air, to punish mankind for his evil ways. The gods tell Uta-napishti to build a boat, which he and his family then board, accompanied by many creatures. After he seals the hatch, a deluge comes. It is so fierce that it 'even frightened the gods who fled to heaven'. All the humans in the world except Uta-napishti and his family die in the floods. When the rains subside, the boat runs aground on the top of a mountain. Uta-napishti releases a dove, then a swallow, then a raven, which does not return, showing that they are near to land. This ancient Sumerian story is, almost word for word, the same as the account of Noah and the Ark in the Jewish Torah and the Christian Bible.

Some experts believe that this story, and similar flood myths in other early religions, originated from the dramatic rise in sea levels that flooded the Persian Gulf at the end of the last Ice Age. They think such stories were passed down through generations by word of mouth for thousands of years, before being written down once the art of writing had evolved.

The Epic of Gilgamesh, and other stories found inscribed on clay tablets, tell us a great deal about how the Sumerian people viewed the world. They give us, for example, some of the first written evidence of religious beliefs.

This Syrian carving from the ninth century BC shows the Sumerian hero, King Gilgamesh, supporting the Sun (Enkidu) with the help of two demi-gods.

Gilgamesh realizes that all humans are just servants of the gods. Gods helped explain to people why unexpected things happened, such as floods, droughts and invasions. A common feature of all human civilizations is that they try to influence nature in order to protect themselves or increase their power. The Sumerians tried to do this by building temples in their cities, each dedicated to a different one of the gods, who it was believed were responsible for everything from love to war and good harvests. Human sacrifices were sometimes carried out because it was thought they would please the gods.

Another source of information about ancient Sumerian religion comes from the remains of clay seals that were stamped on the lids of jars of wine or oil to identify their owners. People believed these seals had magical powers, and could shield them from harm, so they decorated them with pictures of the gods for added protection.

The Sumerians believed that the gods met each New Year's Day to decide what fateful events would happen in the coming year. Their decisions resulted in all manner of disasters, such as droughts and floods, as well as unexpected good fortune like bumper harvests and military success. Aside from these annual fates everything else was, they believed, predetermined by the stars.

The Sumerians and their successors in Assyria and Babylon believed that the world rested on a flat disk, surrounded by water on all sides. Above the sky was a tin roof punctured with small holes through which the celestial fires of heaven could be seen.[2] They studied these holes (the stars), and watched them rotate each night along a predictable path. They discovered that five large stars behaved in a different, unexplained way. They believed these were the stars of the gods – we know them today as the five planets visible to the naked eye: Mercury, Venus, Mars, Jupiter and Saturn. Some of these stars were thought to bring good luck, and others bad. For instance, Mars meant war, and Venus love. The Sumerians dedicated one day of the week to each of the five randomly moving stars – with the sun and the moon, that made seven. The names we use in English, derived from the Latin language later used by the Romans, shows the legacy we still owe to the Sumerian system: Saturday (for Saturn), Sunday (for the sun), Monday (for the moon). The link is clearer in French for the other weekdays: mardi (Mars), mercredi (Mercury), jeudi (Jupiter) and vendredi (Venus).

The Sumerians constructed towers, called ziggurats, so they could be closer to the heavens. These were terraced pyramids built from sun-baked clay bricks. The top of each tower was flat, on which was built a shrine or temple to a god. Only priests were allowed inside them, since these were believed to be the dwelling places of the gods. About thirty-two still survive today, most of them in Iraq. One of the largest, in the city of Babylon, is dedicated to a god called Marduk, and may have been the original inspiration for the story of the Tower of Babel in the Bible.

It isn't just our seven-day week that we owe to these ingenious people. They were also prodigious mathematicians. Amongst the clay tablets found by Layard is evidence of complex arithmetic, with different combinations of vertical strokes and V-shapes used to represent the numbers one to nine. The Sumerians developed a system of mathematics based on the number sixty, because there are so many ways of dividing it up (by 2, 3, 4, 5, 6, 10, 12, 15, 20, 30). They also used the world's first 'true place-value' system, where numbers written on the left have larger values than those next to them, as in our counting system today. We have also inherited from the Sumerians our sixty-second minute, sixty-minute hour, twenty-four-hour day and twelve-month year, as well as the twelve-inch foot, 360-degree circle and 'dozen', meaning group of twelve. Other tablets show that the Sumerians had symbols for division and multiplication, and that they were able to work out squares and square roots.

Their genius for astronomy and mathematics was matched by their inventiveness at making things with their hands. These people are credited with inventing the wheel. These weren't used for carts or chariots, although they were to come soon after. They were wheels for making clay pots – pottery wheels. It didn't take long for the wheel to be adapted as a device for transporting goods. Tamed asses were used to pull the first carts, and

later solid wooden wheels were replaced by spoked wheels that could carry more weight, making them ideal to support chariots of war.

But travelling on a cart was not nearly as convenient for these people as going by boat, since there were few roads, and paths were designed for walking. Instead they made use of the rivers that connected their settlements to the sea. Sumerians designed at least three different types of ship. Some were made from animal skins and reeds, others were stitched together with hair and incorporated bitumen waterproofing and wooden oars. Ships allowed Sumerians to trade with more distant groups of humans who were also beginning to settle into their own cities and civilizations. Sumerian technologies, from the wheel to the word, soon spread throughout the navigable world.

Sumerian craftsmen made precious objects out of soft metals such as silver, gold and copper. During the 1930s, Leonard Woolley, a British archaeologist, unearthed more than 1,800 graves at the royal cemetery of Ur. Inside one tomb he found some of the most extravagant and precious riches of all antiquity. They had been completely untouched for thousands of years. This was the tomb of Pu-Abi, Queen of Ur, dating back to about 2500 BC. Alongside her were buried five soldiers and twenty-three ladies-in-waiting. They had been poisoned so they could continue to serve their mistress in the next world. Alongside, Woolley discovered the most magnificent treasure pile, including a head-dress made of golden leaves, a superb lyre with a gold-encrusted bearded bull's head, golden tableware, a chariot adorned with a lioness's head made of silver, and many golden rings, necklaces and bracelets. Some of these treasures can today be seen in the British Museum, some in the University of Pennsylvania Museum in Philadelphia, and some in the National Museum of Baghdad.

A Sumerian ziggurat at Nineveh. These towers let priests get closer to the heavens, where they painstakingly studied and recorded the movements of the moon, the planets and the stars.

should take over the land and property of dead people was a big deal. What was supposed to happen when a son claimed the land of his father, but his siblings or someone else said they had a prior claim? Or when a person died with no heir?

Typically, either the dispute would escalate into a fight between families, a battle between gangs or, if it got really bad, a war between cities. Alternatively, thanks to the emergence of writing, a strong ruler could impose his will through a series of written edicts and a code of law, backed up by the threat of punishment. Thanks to the invention of writing, people living in towns and cities could develop an alternative to settling disputes by means of violence. Laws could be written down, agreed, publicized and administered by a ruler's bureaucrats. Rulers in each city of Sumeria developed their own laws. By far the most famous was a king called Hammurabi who ruled in Babylon, which grew into a powerful city situated north of Uruk and Ur, between the two great rivers of the Tigris and the Euphrates.

Babylon's rise to power started in around 1900 BC. Hammurabi lived from 1810 to 1750 BC, and it was his code of laws that transformed and stabilized the city, turning it into the most powerful in all Mesopotamia. A copy of Hammurabi's code of 282 laws was prominently displayed on an eight-foot-tall slab of stone in the centre of the city, so that everyone could see it.[3] Ignorance of the law was not accepted as an excuse, a principle that lives on in most societies to this day. Hammurabi had his laws chiselled on to stone so that they were unchangeable: we still use the phrase 'set in stone' to describe something permanent.

Hammurabi's laws were copied by other civilizations, and they set several important principles that are still cornerstones of justice in many parts of the world today. For example, they established the principle that a person is innocent until proven guilty. But to maintain proper order, these laws were necessarily harsh: 'If a man put another man's eye out, his eye should be put out also.' Another one that probably didn't provide a great incentive for people to study medicine went as follows: 'If a patient dies in or after surgery, the doctor's hand will be cut off.'

Woolley's discovery shows that these people had developed magnificent craft-making skills. Their skill in metalworking also allowed them to make weapons and armour. The discovery of how to mould copper, bronze and eventually iron allowed early human civilizations to equip themselves with a range of cutting-edge weapons such as swords, spears and arrows for defence, conquest and invasion.

Sumerian cities like Uruk and Ur that rose up along the banks of the Tigris and Euphrates rivers were constantly at war with each other. Fighting frequently broke out because it was unclear who should inherit the land and property of someone who had died. The concept of private ownership had no place in traditional hunter-gathering societies, since everyone moved from one place to another, and no one owned anything (see page 104). But in a city, where people had built their own houses and dug their own irrigation channels to water their own fields, the question of who

Laws were useless if no one could read them. So for the rule of law to be effective, a stronger emphasis had to be placed on education. Most Mesopotamian cities had public libraries. Men and women were encouraged to learn how to read and write, and a Sumerian proverb had it that 'He who would excel in the school of the scribe must rise with the dawn.' The Epic of Gilgamesh itself was frequently used as a text to be copied out as practice by people learning to read and write.

Like all human civilizations, even the ingenious Sumerians could not survive for ever. In the end, it wasn't war and invasion that led to their decline and fall. Something much more dramatic and unstoppable halted these inventive people. They learned that living a life in one fixed location, rather than moving from place to place as hunter-gatherers do, came at a considerable price. After many generations of intensive farming the land became less fertile, owing to increasing levels of salt, which spread to the fields through artificial irrigation. To start with, the people responded by switching from growing wheat in favour of barley, which could tolerate higher salt levels. But before long even that crop just withered away as the soil turned sour. By about 2000 BC the land around the mouths of the Euphrates and the Tigris had become impossible to farm, and cities like Ur and Uruk fell into permanent decline.

Their misfortune was another's opportunity. The mighty Assyrian king Sargon the Great (ruled c.2270–2215 BC) built one of the world's first empires around Akkad, a city located hundreds of miles further up the Euphrates, where the land was still rich and fertile. A seventh-century BC clay tablet describes how Sargon's mother cast him off as a baby in a basket of rushes. Eventually he was found and cared for by the king's water-drawer, Akki, and reared as his son. This echoes the story of Moses, who it is said in the Bible came from the Sumerian city of Ur.

As the southern Sumerian cities declined they fell victim to Sargon's conquests, becoming part of his enormous new domain that stretched from south-west Iran to the Mediterranean coast. Cultures diffused two ways. While the Akkadian

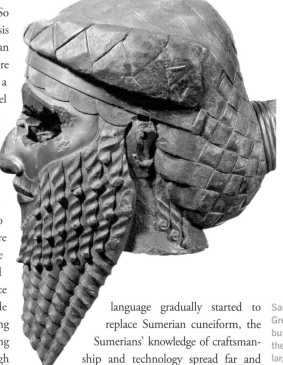

language gradually started to replace Sumerian cuneiform, the Sumerians' knowledge of craftsmanship and technology spread far and wide across the vast Akkadian empire.

Unfortunately for the ancient people of the Middle East, a long period of wet weather (known as the Holocene Climatic Optimum) came to an end in about 3200 BC, making their lives even harder. The land dried up fast, and much of it turned into the desert we know today. From about 2000 BC almost all the cities in Sumeria became uninhabitable because of salt poisoning and increasing droughts caused by climate change.

By evolving a system of writing, the Sumerian civilization allowed recorded history as we know it to begin. The written word meant that knowledge could be transferred, without error or change, from one part of the world to another, and from one generation to the next. It was one of the most potent tools for organizing the construction and administration of man's first artificial worlds.

Sargon the Great, who built one of the world's largest early human empires, based around his capital city Akkad located on the banks of the River Euphrates.

23:59:59

Divine Humanity

How nature's bounty helped some rulers become living gods in the eyes of their subjects, commanding supreme obedience, total devotion and absolute protection – even in the afterlife.

NO HUMAN SO FAR in our story has ever claimed to be a god. Shamans of the hunter-gathering people's caves venerated the spirits and the gods of the earth, sky, beasts and woods, but there is no suggestion that they ever thought that they themselves were part of divinity. Rather, these ancient people were so in awe of the gods that they sensed their presence all around, from the pinpricks in the tin roof of the heavens to the awesome forces of floods, thunder, lightning, sunshine, moon, rivers, woods and war. So, to make the leap from seeing the gods as other-worldly to regarding them as real, living, breathing, walking and talking humans is a big one. What power and magnificence would be bestowed on

the person who managed to convince others that he was a god on earth!

According to one early human civilization a member of *Homo sapiens* could indeed be a living god, possessing that most precious gift so fruitlessly craved by King Gilgamesh – divine immortality. They called him Pharaoh.[1] He ruled a stretch of North Africa which we now call Egypt through a succession of more than thirty dynasties lasting about 3,000 years.

Pharaoh was all-powerful. His people created for him extraordinary monumental buildings in the forms of palaces, temples and tombs – the only survivor of the famous Seven Wonders of the Ancient World is the Great Pyramid of Giza, built as a tomb for one of the earliest pharaohs, called Khufu (known as 'Cheops' in Greek), who died in 2566 BC. This monumental construction originally towered skywards by a massive 147 metres – that's over fifty metres taller than Big Ben. It still contains more than two million blocks of stone, each one weighing more than a pick-up truck. Hundreds of thousands of people worked to build structures like this. Modern experts are still at a loss to explain how the ancient Egyptians could have cut, transported and hauled into place so many huge blocks of stone, punching them upwards into the sky from the flat, sandy desert in defiance of everything natural around. Such power over other men and nature had never been seen before.

The Egyptians were the first example of a human civilization whose rulers amassed extremes of wealth and absolute power over men. Their unprecedented riches and glory were underpinned by a belief that when they left this world they would join the gods in heaven for all eternity. Those who curried sufficient favour could be taken along too, if Pharaoh so chose, entering into a blissful life amongst the reeds of everlasting peace. Why did these people start to believe that man himself could potentially be counted amongst the gods? What caused them to feel so special, so much more important than any other living creatures that they could imagine themselves, or rather their souls, as immortal?

From about 6,000 years ago nature gave these aspiring all-powerful human rulers a big helping hand, in the form of a river and some dramatic changes in the climate. Together they transformed the north-eastern tip of Africa into one of the most fertile and best protected lands on the earth. By the time of Gilgamesh and the rise of the Sumerian cities of Uruk and Ur, a number of large towns and cities had also arisen along the banks of the River Nile. The pharaohs of what is called the Old Kingdom were already at the height of their powers.

It used to be thought that much of what made Egypt great came across the Red Sea from Mesopotamia in the form of trade – seeds, dates, craft-making skills, technology and even writing. Nowadays, historians are not so sure. Did the Egyptians simply borrow and then adapt these hallmarks of civilization from across the seas, or did they develop them independently for themselves? Probably a bit of both. Trade and interchange were established well before the first pharaohs, but what gave the Egyptians a spectacular advantage over any previous civilization was unique to them: their river and the surrounding landscape.

Unlike the rivers of Mesopotamia, the Nile naturally floods once a year. From July to September, as the rainwater flows down from the Ethiopian mountains, the Nile's banks burst, flooding the land for miles around. The floods bring with them a supply of fresh, nutrient-rich soil, earth and sediment – perfect for growing crops.

With a natural supply of nutrients and a fresh deluge of rainwater each year, there was no risk of salt poisoning here. Following the end of the Ice Age 11,000 years ago, North Africa was a verdant land of rolling grasslands dotted with trees and vegetation. Over the years, hunter-gathering tribes

The pyramids of Giza in Egypt are the only ancient wonders of the world still standing – lasting testaments to the power of Pharaohs over men.

This 3,500-year-old tomb painting shows Egyptian sailors taking advantage of the Nile's prevailing winds as they power back up the river, which was blessed with natural two-way navigation.

established themselves near the Nile, settling into small villages and communities. They learned to domesticate the wild cattle, goats and sheep that grazed the savannah, providing them with plentiful supplies of milk, wool and leather. Over time, knowledge about farming crops such as wheat, barley, grapes and flax had reached them via nomadic traders from Mesopotamia, and across the land from people like the Natufians. These river-dwellers were now ideally placed to grow into a rich and powerful civilization.

They also had another advantage. From about 6,000 years ago the land around the upper Nile began to dry out – partly as a result of cyclical changes in the earth's axis that re-directed rainfall patterns and partly because new human activities such as growing crops and herding animals such as camels reduced natural water levels. By 4,000 years ago what was once a landscape full of crocodiles and hippos wallowing in plentiful streams of water, had become the arid land we know today as the Sahara Desert.[2]

Shifting patterns of rainfall account for so much advantage and disadvantage in the story of human history. This time the encroaching desert was good news for these people, because it provided them with an almost impenetrable barrier to invaders. There was no need for defensive city walls, towers, castles or elaborate military installations here. From about 2000 BC the only way other people could disrupt the ancient Egyptians' way of life was either to cross hundreds of miles of barren desert or to come by sea, which was an equally daunting challenge due to a natural defence shield in the form of the boggy, reedy marshlands of the lower Nile delta. Thanks to these natural barriers, the Egyptian people lived in relative peace and security for much of their history, able to develop their own way of life with little outside interruption.

The Nile brought another gift, too – one which helps explain why it was here that such powerful rulers were able to rise up and take for themselves the title of god. The river provided a two-way causeway that allowed easy passage up and down the country. Most river systems are effective as one-way downstream networks. A return journey meant either travelling by dirt track and hauling the boat back upstream, or rowing against the flow. But the

Egyptians found that it was almost as easy to travel up the Nile as down it. The arrangement of tectonic plates means that the prevailing winds across Egypt blow north to south – in the opposite direction to that in which the river flows. A vessel could simply float downstream, then raise a sail for the return journey. What could be better for controlling a kingdom than a well-protected, fertile valley with an easy-to-navigate, two-way river system? Nowhere on earth had as many helpful natural ingredients to aid the growth of an advanced human civilization as did ancient Egypt 5,000 years ago.

Legend has it that in about 3150 BC a king called Menes united the lands of Upper and Lower Egypt. It was he who began the 3,000-year-long reign of the pharaohs, during which time the Egyptian way of life changed remarkably little. After the lands of the Nile were unified the pharaohs of the early dynastic period and the Old Kingdom, which lasted for about 800 years, until around 2200 BC, quickly gained overall control by installing their own local rulers in forty-two separate regions. The king's governors imposed taxes on the people – in the form not of money, but of food. The idea was that if the weather turned bad or the river floods were weaker than expected, there would still be plenty of food kept in a central store to support the needy population. It wasn't hard for a population to worship their ruler as if he were a god when it was he who provided their only insurance in the event of a run of poor harvests.

Pharaoh had to have somewhere to store all this food, hence the need for some of his taxes to be paid in the form of manual labour to build huge granaries and storehouses. The Nile's floods meant that the farmland all round the river was underwater for at least three months a year – usually from late June until the end of September. During these months hundreds of thousands of peasant farmers could sail downstream, construct for their Pharaoh some of the most magnificent buildings ever made by humankind, and then head home on the favourable winds.

All this helped make Pharaoh – in the popular mind – a living god on earth. It therefore followed that everything possible should be done to make sure that when their god departed this world, his soul should pass as effortlessly as possible into the next life. Here Pharaoh could continue to protect the people from other gods and ill fates such as war, drought, famine and disease. To prosper in the afterlife, the Egyptians believed that the souls of the departed, including that of their Pharaoh, revisited their earthly bodies during the night. In this way their souls could be constantly supplied with food and nourishment.[3] For this reason the Egyptians built an elaborate system to allow the souls of the dead to move freely between earth and the next world just as easily as when in life they might have taken a trip up and down the Nile.

Awesome tombs were constructed for the pharaohs, their families and friends, initially in the form of pyramids, the largest of which were just south of modern-day Cairo, at a place called Giza. More than a hundred pyramids were built during the Old Kingdom, but only three massive structures

An ancient papyrus scroll shows the god Anubis weighing a recently deceased person's heart to determine its soul's fate.

survive to this day – the biggest being the one built by the Pharaoh Khufu. This pyramid took twenty-three years to build, and used the labour of more than 100,000 slaves and farmers. Originally, this wonder of the ancient world was cased in brilliant white limestone and topped off with a gold cap. The purpose of this massive monument was to provide an everlasting structure in which to store Pharaoh's body so he could use it again in the next life.

Dead bodies were preserved using a process of mummification, learned over many generations, which typically took as long as seventy days to complete. All the body organs were cut out and placed in a series of canisters called canopic jars, including the brain, which was pulled out of the head through the nose using a special instrument with a hook on the end. The heart was the only organ left in the body, so it could be weighed in the next life by the gods to help judge if the person had lived a virtuous life on earth.

The body was dried out with salty crystals and then stuffed, covered with oils and ointments and finally wrapped in bandages. The completed mummy was packed inside a coffin which, in the case of a Pharaoh, was placed at the heart of the pyramid, in the king's burial chamber. Surrounding the body was everything that the dead Pharaoh could possibly need in the afterlife: food, drink, pets (mummified, of course), games, toys, crowns, tableware, daggers, spears, clothes, books, pictures and magic spells …

The tombs of important people contained teams of servants called *shabti*. These were dolls, sometimes carved out of wood, sometimes of semi-precious stone. Their purpose was to come to the assistance of the dead soul whenever he or she needed help. In some tombs as many as 400 dolls have been found, some of them holding hoes, sickles or ploughs to help the soul perform agricultural tasks in the next world. Before they could do this, a priest had to recite a spell over them:

'Oh Shabti, if the deceased be summoned to do any work which has to be done in the realm of the dead, to make available the fields, to irrigate the land, or to convey sand from east to west: "Here I am," you shall say, "I will do it".'

Relatives frequently visited the tombs of the dead to keep them freshly stocked with food and provisions. Some people had vivid pictures of activities from everyday life – sowing crops,

Shabti figures, designed to accompany and assist souls in the afterlife, were left inside boxes in the tombs of the dead. Priests tried to bring them to life by casting spells.

hunting birds, eating a meal – painted onto their tombs or coffins, or on scrolls made from the stems of the papyrus reeds once common in the marshes of the Nile delta. They believed that these vignettes could come to life by magic in the next world, to keep the souls of the dead constantly nourished and refreshed. These ancient images form some of the oldest and most beautiful works in the history of human art.

Many of the ancient Egyptians' most sacred beliefs were encoded in *The Book of the Dead*, a collection of magic spells and stories, often illustrated with scenes from this world and the afterlife, that were written by the living for the benefit of the dead. Verses from the book were placed on scrolls inside tombs to help the souls of the dead pass through the dangers of the underworld and into an afterlife of bliss. Nearly 200 different spells have survived. Chapter 125 was especially popular. It dealt with the dead soul's judgement by Osiris, who, with forty-two other gods, decided if it should go to heaven or hell. Some people had special brooches, called amulets, woven into their mummies near to the heart. These were inscribed with magic spells whose purpose was to disguise any earthly indiscretions they may have committed, so protecting them from being found out by the gods during this judgement ceremony.

By the time of the New Kingdom (starting in about 1550 BC), the capital of Egypt had been moved further upstream from Memphis to Thebes. Here the ancient art of mummification continued, but with one important difference. Now the pharaohs, along with their families and friends, buried their tombs in secret locations underground. This was to protect against looters, who had taken advantage of the occasional moments in Egyptian history when central power broke down, such as when invaders called the Hyksos came on their chariots from the north and overran the lower part of the country between about 1674 and 1548 BC.

Hundreds of secret tombs have been discovered in the valleys of the Kings, Queens and Nobles, near Thebes. Even though they were buried underground, many have been looted in the intervening years. But, remarkably, some have

survived almost completely intact. Massive temples, built by the rich to glorify the gods, still stand to this day, such as the one at Karnak.

We know a great deal about the ancient Egyptians' belief in an afterlife, thanks to pictures and inscriptions etched on to the walls of the pyramids and tomb chambers, and preserved on papyrus scrolls. These inscriptions were made using a style of writing called hieroglyphics. Did the Egyptians copy the idea of writing from the people they traded with, like the Sumerians? Or, did their rulers' need to store and account for so much grain give them the impetus to develop writing independently? Whichever is true, thousands of ancient scripts chiselled onto walls and etched onto papyrus have survived to this day, and are now scattered in museums throughout the world. No one could read this writing until the decoding of the Rosetta Stone (see page 124).

On 4 November 1922, a British archaeologist who had been searching the valleys for more than fifteen years stumbled across some steps leading

Howard Carter's discovery in November 1922 of the tomb of the boy-pharaoh Tutankhamun made them both famous. The boy, buried in a room full of treasure, is thought to have died of gangrene at about the age of nineteen.

into an unknown tomb. What Howard Carter discovered was the burial chamber of a little-known Pharaoh called Tutankhamun who died when he was only about nineteen. For a long time it was thought he had been murdered, because his mummy shows a mysterious bump on the back of his head. But it is now thought that this young ruler died from gangrene after breaking his leg, probably while out hunting. The discovery of the tomb of this boy-king has transformed our understanding of Egyptian civilization. A huge hoard of treasure was packed inside the chamber; the most famous object of all was found bound into the head of the boy-king's mummified body: his funeral mask, made out of solid gold.

Civilizations with such rich and powerful ruling elites needed groups of people whose jobs were to make things or provide services on their behalf, or to give professional advice. This gave birth to what are now called the middle classes – not farmers, labourers or slaves, but stonemasons, artisans, artists, scribes, lawyers, teachers, doctors, craftsmen, potters, jewellers, architects and metal-workers. A whole community of such tradespeople, along with the ruins of their houses and tools, was discovered in 1904 by an Italian archaeologist, Ernesto Schiaparelli, at Deir el-Madinah, close to the Valley of the Kings, just to the west of Thebes.

In the same place, archaeologists found the tomb of one of the most famous Egyptian queens, Nefertari. She lived from 1300 to 1250 BC, and was married to the Pharaoh Ramesses the Great. Although she was only one of his eight wives, he was clearly devoted to her. Her tomb is the most spectacular yet found in the Valley of the Queens. On the walls is written some of the earliest love poetry of all time.

Nefertari, great wife of Pharaoh Ramesses the Great, was raised to the status of a goddess in her lifetime. Women's rights were well recognized in ancient Egyptian law.

'My love is unique. No one can rival her
She is the most beautiful woman alive.
Just by passing, she has stolen away my heart …'

Nefertari was so loved by her King and her people that she was raised to the status of a goddess in her lifetime. We also know that she was actively engaged in Egyptian politics, after clay tablets were found in Turkey recording how she wrote to the king of a northern people called the Hittites (see page 149) in an attempt to secure a peace.

Nefertari's story is important because it shows that despite ancient Egypt's development of absolute kingship, it was not a society based on unequal rights between the sexes. Women were equal to men in the eyes of the law. They were paid similar wages, could give testimony in court, and could inherit property. Hereditary kingship was valid only for heirs from the female line – a principle called matriliny that became a founding cornerstone for many subsequent monarchies throughout the world.[4]

Ancient Egypt was so well endowed with natural resources and barriers of protection that it had little need to develop military technology in the same way as other nearby civilizations. Why bother protecting yourself when nature has so kindly managed your defences in the form of a surrounding desert and marshes? Why bother going on the attack when staying put along the banks of the Nile provided more than enough natural resources?[5] In the end, this lack of preparation contributed to Egypt's downfall. After 3,000 years of almost uninterrupted dynastic rule a wave of invasions swept over the Empire, starting with the Assyrians in 671 BC (see page 174) then the Persians in 525 BC (see page 177), followed by the Greeks in 332 BC (see page 189) and the Romans in 30 BC (see page 197). By this time ancient Egypt as a separate, distinct civilization had finally reached its own dead end.

Mother Goddess

How veneration for nature's cycles of birth, life and death became a hallmark for some human civilizations that devoted themselves to fertility, femininity and equality.

NOT MANY PEOPLE can genuinely claim to have discovered a lost world. But one day in 1827 a British spy called Charles Masson had the privilege of becoming one of them. He left his army base in Agra, India, site of the world-famous Taj Mahal, and headed westwards with a fellow soldier on some unknown errand – it is even possible he may have been deserting.

On his journey he stumbled across the ruins of an ancient city at a place called Harappa, now situated in north-east Pakistan, which included what looked like a castle on top of a hill. Lying on the ground he found jewels, bangles and arm rings, as well as the remains of three ancient chariots. But he couldn't stay long, as the locals warned him that

he and his friend would be attacked by a plague of stinging gnats. 'Our precautions were in vain against the swarms of our tiny antagonists,' he wrote in his diary, 'and at sunset they so annoyed us, and particularly the horses, which became absolutely frantic, that we had no choice but to decamp and march on throughout the night.'[1]

Before he left, he drew the ruins. He knew he had found the remains of something extraordinary, but it wasn't until a hundred years later that professional excavations revealed the full extent of this lost civilization, hidden for thousands of years beneath the mud, sand and dust. Unfortunately, in 1857 the site was badly damaged when a team of British engineers used bricks from the ruins as ballast to support a railway line being built from Lahore to Multan.

Serious archaeological excavations began in the 1920s. They revealed that Harappa was one of the largest cities in what is now called the Indus Valley civilization. More than 2,500 different sites have since been discovered. These settlements were established at about the same time as the first towns and cities of ancient Egypt and Sumeria – that is, starting from about 3300 BC. Over about 1,700 years, people living here developed what many experts consider was the most advanced and impressive society on earth at the time. Then, quite suddenly, they vanished, seemingly into thin air. To this day, no one knows exactly why or where they went.

The people who lived here arrived from an ancient settlement called Mehrgarh, located near a town called Sibi in modern-day Pakistan. The first signs of farming and agriculture in this area – including traces of crops such as wheat and barley, and the bones of domesticated sheep, cattle and goats – date back as far as 7000 BC. The people here lived in simple mud buildings, and like the Egyptians they left many items in their burial sites which have helped modern historians piece together what their lifestyle was like. Jewellery, baskets, bone tools, beads, bangles and pendants have all been recovered from their graves. The most intriguing finds consist of hundreds of small, simple statues of women, often decorated with red ochre paint and featuring different types of hairstyles and ornaments.

From about 2600 BC the landscape dried up as the climate changed, and these people moved northwards to the more fertile river valley of the Indus, taking their expertise in making things with them. By the end of the second millennium BC, people living in the Indus Valley had built a number of stunning cities. They contained many of the features we associate with modern living, making them unique in the world at that time.

These people were brilliant town planners. Their streets were designed in convenient, well-measured grid patterns, like a modern American city. Each street had its own sewerage and drainage systems which were, in some people's opinion, more advanced than many found in modern-day Pakistan and India. Excavations have unearthed a series of large public buildings, including assembly halls and a meeting place for up to 5,000 citizens. Public storehouses, granaries and bath houses were surrounded by colonnaded courtyards. Indus Valley builders even used a type of natural tar to stop water from leaking out of what is almost certainly the world's first ever artificial public swimming pool.[2] Underneath one house there are remains of what looks like an under-floor heating system, pre-dating the famous Roman hypocaust system introduced more than 2,000 years later.

Each house had access to a well, and waste water was directed to covered drains which lined the main streets. Some houses opened on to inner courtyards and small lanes, and for the first time houses were built on more than one level. The people wove cotton, fired exquisite pottery and crafted copper and bronze for making jewellery and statues. Metal-workers lived here in abundance. Evidence of at least sixteen copper furnaces have been uncovered in Harappa alone.

But unlike in Egypt and Sumeria, there is a noticeable absence of royal tombs. There are no ziggurats, pyramids, temples or big palaces characteristic of a rich, dominant ruling class. What makes the Indus Valley civilization so interesting is that it suggests a way of life which was organized and efficient, but above all egalitarian. Most people, it seems, shared their wealth and lived in comparative equality.

This civilization was based on trade, because it needed access to raw materials such as copper and tin from other places. One Indus Valley city, called Lothal, featured a massive artificial dock with a dredged canal and loading bays for filling and emptying ships.

Because of their expertise in trade and their proficiency as builders and craftsmen, these people developed great skill in precise mathematical measurement and accuracy in weights. They followed a decimal system, and their building bricks were perfectly proportioned. They even invented an instrument that could measure whole sections of the horizon, to help them study tides, waves and sea currents. More recently, archaeologists examining skulls recovered from Indus Valley sites have found evidence that these people practised the world's first dentistry, using their jewellery-making drills to repair broken teeth.

Seals engraved with animals and pictures of dancing women are some of the most common objects recovered from these ancient sites. Some are inscribed with a mysterious script. More than 400 different written symbols have been found, but so far no one has successfully deciphered them. There is as yet no Rosetta Stone to help unlock their written code.

One discovery stands out from everything else. It is the bronze figurine of a naked young girl in a dancing pose found in the ruins of the city of Mohenjo-daro and now kept in the Indian National Museum in New Delhi. Cast about 2500 BC, and only eleven centimetres tall, her uniqueness lies in the advanced and technologically sophisticated way in which she was made. She was cast using a technique called the 'lost wax' process. First she was modelled in wax, which was then covered with layers of clay to make a mould. This was then heated to make the wax centre melt and run out. The hollow middle was then filled with molten bronze. Once it had cooled, the clay mould was chipped off to reveal the sculpture underneath. This advanced way of moulding bronze was not properly described in Europe until about 1100 AD – that's more than 3,500 years later.

But it's not just the advanced skill and craftsmanship shown in the figurine of this young dancing girl that are so striking. Everything about this world of the Indus Valley seems to have been far before its time – from the sanitation of its streets and the central heating of its houses to the fabulous dockyards and meticulous works of art. Craftsmen and women were on a par with farmers, tradesmen – even priests. They all seem to have worshipped what is known as a mother goddess, which accounts for the hundreds of female figurines, including the small bronze dancing girl, found in sites throughout the region.

But was this highly advanced, artistic and largely peaceful early civilization in the Indus Valley unique? Beyond the big, ostentatious urban civilizations of Egypt and Sumeria, it would seem that the Indus Valley people were far from alone in their peace-loving ways. Plenty of evidence suggests that until about 4000 BC, and in some areas until 1600 BC, much of Europe and the Near East lived in a similar way.

Remarkable remains have been found buried underground in the graves of Neolithic farming people across Europe who lived from about 8000 BC to 3000 BC, which is when the first bronze tools and weapons appear, and what is known as the Bronze Age begins. Mostly, people were buried together, men and women equally, in large communal graves called barrows. Studies of their bones have shown that these people did not generally die as a result of violence.

More than 10,000 tombs and barrows are known about in western Europe alone. They are called 'megalithic' because huge structures are often to be found near these graves, usually built from large blocks of local stone. Many were set upright in circles, like the sites in England at Stonehenge and Avebury. Elsewhere they were constructed as temples, with altar tables at one end or in the centre. Famous examples such as Hagar Qim and Mnajdra survive on the island of Malta.

Some of these constructions are so large and so skilfully made that they required similar levels of skill, organization and craftsmanship as the Giza pyramids of Egypt. The enormous Newgrange megalithic temple and burial chamber in County Meath, Ireland, recently restored, was built 500 years *before* the pyramids. It is thought to have taken 6,000 man years to complete. As a testament to the skill of its builders, its roof has remained intact and waterproof for more than 5,000 years.

Many thousands of megalithic structures still stand as monuments to farming people

This small, bronze and perfectly formed dancing girl from the Indus Valley civilization shows highly advanced skills in casting and craftsmanship.

23:59:59

strong. Although only part of it has been excavated, a series of stunning wall paintings, dating back 8,000 years, shows a goddess in the form of a vulture consuming dead bodies – an ancient practice known as excarnation. Neolithic people all over Europe put out the dead bodies of their loved ones so they could be eaten by birds of prey and other animals as part of a continuous cycle of life, death and regeneration. Once the flesh had been eaten the bones would be buried, often following a ritualistic ceremony that may have taken place at megalithic sites such as Stonehenge. Such ceremonies expressed these societies' deeply held reverence for the cycle of life, death and renewal, with bones representing the seeds of new life.

Dozens of clay figurines of naked women, often looking pregnant, have also been found at Çatal Hüyük. One, discovered as recently as 2005, shows a female deity on the front, but on the back is a skeleton with depleted bones, symbolizing the process of life and death joined together in a continuous cycle.

As these people wandered with their domesticated flocks across Europe they took their culture with them. Female goddesses just like those found in the far-off Indus Valley or Çatal Hüyük have also been unearthed in Greece, at a place called Achilleion. Archaeologists have recovered more than 200 clay statues from this site, many of them buried next to bread ovens. Here, perhaps, we can see the beginnings of the belief in the association of bread and divinity – the bread of life, the bread of heaven.

These people deliberately started to change the natural environment around them to support their new agricultural lifestyles. Between 6000 and 3000 BC millions of trees were cut down all over Europe to make way for fields. Large areas of open moorland such as Dartmoor and Exmoor in the West Country of England were formerly ancient forests, cut down by Neolithic farmers to provide open areas to grow crops and graze livestock. They needed to do this so they could settle permanently in small villages and towns – with some communities growing as large as 500 inhabitants.

An 8,000-year-old mother goddess figurine found at Çatal Hüyük, Turkey, gives birth between lions, showing how fervently these early farming and tradespeople venerated fertility, femininity and the ongoing cycle of birth, life and death.

who spread out from the Near East with their domesticated animals and seeds from about 7000 BC, gradually replacing or absorbing old established hunter-gathering tribes wherever they went.[3] Mostly they travelled along the coasts by sea, and up river valleys such as those of the Danube and the Rhine, fertile areas with rich soil and plentiful moisture for their crops. Many skirted around the Mediterranean coasts, establishing themselves on islands such as Malta. They then travelled north, settling in Portugal, northern Spain and Brittany before reaching England, Ireland, Wales and as far up as the Orkney islands off the coast of Scotland, where some of the best-preserved structures and stone houses still remain.[4]

These people all seem to have shared a common set of religious beliefs, as exemplified by an ancient Neolithic site called Çatal Hüyük, discovered in 1958 in southern Anatolia, Turkey. In about 6000 BC this was a bustling trading city, like Jericho, in close proximity to the Fertile Crescent where farming communities first emerged. Trade in obsidian helped this impressive city grow big and

There is no evidence to suggest that these people were violent. As at Harappa, there is no trace of a dominant ruling class, from Çatal Hüyük in central Turkey to Malta, Britain or Scotland. Objects buried alongside dead early Neolithic people were typically figures of goddesses, not axes, arrowheads or spears. The absence of violent deaths, fortifications and weapons of war suggest that these were peaceful times. Villages were built along fertile valley floors, not at the tops of hills, implying that territorial aggression, invasion and terror were little known.

Like the population in the Indus Valley, early European Neolithic farmers were sophisticated and technologically advanced. They too had a distinctive form of written symbolism. Objects covered with swirling whorls and spiral shapes have been found at more than a hundred megalithic sites across Europe. Unlike the cuneiform script of Mesopotamia or the hieroglyphics of ancient Egypt, no one has yet worked out exactly what they mean. What is clear is that this form of writing did not originate out of trade and commerce. Rather it is found on temples, tombs and graves, and on religions objects such as female figurines. It is likely that these inscriptions were some form of communication between the people and their gods and goddesses in the world beyond.

Some experts believe megalithic societies were matrilineal, with women placed at the apex of the civilization – not as rulers, but as birth-givers. Perhaps a line can be traced from the Natufian women of Lebanon (see page 115), or even as far back as the Venus of Willendorf, that 24,000-year-old statue of a pregnant woman found in Austria (see page 102). After all, women were the original seed-gatherers while men went out to hunt. It was they who probably developed the most intimate expertise in agriculture, using a mixture of instinct and common sense to select the best seeds for the next year's crops, unwittingly instituting what we now call artificial selection.

Swirling whorls and spiral shapes are common inscriptions at megalithic sites like this one in Brittany, France, but no one is sure of their exact symbolic meaning.

23:59:59

The mother goddess took a variety of different forms. Sometimes she was a snake, or a vulture, or the moon. Each symbol represented a cycle of death, birth and regeneration: the snake hibernates, then wakes up and sheds her skin; the vulture recycles dead flesh by eating it; and the moon dies and is reborn every twenty-eight days, mirroring the feminine menstrual cycle.

Moon worship was very highly advanced in megalithic times. It has recently been recognized that temples such as Stonehenge were originally built to glorify the moon as well as the sun. Every month shafts of moonlight line up perfectly with gaps in the massive stones, the architects having positioned them precisely to accommodate the subtly shifting patterns of the moon's varying rising and setting cycles, that repeat themselves exactly every 18.6 years. The full moon has had historic and religious significance going back thousands of years, since it was by the light of the full moon that many hunter-gathering tribes hunted, providing the best opportunities for a good catch.

Europe's mother goddess culture grew to its climax on the Mediterranean island of Crete in the second millennium BC. Here it also survived longest. Crete thrived on trade routes that linked the Mediterranean with the rest of megalithic Europe and North Africa. The flowering of the island's Minoan civilization coincided with the growth of the Indus Valley civilization, from c.3300 to 1700 BC. Homer, a Greek poet who wrote in the eighth century BC, claimed there were as many as ninety cities on Crete, and archaeologists have found a number of 'palaces', including the largest of all at the island's capital, Knossos.

The discovery of this ancient island civilization was chiefly the work of Sir Arthur Evans, an eccentric but meticulous Victorian archaeologist. As soon as he set foot on Crete in 1894, Evans rigorously pursued the mystery of the mythical King Minos, who, legend has it, ruled from a fabulous palace at Knossos which housed an appalling monster, the minotaur. Half-man, half-bull, this beast lived in an impenetrable maze and feasted off the flesh of still-living virgins.

Myth has it that King Minos's palace and its maze were built by an ingenious architect called Daedalus. The stories of how the Greek hero Theseus defeated the minotaur, and how Daedalus escaped from imprisonment in a tower with his son Icarus by creating the world's first artificial wings, beautiful as they are, are fancy not fact. But these were myths that inspired Sir Arthur Evans to invest more than £250,000 of his own money excavating and restoring the palace at Knossos.

Minoan Crete was like a heart pumping at the centre of the Bronze Age trading system. Its trade links stretched as far as Mesopotamia in the east, to Spain in the west. Tin and copper were imported and exported for smelting into bronze, while

A 3,500-year-old mother goddess figure clutching serpents that represent new life was found by Arthur Evans at the palace of Knossos in Crete.

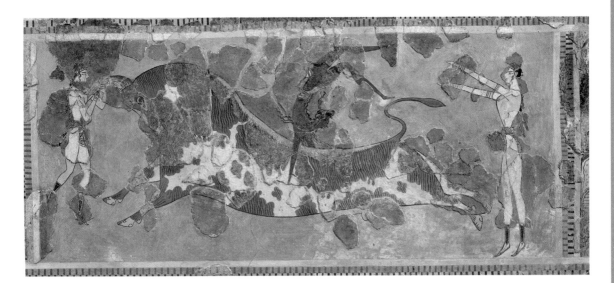

luxury crops such as bright yellow saffron were grown in the island's fields and exported as flavouring for food.

Evans discovered that the people who lived on ancient Crete followed the megalithic tradition. Women and men had equal rights. Wall paintings from the palaces of Knossos and Phaistos show that women were able to express themselves freely. They are depicted as bare-breasted, wearing short-sleeved shirts open to the navel and long, flowing, layered skirts. When they were first discovered, these frescoes took experts completely by surprise. Like the Indus Valley dancing girl, they seemed far too modern for their age. Here's how Evans described the women in his account of the excavations:

> 'They are fresh from the coiffeur's hand with hair frisé and curled about the head and shoulders. The sleeves are puffed and the constricted girdles and flounced skirts equally recall quite modern fashions … These scenes of feminine confidence … take us far away from the production of Classical Art in any age … They bring us quite near to modern times.'[5]

Statues, vases and wall paintings show images of sporting contests where women competed equally alongside men. The island's favourite sport was the impossible-sounding bull-vaulting. An acrobat (sometimes female) would grab the horns of a bull and somersault on to its back. Then, in a second somersault, she would leap off its back and land upright, with her feet back on the ground. No wonder Minoan women were the first people known to have worn fitted garments and bodices – essential prerequisites, you would think, for a sport like this.

Women did not dominate society, but they did oversee it. Frescoes at the palace of Thera, on the island of Santorini, a hundred kilometres north of Crete, show women standing on balconies overseeing processions of young men who are carrying an animal for sacrifice. Most priests on Minoan Crete were female. In Minoan law, women retained full control of their property. They even had the right to divorce at pleasure. It was a tradition, too, that a mother's brother was responsible for bringing up her children. Customs such as these, which seem strange to us today, lingered long in the Mediterranean mind.

As in the Indus Valley and all across Early Neolithic Europe, these people worshipped a mother goddess. Evans found painted vases of her depicted as a snake. Figurines of naked pregnant women dating from as far back as 3000 BC have been found not just on Crete, but on islands all around. Most scenes in Minoan art show female goddesses – sometimes they are represented as serpents or birds (as in Çatal Hüyük), or sometimes wearing masks. More than 300 sacred caves have

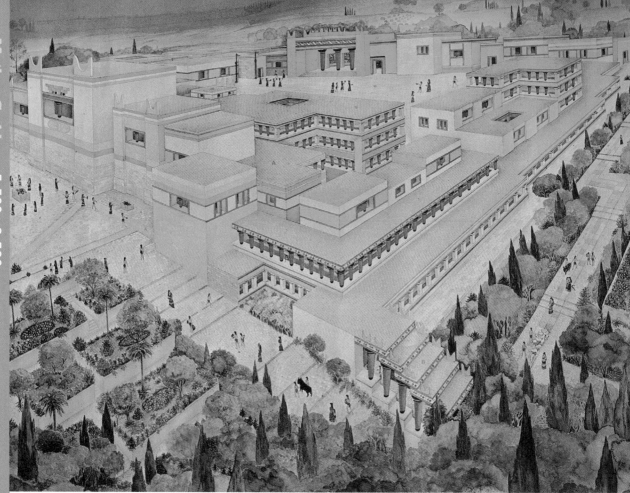

An impression of the maze-like Minoan palace at Knossos, more of an administrative centre than a grand royal residence.

been found. Many are believed to have been centres of goddess worship.

Minoan palaces were not mighty and dominant like those in Egypt or Sumeria. Rather, they functioned as the region's communal administrative and religious centres, providing a place of work for craftsmen, storage spaces for food and temples for goddess-worship. One look at a model reconstruction of the palace at Knossos and you can understand why Greek invaders might later imagine that the corridors and irrigation channels resembled an impenetrable maze.

Like the traders of the Indus Valley and other European megalithic people, the Minoans had their own form of symbolism which shows that their civilization was culturally and technologically advanced. In 1903 archaeologists excavating the palace of Phaistos, on the southern side of the island, made a discovery which has had historians

baffled ever since. Buried inside an underground temple they found an object which many believe is a product of the world's first printing press.

The Phaistos Disc (see opposite), currently on display at the archaeological museum in Herakleion, Crete, is thought to date from some time between 1850 and 1600 BC. It contains forty-five unique symbols arranged in a spiral shape, resembling the swirls found on vases at Knossos, or even in European megalithic tombs such as that at Newgrange in Ireland. They were made by impressing a set of re-usable seals into the disc when it was soft clay, while it was rotated clockwise. The disc was then fired to make the symbols permanent. This early printing technique (called moveable type printing) pre-dates the invention of the modern printing press by more than 2,000 years.

No one really knows who made the disc, or what the symbols mean, but it does show that the people of Minoan Crete were artistic, prosperous and, like Daedalus in the legend, highly ingenious.

Following excavations at a site called Akrotiri in 1967, the Minoans are now known to have spread to the island of Santorini. There, archaeologists have discovered the remains of a vast, ancient island city which had been buried for thousands of years under thick layers of volcanic ash. Although only the southern tip of the town has so far been examined, houses three storeys high have been unearthed with fine wall paintings, stone staircases, columns and large ceramic storage jars, mills and pottery. Minoan Akrotiri even boasted a highly developed drainage system, featuring the world's first known clay pipes with separate channels for hot and cold water supplies. Frescoes show women gathering saffron, offering flowers to a seated goddess. Another shows a group of fishermen bringing back their catch, and a flotilla of boats accompanied by leaping dolphins. The scene is overlooked by finely dressed ladies watching them from under the cooling shade of a canopy.

A distinct pattern is discernible from the evidence that has been left by these early civilizations.

Stretching from the ancient Indus Valley, right across the mountains of Anatolia, to the islands of the Mediterranean and as far as the topmost island of Orkney in Scotland, what emerges is a series of like-minded civilizations whose temples and graves bear witness to a lifestyle of peace and a veneration for mother nature. Their common belief in the continuous cycle of birth, death and regeneration is personified by their worship of a mother goddess in all her forms: snake, vulture, pregnant woman or moon. Excellence in craftwork, technical skill and exquisite art are some of their legacies, along with a spirit of natural equality.

It was not to continue. During the second millennium BC, the last of these early civilizations fell. New power in the form of military might was in the process of sweeping across Europe, the Middle East and Asia. Warriors, like those soldier ants that evolved in another of nature's civilizations millions of years ago (see page 68), had by now worked out how to prey off the profits of others, ushering in an age when human elitism, ruthlessness and terror had their true beginnings.

The 30cm-wide Phaistos Disc may be the world's oldest ever printed document, but no one has yet been able to decipher its symbolic text.

23:59:59

Triple Trouble

How a trilogy of domestic horses, wheeled chariots and Bronze Age weapons fanned out across Asia, Europe and North Africa, creating waves of destruction, conquest, and inequality.

AS WE PASS THROUGH the final tenth of a second to midnight on the twenty-four-hour scale of all earth's history, it now becomes quite impossible to disentangle the natural history of the world from the story of the rise and fall of human civilizations. Sometimes the planet's natural systems – like the climate – shuffled packs of people like a deck of cards between games. Sometimes it was humans that took the lead, by harnessing the earth's other life forms, be they animals, crops or forests, for making communities of houses, farms and fields. The history of *Homo sapiens* makes sense only when it is seen as having an umbilical connection to the natural world.

From about the beginning of the second millennium BC – that's 4,000 years ago – the lives of most people living in the orbit of new civilizations became a great deal more stressful, violent and aggressive. The causes were sometimes natural and

unavoidable. Global catastrophes such as the super-volcano that is thought to have caused the Permian Mass Extinction 252 million years ago are very rare (see page 52). But regional disasters such as the Category 7 volcano that erupted in the middle of the Mediterranean Sea in about 1627 BC happen a lot more often. Although on nothing like the scale of a super-volcano, this eruption still devastated the fabric of several early human civilizations living nearby.

Recent research has shown that this single eruption, close to the Greek island of Thera (now Santorini), threw some sixty-one cubic kilometres of boiling-hot rock into the atmosphere, plunging the surrounding region into a sudden dark age. There's no lack of evidence of the eruption today. Cliffs of dry volcanic sediment more than sixty metres thick are encrusted on top of the island. This blast is thought to have been one of the two largest volcanic eruptions to have taken place anywhere on earth in the last 5,000 years. (The largest is believed to have been the eruption of Mount Tambora, in Indonesia, in 1815. It released about a hundred cubic kilometres of magma into the atmosphere.)

It wasn't just the size of the eruption near Thera that mattered – it was also its location. With some of the world's earliest human civilizations clustered around the Mediterranean, the impact of such a large explosion was bound to be significant. A series of huge tsunamis, some perhaps as much as 150 metres high, ripped across the seas, smashing into the trading empire of the Minoans on the island of Crete, a hundred kilometres south of Thera. This eruption explains how that civilization, so advanced for its time, was quite literally swept away.

An ancient scroll, called the Ipuwer papyrus, has been dated to about this time. It describes in verse a period of chaos that struck Egypt in this era, thought by some to have been caused by the aftershocks of this mighty eruption. The country was being spun around 'like a potter's wheel'. Its towns were destroyed, the government had collapsed and all the surrounding land was 'transformed into an empty waste'.

> 'Everywhere barley is perished and men are stripped of clothes ... Scribes are killed and their writings are taken away, the laws of the council chamber are thrown out. The king's storehouse is the common property of everyone and the entire palace is without its revenues. The robber is a possessor of riches and the rich man is become a plunderer. Towns are destroyed and Upper Egypt has become an empty waste ...'

At about the same time as Thera blew its top the Indus Valley also fell to the forces of nature. A series of violent earthquakes in the Himalayan mountains were probably what disrupted the all-important Indus Valley river system near its source, diverting its waters further to the east, towards where the Ganges flows today. Without nature's gift of free-flowing fresh water, this sophisticated civilization found itself stranded next to nothing but dried-up riverbeds. The creators of the remarkable dancing girl evaporated from the historical record almost as rapidly as the water they once relied on.

Not all human civilizations were as vulnerable to nature's disruptive fits. While some grew up as fixed communities next to rivers such as the Nile and the Indus, or beside the ocean, like the Minoans, others stuck to a more nomadic way of life. These people didn't depend on any fixed source of water, like a river, nor did they live in settled communities that could be swept away by the sea. Theirs was a lifestyle closer to the original human hunter-gathering state of nature, but with one important difference. Rather than always relying on hunting, they took with them domesticated animals such as sheep, goats, pigs and cows as a regular, dependable supply of food, drink and transport.

People who wander from place to place living off domesticated flocks of animals that travel with them are called nomadic pastoralists. If disaster struck in one region, they simply moved on to another where conditions were less volatile. These people found they could make a handsome living as traders by wandering with their herds from place to place, carrying goods from one civilization to the next. Nomadic pastoralists formed the overland backbone of the ancient world's transportation network.

Over time, their wandering lifestyle brought these people into a new relationship with nature that dramatically altered the course of human and natural history. Changes began when they discovered how to harness horsepower, allowing them to travel more quickly and to carry greater loads. Then they learned how to smelt tin and copper into bronze – which was ideal, they soon found out, for making strong, robust and lethal weapons of war.

There is no fixed date from when the Bronze Age can be said to start. It began at different times in different places. The earliest known bronze artefacts may have been made as early as 4000 BC in Mesopotamia, where the alloy was used by artisans for making objects such as the Indus Valley dancing girl. From about 2000 BC, nomadic chiefs living in central Anatolia (Turkey), an area rich in

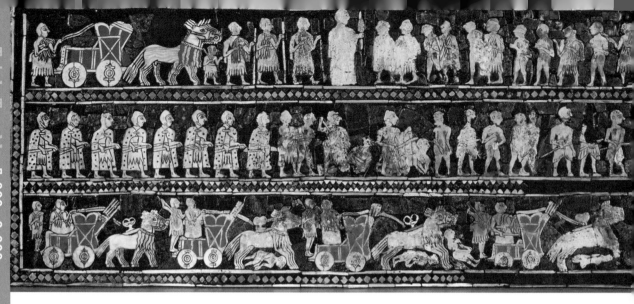

One side of the beautifully painted Standard of Ur (c.2600 BC), found by Leonard Woolley in the 1920s. Soldiers are shown marching to war using chariots pulled by animals.

copper and tin deposits, started to adapt bronze for making armour, shields and weapons in the form of axes, swords and spears. Bronze is hard. People who could craft bronze into weapons and armour were at an immediate military advantage over those with just stone implements or wooden clubs.

Nomads transported precious raw materials like tin and copper from places as far flung as Cornwall and Wales in the west and Spain in the south to the Rhine Valley of Germany, the Caucasus mountains in Asia Minor (modern-day Turkey) and beyond, into the Middle East and south towards India. Once these traders became expert at making bronze weapons it was a simple step to invade, conquer and subjugate the many settled farming communities along their various trading routes between the markets and civilizations of Europe, the Middle East and Asia.

Just as important as learning how to harness some of the earth's natural resources into a powerful new type of hardware was the conquest of a wild animal that revolutionized the nomad's ability to travel long distances. From as far back as about 4000 BC, nomadic people roaming around a region of southern Russia called the Pontic Steppe had begun to discover how to domesticate the wild horse. While women were perfectly placed to study and understand the science of agriculture by choosing and then sowing the best-looking seeds, men were better suited to the task of taming wild beasts. Energy, determination and brute force were

the ingredients required to bring these powerful animals under control.

Wild horses called tarpans roamed the lands of southern Russia, near the edges of the forests. These herd animals were tamed, bred and broken in as slaves for humans. Originally more the size of ponies than thoroughbreds, tarpans are now extinct: the last known specimen died in a Moscow zoo in 1875. No one knows exactly when the first tarpans began to be bred in captivity, nor who first managed to mount one, but since they were small horses the first breeds were better suited to pulling carts and carrying loads than transporting people.

Chariots had begun to appear in Mesopotamia by 2600 BC, according to a beautiful painted box called the Standard of Ur discovered by British archaeologist Leonard Woolley in the 1920s. This box, now on display at the British Museum, includes a graphic image of 'War', with Sumerian spearmen following a four-wheeled chariot pulled by what look like horses, cows or oxen.

Nomadic traders from the Russian steppes were probably the first to possess the triple combination of domesticated horses, wheeled chariots and bronze weapons. They also learned to increase the loads that their horses could pull, by using spoked rather than solid wheels, making their chariots lighter and more manoeuvrable in battle. The first military chariots with spoked wheels have been found in the graves of nomadic people from about 2000 BC in Russia and Kazakhstan. Their heavily fortified settlements were built on hilltops for best defence, and they used their expertise in bronze metalwork to make armour and weapons.

Evidence suggests that these people depended heavily on their horses. Bone bits and horse remains have been found together in human graves dating from as far back as 4000 BC in the steppes of Russia. Gradually from about 3500 BC their way of life spread across Europe, eventually reaching Britain, Ireland and Spain, superimposing a new world order ruled by a more aggressive, male-dominated society built on controlling horses and making metal weapons. Traces of this cultural shift can be seen in fragments of a style of pottery found all over Europe, originating from the Pontic Steppe. These pots were made in the shape of an upturned bell, and were probably used by well-off families to contain mead or beer. This is why theirs is known as the Bell Beaker Culture.

Graves of Early Neolithic farmers all over Europe gradually changed from being communal barrows located in fertile lowland areas to individual burial mounds in fortified sites, in which can be found the remains of once-proud chiefs laid to rest alongside their own splendid weapons, chariots and armour.

Speed, height and firepower gave people with domesticated horses a huge military advantage over those without. Horses provided a means of transport at least five times quicker than any other known at the time. Reconnaissance, fast communications and the precious element of surprise were now at a rider's discretion, powerfully exacerbating the potential for terrorism, blackmail, subjugation and war.

Once people saw the huge advantage of domesticated horsepower, everyone wanted it – if not to threaten other people, then at least to protect themselves. From about 2000 BC civilizations, both nomadic and settled, became militarized. The empires of the Hittites, Hurrians and Mycenaeans were all forged on bronze weapons, spirited horses and loaded chariots that probably originated from the Pontic Steppe around the north, east and west flanks of the Black Sea.

The Hittites, who emerged by about 1800 BC, built their capital city at Hattusa in central Turkey.

This warrior race was well known for its brilliance at using horses and chariots in warfare. Their expertise in metallurgy helped them discover how to mine and smelt iron, giving them an even bigger range of more powerful weapons. By 1400 BC the Hittites had become so mighty that they successfully conquered all of Lebanon, Syria and Canaan, as far as the border with Egypt.

Skills in how to train horses were usually passed on from one generation to the next by word of mouth. The first known written training manual dates back to about 1350 BC. It was the work of a Hittite horsemaster called Kikkuli, and it detailed how to train a horse for chariot warfare. It described a series of intricate disciplines, many of them similar to modern dressage techniques using 'interval training', which were designed to improve a horse's stamina and increase its agility.

The Hurrians, like their neighbours the Hittites, came from the north, where wild horses were first tamed. Ancestors of today's modern Kurdish populations, these people ruled much of the Middle East during the second millennium BC. With their warrior society came a new view of the gods. Now it was a male god, Teshup, who became

A king, mounted on a horse-drawn chariot, strikes down his enemies in this stone carving from central Turkey, where the Hittites ruled from about 1800 BC.

23:59:59

the most powerful of all, supplanting the female, agricultural mother goddess of the moon and fertility. Teshup controlled the weather. His wife, Hepa, represented the sun. Historians believe that large parts of ancient Greek mythology originated from the Hurrians' gods, and that Teshup was probably the inspiration for the later stories of Zeus.[1]

By about 1180 BC both the Hittites and the Hurrians were themselves subjected to invasion and conquest by a confederation of other warrior tribes called the Sea Peoples. In fact, thanks to the unsavoury combination of wheels, horses and bronze, from about 1800 BC until 500 BC Mediterranean Europe and the Middle East were in constant turmoil. Wave upon wave of horsemen wielding bronze and iron weapons availed themselves of each and every opportunity to conquer and invade.

Sometimes opportunities presented themselves in the wake of natural disasters like the tsunamis that swept away Minoan Crete. By about 1600 BC the Mycenaeans, who also originated from near the Black Sea, had taken over large parts of southern Greece. They built a splendid capital at Mycenae where large beehive tombs have been found, containing daggers, masks, armour and jewelled weapons.

By 1400 BC the Mycenaeans had reached across to Crete, invading and occupying the ruins of Knossos and other deserted Minoan cities. Although many influences from the Minoan civilization rubbed off on Mycenaean art and pottery, the differences between the two cultures are marked. Wall paintings reveal very different attitudes towards the natural world. Minoan pictures show a delight in animals, such as the leaping dolphins painted on their palace walls. Mycenaean works of art celebrate animals only as the freshly slaughtered victims of a violent hunt.

The period between 1400 and 1100 BC was the stage for what was once thought the most epic military struggle of all time – the Trojan Wars, fought between a confederation of small Greek states under the overall command of Agamemnon, king of Mycenae, and the people of Troy. Accounts of the wars are contained in two epic poems written some 500 years after they supposedly took place. Homer's *Iliad* and *Odyssey* are full of divine intervention, with the gods influencing the affairs of their human favourites. After ten years of war, military victory was finally secured by the Greeks after they left a giant wooden horse outside the walls of Troy. The curious Trojans pulled the horse inside their city, not knowing that it contained dozens of enemy crack troops who, after nightfall, crept out of the enormous structure, opened the gates and let the Greek army come flooding in to sack the city.

Heinrich Schliemann (1822–90) was a German treasure-hunter who, after making a fortune in the Californian gold rush (see page 350), dedicated his life and his riches to trying to prove that Homer's stories, which his father had read to him as a child, were based on fact, not fiction.

His first quest was to discover the site of the ancient city of Troy, which historians then believed was a mythical place. After two years of exhaustive excavations at Hissarlik, in Turkey, he announced to the world that he had discovered the great Trojan King Priam's treasure, in the form of a cache of gold objects including diadems, bracelets, goblets, necklaces and earrings, as well as vases of silver, copper and weapons of bronze. The level in which he found this treasure is now known to have dated several hundred years before the alleged Trojan Wars took place and exactly how genuine these artefacts are is still a matter of dispute.[2]

When Schliemann double-crossed the Turkish authorities and smuggled the hoard out of the country, they banned him from further excavations for an indefinite period, although, thanks to friends in high places, he gained permission to excavate at Hissarlik again a few years later.

Today most historians believe that Hissarlik is quite probably the remains of Troy. According to the Greek historian Eratosthenes (276–194 BC), the Trojan Wars took place between 1194 and 1184 BC. These dates tie in with archaeological evidence uncovered by Schliemann which shows that the city was razed to the ground by fire at about that time. But there is little or no

archaeological evidence of a ten-year-long struggle between Greeks and Trojans, although, given the turmoil in the region at that time, the possibility of a war of some sort remains highly plausible.

By 1876, Schliemann turned his attention to excavating the city of Mycenae on the Greek Peloponnese. This time it seems his finds were genuine. He dug up a series of five royal graves, inside which were the remains of nineteen people, many of them wearing gold masks. Buried next to the men were swords, daggers and armoured breastplates. Despite Schliemann's hopes, these people were not victims of the Trojan Wars, because they were dug up from a level in the ground that dated back some 500 years before, to about 1600 BC, when the Mycenaean Greeks first established themselves as rulers in the region. These were rich warriors who lived in a highly structured society, with kings and a ruling class that celebrated their military conquests with glorious artefacts and weapons of war.

None of which means to say that Homer's epic poems aren't at least partly based on fact. If nothing else, they can be enjoyed as great stories. Many scholars regard Homer's epic poems as surpassing even the plays of Shakespeare, because of their swift-flowing movement and their clearness of thought and style. Tradition has it that Homer was blind, but in fact there is no hard historical evidence that he was even a real person. One theory suggests that his name is derived from an ancient society of poets called the *Homeridae*, which means 'of hostages'. These were prisoners who were entrusted with inscribing epic accounts of the past that had never before been written down.

While Homer's poems provide a vivid account of the chaos of the late Mediterranean Bronze Age, another, even more famous piece of literature purports to tell us what was happening further south, in Egypt, Jordan and Israel, at just about the same time.

Like Homer's poems, the first five books of the Bible were written down many hundreds of years after the events they describe are supposed to have taken place.[3] These religious texts are sacred to Jews and Christians, some of whom believe every word

they contain to be literally the true word of God. As a piece of history, however, they are as confusing as they are vivid. The fact that they were passed down orally through many generations before they were written down means that they are, like Homer's poems, hard for historians to interpret.

The most dramatic historical events in these early books are the exodus of the Israelites out of Egypt and the plagues sent by God through Moses to punish the Egyptian Pharaoh for enslaving his chosen people. No one really knows quite when these events are supposed to have happened. Estimates range from 1650 BC, at about the time of the Thera eruption, to about 1200 BC, which is when the first archaeological evidence of the Israelite occupation of the Promised Land of Canaan (now Israel) can be traced. This later date tallies with the first known mention of Israel as a real place, in an Egyptian inscription dating to 1208 BC called the Merneptah Stele, chiselled during the reign of Pharaoh Merneptah (ruled 1213–1203 BC) to celebrate his victories over the Libyans and the Sea Peoples.

Sigmund Freud, a man sometimes dubbed the 'father of modern psychology', put forward an intriguing theory in a book he wrote in 1939 called *Moses and Monotheism*. He linked the Jewish Exodus as told in the Bible to the reign of the maverick Egyptian Pharaoh Akhenaten (ruled 1353–1336 BC). Along with his beautiful chief wife Nefertiti, Akhenaten turned the ancient Egyptian religious world upside down by declaring that there was only one God. According to him, the sun god (Aten) was the sole deity, the source of all life, and Akhenaten was his representative on earth. All other gods traditionally worshipped by the Egyptians (like Horus, Amun and Osiris) were banned. The Pharaoh and his wife had a new capital built at Amarna, on the east side of the Nile, where the sun rose. He ordered that Aten be worshipped in open sunlight, not in dark, gloomy temples – which, if they were dedicated to other gods, he banned or had defaced.

But the people of Egypt weren't ready for such a dramatic change in their beliefs. Soon after Akhenaten's death the powerful priests of Amun restored the *status quo ante*, and removed almost every reference to the revolutionary Pharaoh from the historical record. Freud believed that Moses' life in Egypt dated from this time, and that it was from Akhenaten that his belief in a single, all-powerful God originated. The Israelites represented an enslaved faction of Egyptian society who rebelled when Akhenaten's reforms were reversed, which would date their Exodus to about 1350 BC. The world's first known influenza pandemic is now thought to have occurred at this time, providing a possible natural explanation for the distresses depicted in the biblical account of the plagues of Egypt. Another theory has recently been put forward by Simcha Jacobovici, who ascribes the plagues to the chaos caused by the eruption of Thera and the invasion of the Hyksos (1648–1540 BC).

Despite the lack of hard archaeological evidence for a ten-year war between Greece and Troy, or a mass migration out of Egypt, such oral stories, later captured in writing by Homer and biblical scribes, paint an unmistakable historical picture. They tell of natural disasters, as well as the stress, violence and cruelty that came with the noxious blend of horses, wheels and metal weapons that beset the Bronze Age world of the Mediterranean, Europe and the Middle East.

But however bad things get for most people, there are usually some who find a way of living well. In this case, it was the Phoenicians, a seafaring empire based on the Mediterranean coast, in modern-day Lebanon. Despite the mayhem all around, from about 1200 BC to 800 BC their civilization established a fine trade based on making a precious purple dye, highly prized for colouring rich and exotic textiles, from the crushed shells of sea snails that were harvested from the sea floor near the coast. They also traded and transported cedar wood to Egypt for building ships, pioneered a technique for producing clear, transparent glass, and transported tin, silver and copper from as far away as Spain in the West to Mesopotamia in the Middle East. Wood and metal were vital supplies for the arms race that was now tearing so many other societies apart.

Given their maritime prowess, it should come as no surprise that the Phoenicians were experts at shipbuilding. They invented designs for warships such as triremes and quinqueremes that became the industry standard for more than 800 years. Phoenician trading outposts spread across the Mediterranean to Sicily, Cyprus, Sardinia, Carthage, on the North African coast, and Cádiz in southern Spain. These settlements were a good insurance policy. They meant that the Phoenicians could transplant themselves and their culture if ever they were threatened by invaders back at home. Indeed, in 539 BC Persian armies under Cyrus the Great overran the Phoenicians, who then fled to their settlement at Carthage (see page 176).

The Phoenicians' biggest legacy was their conversion of primitive symbolic writing into a more flexible alphabet that consisted of a limited number of shapes (letters), each representing

Phoenicians loved to make boats, as seen on this carved tomb dating from about 200 BC.

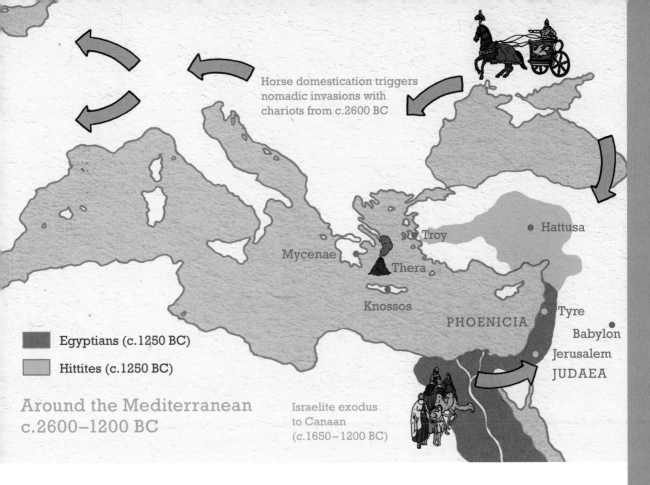

Horse domestication triggers nomadic invasions with chariots from c.2600 BC

· Hattusa

Troy

Mycenae

Thera

Knossos

PHOENICIA

· Tyre

Babylon

Jerusalem

JUDAEA

Egyptians (c.1250 BC)

Hittites (c.1250 BC)

Around the Mediterranean
c.2600–1200 BC

Israelite exodus
to Canaan
(c.1650–1200 BC)

small, individual oral sounds. This system, first developed around 1200 BC to make trading transactions easier to record, later became the basis for the ancient Greek alphabet of Homer (*alpha*, *beta*, *gamma*, etc.) as well as the writing systems of the Middle East and India. In fact, all alphabetically based languages in use today are ultimately derived from this radically improved way of writing.

From about 2000 BC, horses, metals and wheels rapidly transformed the ancient civilizations of Europe, North Africa and the Middle East. Together, they unlocked a voracious human appetite to experiment with exploiting the natural world to advance military and social advantage. Anyone wishing to trace the roots of economic inequality, social repression and military competition can find it all right here.

Volcanic eruptions, Bronze Age invasions and mass emigrations created a trilogy of troubles for some early societies.

Dragon's Lair

How a powerful and enduring human civilization arose from the East thanks to an abundance of natural riches in the forms of rice, silk and iron.

IF AN OLYMPIC GOLD medal were to be awarded to the largest, most robust human civilization ever to have existed on the earth, there would only be one serious contender – China.

Modern China is awe-inspiring. It is home to 1.3 billion people, more than a fifth of the world's population. It has the fastest-growing economy in the world, and can arguably take the credit for cradling more inventions and discoveries that have made a real difference to people's lives than any other country in history. The list includes the blast furnace, paper, gunpowder, the compass and printing, not to mention competitive examinations (see page 245).

Most impressive of all is its age. It is as ancient as any of those early civilizations that grew up around the Fertile Crescent of the Middle East, all of which have long since collapsed or been subsumed into other encroaching cultures and empires. Yet the foundations of modern China, both politically and culturally, were laid down more than 3,000 years ago. This land of fiery dragons and giant pandas (there are still a few left) is humankind's most remarkable survivor, and, quite possibly, it is now this civilization more than any other that holds the key to the future of both mankind and the overall health of the planet earth herself (see page 377). What was it that made this great power so different and so special, and has enabled it to survive to this day?

China's story pivots around key relationships with three products of the natural world which were, from early on, relentlessly exploited as nowhere else on earth: rice, silk and iron.

By about 2000 BC, two distinct civilizations were emerging in China along the banks of its river systems, the Yellow River to the north and the enormous Yangtze, located further south. Roaming hunter-gathering tribes probably began cultivating rice as early as 7000 BC along the banks of the Yangtze, the greatest river system in all east Asia, and the fourth longest river in the world, after the Nile, the Amazon and the Mississippi. With its source in the glaciers of Tibet, this mighty river flows west to east, then, after twisting and turning over the course of some 4,000 miles, it finally spills its muddy load into the East China Sea. Today heavy river traffic and intensive dredging of the Yangtze's riverbed are driving some the world's rarest river animals into extinction; they include the Chinese river dolphin, officially declared extinct in 2006, and the finless porpoise or river pig, of which only about 1,400 are thought to be left alive.

By about 3000 BC, though, frequent flooding of the river and its 700 tributaries made this an ideal place for growing rice, an almost magically productive and nutritious source of food which has the best record of any crop on earth for supporting large, intensive human populations. Today India and China are the most populous human civilizations largely thanks to their early production of rice, which can feed more humans per hectare than any other staple agricultural crop.

Rice is highly nutritious and remarkably resistant to pests. Flooded paddy fields provide ideal habitats for water-loving creatures like frogs and snakes, that feed off insects which could otherwise spoil the crops. Water cover is also perfect protection from the threat of self-sowing weeds, significantly increasing the chances of a successful crop. Nutrients flow freely around the waterlogged soil of paddy fields, renewing them just as the nitrogen-rich mud revitalized lands flooded by the Nile each year, giving the ancient Egyptians their head start thousands of miles to the west.

But growing rice is hard work. Each plant has to be sown individually, and expertise in irrigation is necessary to ensure flooding at the right times of the year. Thankfully, the abundance of a suitable natural workforce in the form of water buffalo meant that people here learned early on how to harness animal power for ploughing, puddling and harrowing the fields.

The rich allure of rice didn't go unnoticed by the people living further to the north, along China's other river system. The Yellow River Valley wasn't suitable for cultivating rice, because the climate there wasn't wet enough. Instead, the people there had their own staple crop, in the form of millet, which they made into noodles. In 2005, the world's oldest intact noodles, estimated to date back 4,000 years, were found by archaeologists at a site called Lajia in north-west China. Although nutritious, millet seed cannot be made into leavened bread, so it never really caught on in Western civilizations, which have tended to use it as birdseed instead of eating it themselves.

The Yangshao people, who lived between 5000 BC and 2000 BC along the Yellow River Valley, are thought to have been the first ever to have practised China's most lucrative long-term secret – silkworm cultivation.

Silk is extraordinary stuff. It is entirely natural. It reflects the light, making it look shiny and glamorous, and above all it is amazingly strong. In fact, silk is the strongest natural fibre known to man. Insects manufacture silk as a kind of miniature binding rope to protect their larvae. Gradually, as the larvae hatch, they gnaw through the silk rope to emerge out of the cocoon as caterpillars, eventually metamorphosing into butterflies or moths. Some adult insects, such as spiders, use silk for other purposes, such as making webs for catching prey, such as flies.

Quite how the ancient people of the Yellow River Valley in China discovered that they could harness the silk from a certain type of moth larva that feeds off the leaves of mulberry trees is a mystery that will probably never be solved.

Legend has it that the magical properties of silk were first discovered by Leizu, wife of the Yellow Emperor, who reigned from 2697 to 2598 BC. While out for a walk she is supposed to have noticed something wrong with the Emperor's mulberry trees. She found that thousands of caterpillars were munching their way through their leaves, causing a great deal of damage. She collected some of the cocoons from which the caterpillars came, and then sat down to have a cup of tea.

While she was taking a sip, a cocoon accidentally dropped into the steaming water, and a fine thread started to appear, unwinding itself from around the cocoon. Leizu found she could wrap this fine, strong cord around her finger. Inspiration struck. She persuaded the Emperor to plant a whole grove of mulberry trees, following which she worked out how to harness silk by reeling it into long threads that could then be woven into shiny pieces of precious cloth.

The practice of what is now called sericulture, which is the deliberate farming of a type of caterpillar called *Bombyx mori*, brought enormous wealth and prosperity to China. For as long as 3,000 years, Chinese farmers and tradespeople have profited by trading silk with other civilizations who marvelled at its shimmering appearance. Man-made alternatives, in the form of satin, nylon and acrylic, weren't concocted until the Second World War (see page 342).

Silk was the primary cause for the development of a series of overland trade routes that later become known as the Silk Road, a term coined by the German geographer Ferdinand von Richthofen

in 1877. By the time of the Roman Empire (44 BC–476 AD), silk was in huge demand by Mediterranean people. Its fine texture and semi-transparent shimmer made it one of the great luxuries of the ancient world. During the reign of the Emperor Tiberius (14–37 AD) the Roman government tried in vain to ban the wearing of the material on both economic and moral grounds, by issuing a series of edicts. Huge quantities of gold were flowing out of the Empire along the trade routes in exchange for silk, making China very rich. The Romans had absolutely no idea how silk was produced. Pliny, one of Rome's most celebrated historians, thought it was grown on some type of mysterious tree.[1]

It wasn't until about 550 AD that people in the Mediterranean learned the art of making silk by cultivating the right kind of caterpillars and feeding them mulberry leaves. Even then, so precious was the secret that the Byzantine Greeks kept its production limited to the imperial court inside Constantinople (Istanbul), in a bid to protect its mystique and keep the price high.

According to Greek historian Procopius of Caesarea, the knowledge of silk-making spread westwards only after two monks smuggled silkworm eggs from central Asia back to Constantinople hidden in bamboo rods which they presented to the Emperor Justinian I in c.550 AD.[2] From then on silk manufacturing in Europe became a preciously guarded Byzantine secret until the Crusades when the Normans of Sicily attacked Corinth, then an important Byzantine centre of silk production, and transferred the know-how to Sicily. Later, silk artisans left Constantinople and settled in Italy and Avignon following the sack of the city by Crusaders in 1204 (see page 264).

The third cornerstone of China's enduring success came from her great expertise in casting iron. The Chinese probably weren't the first to discover that iron was a cheaper and more effective metal than bronze for making tools and weapons. The Hittites of central Turkey are known to have mastered the technique of smelting iron ore, and were hammering on the first blacksmiths' anvils by about 1400 BC, which is when the European and

The ancient art of silk-making (sericulture) can be seen here in this seventeenth-century Chinese drawing.

Mediterranean Iron Age is said to begin. The iron they first used probably came from meteorites, although they soon discovered how to source iron from the ores they found in the rocks all around.

But it was Chinese people, from the region of Wu on the banks of the Yangtze, who worked out how best to harness iron found inside the ores of the earth's rocks. The manufacturing techniques they developed for casting iron were so advanced that they weren't matched in Europe for another 1,500 years.

Iron is a natural gift of the earth that is almost as essential to the development of modern human civilizations as oxygen is to animal life. Iron is by far the most common metal in use today – about 95 per cent of all metals used today are based on it. Without it, modern civilization would be very different indeed. Iron and steel, its derivative, are the materials of choice for making everything from cars to ships, pipes to forks, and computer disk drives to guns and skyscrapers.

Iron can be found almost everywhere. It is the fourth most common element on earth. Cast your mind all the way back to the catastrophe of 4.5 billion years ago, when the young planet Theia smashed into the earth, forming the moon (see page 17). Here lies the origin of how this precious substance was scattered deep, far and wide across the whole world.

Unlike copper, though, iron is not found in a pure form. Other elements like to react with it – for example, oxygen – making compounds such as a red iron oxide. To get the iron out requires effort and a little know-how, something the Chinese had attained by about 500 BC, when they built the world's first ever blast furnace. When iron ores are heated to about 1,450°C a molten liquid is formed. It can then be poured into moulds to make implements of any shape and size. As it cools, the metal becomes strong and rigid. The first iron implements were almost all used in agriculture. Iron ploughs were a magical leap forward, because they could cut through the hardest of clay soils, turning huge areas of land from scrubby waste into high-yielding fields of rice. The more food, the more people could be fed. The more people, the stronger a government could

become by creating well-nourished, easily supplied permanent standing armies.

Southern China, around the Yangtze, was attractive not just because it was a land of rice, but also because it was from here that cast iron first came. Knowledge of how to smelt iron spread rapidly to the north. A mass grave in Hebei, near Beijing, dated to about 300 BC, contains several soldiers buried with their weapons and other equipment. Almost all are made from cast iron, except for a few ornamental pieces of bronze. By the time of the Hàn Dynasty (206 BC–220 AD), Chinese metalworking had become established on a scale not reached in the West until the eighteenth century. The Chinese government built a series of large blast furnaces in Henan Province, each one capable of producing several tonnes of iron a day. It was also at about this time that the Chinese worked out how to combine various types of iron to make steel, an even tougher iron alloy.

While iron and rice were initially the preserve of the southern Chinese people, those from the silk-weaving north were determined not to be left behind. It was they who provided the impetus to centralize, consolidate, conquer and combine the whole area into a single civilization. For rice, silk and iron read food, wealth and war. It is not hard to see why such deep knowledge of how to exploit the natural world to human advantage became a magnet around which a single, powerful civilization arose, uniting the people of the two great river valleys.

The Shang Dynasty (1766–1050 BC) was the first set of Chinese rulers to leave tangible archaeological evidence in their wake. Before them the country's history is a wonderful *potpourri* of magic, myth and legend, dominated by the Three Sovereigns and Five Emperors, mythological rulers of ancient China between c.2852 and 2205 BC.

Their stories are known thanks to a surviving literary source for the early history of China called the *Bamboo Annals*, found in the grave of the King of Wei, who died in 299 BC. Another source is *Shiji*, a magnum opus written in 130 volumes between 109 and 91 BC by a single scribe called Sima Qian.

According to these records, the third Sovereign, Shennong, was the father of Chinese agriculture, and is credited with the discovery of tea. His close relation was the first Emperor (the Yellow Emperor), whose wife Leizu demanded her mulberry grove. He is said to have been the father of Chinese medicine. Their rule was followed by the Xia Dynasty (2070–1600 BC) but again, no material evidence yet exists to verify that they really existed. Their founder, Yu the Great, was responsible, apparently with a little help from the goddess Nuwa, for teaching people flood control techniques, and for overseeing the building of numerous irrigation dykes and channels, involving the labour of some 20,000 workers.

These stories include creation myths and legends about dragons, the most important mythical creatures in Chinese folklore. Dragons were considered the most powerful of divine creatures, and were believed to control the flow of water. They were even thought to be able to create clouds with their breath. At times of drought, Chinese people still pray to one of the most famous dragons of all, Ying Long, their god of rain.

The first historical evidence of Chinese rulers dates to the Shang Dynasty beginning in about 1600 BC. Archaeologists excavating a site called Yin in the 1920s uncovered eleven royal tombs and the foundations of a palace. Tens of thousands of bronze, jade and stone artefacts were found. They show that this was a highly advanced culture, with a fully developed system of writing, ritual practices and impressive armaments that helped its people conquer and rule lands for miles around. Human sacrifices were common. Like Queen Pu-Abi in Mesopotamia (see page 127), many members of the Shang royal family were buried with a complete household of servants, including chariots, horses and charioteers – all thought essential for protection in the afterlife.

One remarkable royal tomb is the grave of Fu Hao, discovered in 1976. She was wife of one of the most charismatic Shang rulers, called Wu Ding. Inscriptions found on bones in her tomb say she led more than 13,000 men into battle. Over 1,600 objects were discovered inside her grave, including eighty-nine daggers and a complete model army carved out of green jade.

The mastery of metal weapons, chariots and horses in China mirrored what was already common in the Mediterranean world 5,000 kilometres to the west. Did this Chinese lust for war and violence stem from abroad, or did it emerge independently? The chronology suggests that early trade links over the Asian grassland steppes may have been responsible for bringing these triple troubles to China, since the earliest evidence of chariot warfare and bronze weapons there dates from at least 200 years after the first Mediterranean finds. This would allow plenty of time for knife-wielding charioteers to gallop across the grasslands and barren deserts of the Asian steppes, stopping off at one or more of the many oases along the way. The problem is that to date no actual evidence that this migration happened has yet been found, so it has been left to conjecture.

Fighting wasn't the only concern of these early Chinese kings of the north. Just as important was their strongly held belief in being the sole link between the gods and mankind. Unlike any other rulers we have seen so far, these kings took it upon themselves to perform detailed and highly technical fortune-telling rituals. They needed no priests or holy men like shamans. It was the king's job to make contact with the heavens above.

This was done in a most bizarre and ingenious way, using turtle shells or the bones of an ox. A heated rod would be pushed into the shell or bone, causing it to crack. Like a modern-day palm reader, the king would interpret the length and direction of the lines to reveal the answers to questions which were important to him and his people. *When will it rain? Will we win the next battle? Will we have a bumper harvest this year?* Sometimes these questions would be inscribed on the shells themselves, using a form of symbolic writing that has been easy for modern historians to decipher. Unlike cuneiform script or Egyptian hieroglyphics there is no need for a Rosetta Stone or Behistun Inscription to help decipher these words, since they resemble modern Chinese writing so closely. This in itself is testament to how today's China has its roots sunk deep in the past, and that its culture represents by far the longest surviving pattern of continuous civilized human behaviour.

Hundreds of thousands of oracle bones with ancient writing on them have been found in recent times. They used to be advertised in Chinese markets and sold as medicinal 'dragon bones'. Their significance was only realized in 1899, when a scholar called Wang Yirong who was suffering from malaria acquired some medicinal bones from a local market. Just as his friend Liu E. Wang was about to start grinding them into powder, he noticed that they were inscribed with a curious type of ancient writing.

Most oracle bones have now been traced back to the tombs at Yin, where more than 20,000 were found during the excavations of the 1920s and 1930s. They form the earliest significant body of Chinese writing yet to have been discovered.

The first thousand years of recorded Chinese history, from about 1200 to 200 BC, is a story of consolidation and conquest, largely thanks to the combined impact of rice, silk and iron. The Kings of Shang, and then the Zhou, who took charge in 1046 BC following victory at the Battle of Muye, used the Yellow River as their main corridor of power. They claimed that their power came directly from heaven. They meant it. Their style of rulership was symbolized by an axe, usually embellished with hungry smiles and devouring teeth.

The Kings of the Zhou also masterminded a series of campaigns southwards in the hope of capturing the agricultural and metallurgic riches of the Yangtze, crossing a series of mountains that lay between the two great river systems. From about 800 BC they increasingly devolved power in their conquered territories to nobles and trusted family members, while threats from the north diverted their attention, making effective and peaceful government impossible.

In the end, their capital Hào (near the present-day city of Xi'an) was sacked by barbarian invaders, and in 722 BC the Zhou had to move their headquarters further east to Luoyang (in the present-day Henan Province). Central Zhou power rapidly fell apart, and a series of smaller states, some with rulers calling themselves kings,

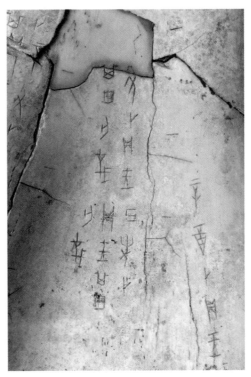

Oracle bones: Chinese fortune-telling began thousands of years ago with kings asking the gods questions inscribed on bones. Answers came in the form of cracks made by heated rods.

emerged to fill the gap. By 500 BC these states had been consolidated into seven major powers, each vying for the prize of a united China, with its promise of almost infinite supplies of food, wealth and power. Over the next 300 years it was the military struggle for supremacy between these states, known as the Warring States Period, which finally constructed the platform for uniting all China.

During this period a number of different philosophies evolved, called the 'Hundred Schools of Thought'. Wise men and thinkers wandered from court to court, advising kings and nobles on how they might live justly, rule wisely and advance the progress of their kingdoms. One such man was Kongzi, later known to the Western world as Confucius. Tradition says he lived from about 551 to 479 BC. His legacy lives on in societies across the Far East, from Japan and China to Korea and Vietnam.

Kongzi was a minister of justice in the state of Lu. One day, aged about fifty-five, he decided to quit his job and go on a trek around the kingdoms of northern China, to preach his message of the

23:59:59

right way to lead a virtuous life and the best way to rule a kingdom. Confucius sought a system for living that could restore unity, because he thought that the world was descending into an abyss of internal power struggles and military confrontations. He taught that obedience, correct behaviour and good etiquette were ways in which order in society could be restored. A good king would set a good example to his people, and good subjects were bound to obey. What have been passed down as the teachings of Confucianism are not necessarily what Confucius taught in his day. His followers, especially Meng-tse (also known as Mencius) and Xunzi, developed his sayings into more complete but different philosophies.

Meng-tse emphasized the potential for individuals to realise the inherent good in themselves, and said that the king could not rule without his people's tacit consent. If he lost their support, he would also lose the mandate of heaven, providing a possible justification for civil disobedience. Xunzi, however, advocated that the state should be strong, and must always take responsibility for the control of people's individual actions, since human nature was innately selfish and evil.

What Confucius didn't concern himself with is almost as revealing as what he actually taught. His philosophy has no place for gods, no afterlife, no discussion or consideration of a divine soul or spirit. In a way, Confucius developed the first godless theory of personal and political behaviour. Family loyalty, respect for older people and reverence for the past were his three pillars of social virtue.

A flavour of his philosophy can be captured by some of his most famous sayings. He hated war and confrontation, had a love of history, and was always pragmatic:

'Before you embark on a journey of revenge, dig two graves.'

'Study the past as if you would define the future.'

'The only constant is change …'

A number of great scholarly works are attributed to Confucius, although it is far from clear whether he actually wrote any of them. For almost 2,000 years Chinese civil servants, lawyers, military officers and other officials were required to study these texts, called the Four Books and Five Classics, in order to qualify to serve the state. This emphasis on education, teaching, conformity and obedience is still a hallmark of the enduring society that is China today.

Confucius's concern for order and peace threatened to become lost in the noise of war and battle that overran Chinese life until the year 221 BC, when the country was unified by the supreme triumph of the state of Qin. The story of the rise of Qin (from which the name 'China' comes) is as bloodcurdling as it is brutal.

Qin was a kingdom in the north-west corner of China, a land of horse-rearing and bounty-hunting. Selective breeding meant that larger horses were now available, allowing soldiers to ride into war on horseback, liberating them from expensive, unwieldy chariots. Anyone with a horse could now charge into battle, giving immediate superiority

An eighteenth-century portrait of Qin Shi Huang, First Emperor of China, who consolidated power after a protracted period of civil war.

over those standing with a bow and arrow on a moving chariot that needed at least two, if not four, horses to pull it. Cavalry warfare gave some rulers a huge military advantage.

The military might of the Qin was matched only by its brutality. One famous general, called Bai Qi, is reputed to have killed more than a million soldiers and seized more than seventy cities. In 278 BC he led the Qin army to victory against its biggest rival from the Yangtze south, the Chu. He then went on to defeat the nominal kings of China, the Zhou, at the Battle of Changping (260 BC). After this battle he had more than 400,000 prisoners of war slaughtered by burying them alive.

Civil administrators were no less harsh. One such, called Shang Yang, is credited with reforming the running of the Qin kingdom, turning it from a disorganized tribal power into a slick, effective, military machine. With the support of the ruler Qin Xiaogong (381–338 BC), Shang Yang was able to put into practice his belief in the absolute rule of law. For him, loyalty to the state was always superior to loyalty to the family. His reforms included stripping nobles of their lands and giving them

instead to generals as a prize for victory in war. He put great emphasis on agricultural reform, so that the land could support more people and feed more soldiers. Farmers who met government quotas for supplying food were rewarded with slaves.

Shang Yang's reforms were later codified into a book of law called *The Book of Lord Shang*. Soon Qin was by far the strongest state, and it rapidly became pre-eminent amongst the Seven Kingdoms. The climax came with the rise to power of Ying Zheng as ruler of Qin. After defeating the last independent Chinese state, Qi, in 221 BC, Ying became the first Emperor of all China (ruled 221–210 BC), renaming himself Qin Shi Huang, after the divine rulers of Chinese mythology.

With the assistance of his Prime Minister, Li Si, Qin Shi Huang rewired China into an awesome centralized powerhouse. Regional rulers were sacked, and in their place he appointed loyal civil governors to each of thirty-six new civil regions. Alongside them military governors were appointed, and a team of inspectors roamed the country to ensure none of them overstepped the mark. Governors were rotated every few years to prevent

This water-colour painting, made on silk, shows how precious books were burned and scholars executed by the paranoid but powerful ruler, Qin Shi Huang.

any one of them building up a regional power base. All this was an extension of what Shang Yang had implemented across the kingdom of Qin more than a hundred years before.

In 213 BC Qin Shi Huang ordered what is called the Great Burning of Books, suppressing freedom of speech in an attempt to unify all thought and political opinion. Hundreds of thousands of books were burned, many of them originating from the philosophies of the Hundred Schools of Thought. All books were banned, except for the legal works promoting the supreme control of the state. Anyone found discussing illegal books was sentenced to death, along with his family. Anyone found with proscribed books within thirty days of the imperial decree was banished to the north to work as a convict building the first Great Wall of China.

A massive canal, begun in Qin Shi Huang's father's reign and built by a brilliant engineer called Zheng Guo, was completed in 246 BC. It unlocked opportunities for rice-growing further north, and provided the Qin Dynasty with an almost limitless supply of food with which its armies and people could gain an unassailable position of strength.

Finally, the new imperial government standardized just about everything that could make running a large centralized empire easier – from the characters used in handwriting to the width of axles for carts,

so they would run more smoothly in the ruts of imperial roads. Edicts, some of which survive, were inscribed on the sacred Mount Taishan in Shandong, to let heaven know of the new unification of the earth under a single, all-powerful Emperor.

Towards the end of his life, like King Gilgamesh (see page 124), Qin Shi Huang became obsessed with finding an elixir that would make him immortal. He sent ships full of servants southwards in search of the Pengali mountain, where it was believed immortal gods lived. They never came back, for fear of losing their lives if they returned empty-handed. Legend has it that it was to the Japanese islands that these people travelled, giving an early injection of Chinese culture to what was still a largely hunter-gathering society.

So paranoid was Qin Shi Huang about being assassinated (several attempts were made on his life) that when on tour he took with him a series of doubles to reduce the chances of his falling victim to a vengeful blade. He eventually died during a tour of eastern China in 210 BC, after swallowing mercury pills which his advisers believed would give him everlasting life.

For 2,000 years no one knew where he was buried. Then, one day in 1974, some well-diggers struck an unusual object buried several feet underground. What they found led to one of the

Lines of terracotta soldiers prepare to defend the First Emperor in the afterlife as seen in this archaeological wonder found by Chinese water workmen in 1974.

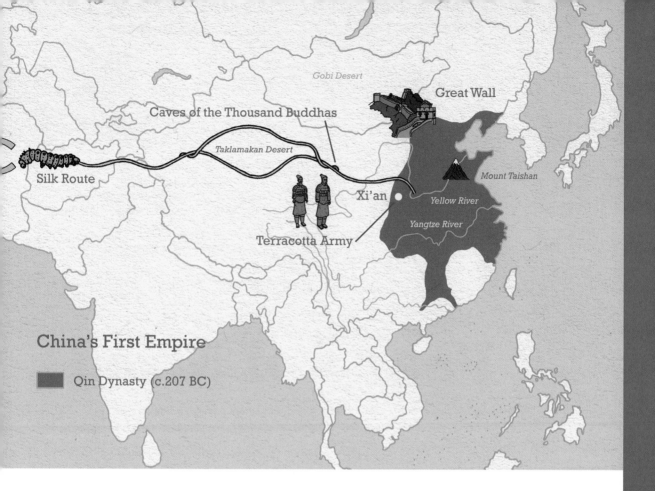

China's First Empire

Qin Dynasty (c.207 BC)

most incredible archaeological discoveries of all time. It was an enormous royal tomb, some three miles across, containing a terracotta army of more than 8,000 life-size soldiers, designed to defend the Emperor Qin Shi Huang in the afterlife.

More than 700,000 workers were involved in its construction. Each soldier is an individual, hand-crafted work of art, originally equipped with bronze spears and bows and arrows. The army is arranged in battle formation, supported by 600 clay horses and more than a hundred life-size working wooden chariots. The main tomb, containing the Emperor himself, was sealed with molten copper, and has yet to be opened. Sima Qian, the Chinese scribe writing less than a hundred years after the Emperor's death, claimed that it contained rare jewels and a map of the heavens with the stars represented by pearls, and that on the floor was a map of China with the rivers and sea represented by flowing mercury.

Magnetic scans of the tomb indicate that a large number of coins are buried there, with unusually high concentrations of mercury.

Although Qin's dynasty crumbled only a few years after his death, thanks to the hatred and vitriol that accompanied his life's work, his achievement was complete. He didn't just bring about the unification of seven warring kingdoms into the largest empire on earth, he created a top-to-bottom model for imperial administration, from the principles of its ruling culture down to the nitty-gritty plumbing needed to make it all work in practice.

Rice, silk and iron provided both the appetite for expansion and the means of conquest, to create the largest and most enduring human power on earth. Supreme command over nature turned these ancient people into an ingenious and unassailably robust civilization for thousands of years to come.

China's two great river valleys were eventually united by the Kingdom of Qin (c.207 BC) and were made powerful through rice, silk and iron.

23:59:59

Peace of Mind

How a particular civilization re-discovered that humans could live in harmony with nature and set about trying to spread its enlightened message.

GLANCE BACK fifteen minutes before midnight on our twenty-four-hour journey, and we find ourselves on the threshold of a momentous geological event that has dramatically altered the course of recent human history. Mountains made by the crashing of the Indian continental plate into the southern flank of Asia, starting about forty million years ago, may seem irrelevant to the story of man and his attempts to master nature over the last few thousand years. But historically the Himalayas have a lot to answer for.

Scientists think that the enormous height of this dramatic range has played a vital part in controlling the earth's temperature. For millions of years massive monsoons have been formed by the cooling of moisture-laden air as it passes over the towering Himalayan peaks (see page 84). In the process, billions of tonnes of carbon from the air have been dissolved by the rainwater and transported down rivers to the sea, where living creatures use it as material for building cells. When they die, their carbon-rich shells and bodies fall to the ocean floor, eventually to be buried deep in the muds of the earth. In this way the quantity of carbon in the atmosphere is regulated through the interaction of the forces of nature and living things, which together have kept global temperatures cool despite the ever-increasing heat of the sun.[1]

Perhaps this ancient partnership between life and the earth explains why, just beyond the foothills of this massive range, a human society developed its own very distinctive relationship with nature over the course of some 2,000 years.

The Himalayas certainly helped protect the hunter-gathering people living in what we now call India from the centralizing, conquering and consolidating forces of China. But their effectiveness as a barrier decreases to the north-west, where passes permit the passage of people travelling by foot, horse and chariot.

Several waves of invaders came from the north. Some of them probably originated from the steppes of central Asia, from where, having passed through Mesopotamia, they swept into the Ganges plain in northern India, stopped only by the towering Himalayan peaks. Surprisingly few archaeological remains have been found so far to help piece together the early history of these invasions. Most surviving evidence is literary.

Sacred texts called the Vedas were written in a language called Sanskrit that originated in the Middle East. They tell stories that weren't written down for hundreds or perhaps thousands of years. The Vedas were used as handbooks for instructing priests (called *brahmanas*) on how to perform sacrifices to the gods. These texts describe life between c.1700 and 1100 BC – although some experts believe they date from a much earlier period, perhaps as far back as 4000 BC. They paint a vivid picture of the tools brought by these early invaders in the form of horses, wheels and metal. Their verses talk of noble archers engaged in duels with rival heroes, exchanging volleys of arrows while galloping across fields of battle in horse-drawn chariots. They also describe the use of tools, which were employed to clear the jungle around the Ganges Valley. This was a good place to settle. Heavy rains made for lush vegetation, allowing the growth of rice, which could sustain armies.

A superficial reading of some of the ancient texts of India might lead to the impression that these people's destiny was to be as violent as those of the emerging societies to the north, east and west. At the heart of the ancient Indian religion, Hinduism, is the *Mahabharata*, one of the most famous sacred poems ever written. It is also the longest epic poem of all antiquity, far longer than Homer's tales of Troy, stretching to 74,000 verses and containing more than 1.8 million words.

It tells the story of an epic struggle for the throne of the kingdom of Kuru between two rival branches of a dynastic family, the Kauravas and the Pandavas. The tale's climax centres around what is

This carving shows soldiers, horses and chariots charging to war in the eighteen-day Battle of Kurukshetra, described in the Mahabharata, a sacred Hindu poem.

purported to the biggest and bloodiest battle in all history – the eighteen-day-long Battle of Kurukshetra, in which the Pandavas were ultimately victorious. The most sacred section, possibly added in about 550 BC, tells of a debate between the leader of the Pandavas, Arjuna, and the god Krishna, who had incarnated himself into human form to serve Arjuna as his personal charioteer. On the eve of battle Arjuna urgently seeks Krishna's advice as to whether or not to wage war. He has a dilemma: he knows war will mean having to kill various members of his own family, who were obliged to fight against him owing to previous oaths of allegiance.

In this part of the story, called the *Bhagavad-Gita*, or *Gita* for short, Krishna reveals the mysterious philosophy that still binds Hindu people together and defines their reverence for nature and all living things. He explains to Arjuna that despite the inevitability of war, there is no need to lament those who die in battle, because the spirit of the self (called *atman*) is indestructible. Fire cannot burn it, water cannot wet it, and wind cannot dry it, he says. This self, says Krishna, passes

23:59:59

from one body to another, like a person taking off worn-out clothes and putting on new ones.

Reincarnation is the belief at the heart of what makes Hinduism different from other religions. Each living thing possesses an individual spirit (*atman*), which is part of an über-spirit (*brahman*), the universal force that binds together all life. The goal of all individuals is to liberate the *atman*, freeing it to join the *brahman* in eternal bliss. Its destiny is to be recycled again and again in any living thing, plant, animal or human, until it reaches a sufficiently advanced state of development to attain enlightenment (*moksha*) and eternal liberation.

Individual spirits can be freed through the practice of meditation. In the *Gita*, Krishna goes into exquisite detail explaining to Arjuna precisely how an individual can free his or her spirit by stilling the mind of selfish desires, using as many as four different types of yoga.[2]

Krishna's advice has had a profound influence on millions of people throughout history. More than 2,000 years after it was first written down, Mahatma Gandhi, the pacifist Indian leader who led the non-violent revolt against British rule in India (see page 365), identified the *Gita* as his greatest source of inspiration:

> '*The Gita is the universal mother. When disappointment stares me in the face and all alone I see not one ray of light, I go back to the* Bhagavad-Gita. *I find a verse here and a verse there, and I immediately begin to smile in the midst of overwhelming tragedies – and my life has been full of external tragedies – and if they have left no visible or indelible scar on me, I owe it all to the teaching of* Bhagavad-Gita.'

The doctrine of reincarnation and the idea that anyone can practise yoga in their quest to liberate their soul into a union with the universal spirit in everlasting bliss has proved a very popular and appealing concept, in both ancient and modern times. Its philosophy helped ancient Indian civilization come to terms with the wave upon wave of migration, invasion and violence that have sporadically plagued the south Asian subcontinent.

A divisive social structure, known as the caste system, evolved as a response to the increasing complexity of Indian society as different cultures and traditions all piled on top of each other into one space. Rather than each new culture blending in to create diverse social groups, the Indian way of life evolved as a kind of multi-layered cake of people who preferred not to mix.

The idea probably originated from the first horse-and-chariot-borne and bronze-wielding invaders who charged in from the north and west from about 1500 BC, bringing with them their priests, writing (Sanskrit) and a belief in many different gods, such as Indra, the god of war and thunder. Originally there were only four castes. *Brahmans* were priests who prayed; *kshatriyas* were the soldiers who fought; *vaisyas* the farmers and artisans who worked; and finally *sudras*, at the lowest end of the scale, dealt with everything that was 'unclean'.

These slave-like people were consigned to the sewers of society and came to be known as 'untouchables'. Mixture between these classes was never encouraged; each one therefore maintained its own identity and culture. However, the idea of reincarnation gave these people some hope that, in the next life at least, there was a prospect of joining the ranks of the higher castes.

The Indian caste system has become a great deal more complicated over time, but it has endured to this day. The concept has even been adopted by other societies – some close by in Korea and Japan, others further away in Africa and even South America. Organizing a human civilization into castes has stood the test of thousands of years because it has provided an effective way of dealing with generations of immigration without causing existing cultures to feel threatened by the dilution or extinction of their own distinctive ways of life. Each time a major new culture arrives, a new caste comes into being, finding its place above or below those already there, without any need for radical adjustments of existing customs or habits. This system has ensured that ancient cultures and beliefs have been preserved longer in India than in most other parts of the world, which explains why Hinduism is the longest-surviving religion in all human history.

Veneration for nature, rather than violence towards it, is a distinctive characteristic of Hindu thought. The *Upanishads* are a collection of ancient Hindu texts, first written down in about 500 BC, designed as commentaries to help interpret the Vedas. In them, *ahimsa* is first mentioned. It is a vow that many Hindus take to be non-violent towards nature.[3] Vegetarianism is part of this philosophy, and for this reason as many as 40 per cent of all Indians are vegetarian to this day – that's about 300 million people. Even those Hindus who do eat meat hardly ever eat beef, since the cow is venerated above all other animals as a gift from nature providing milk to drink, power for pulling ploughs and manure for nourishing the soil. For Hindus the sacred cow is a symbol of unselfish natural giving, and cow-slaughter is still banned in almost all the states of India today.

This respectful, peace-loving relationship with the natural world was given a huge boost by four men who, between them, deeply influenced how *Homo sapiens* found out that it was possible to adapt its civilizations to live in harmony with natural world. Two of them founded religions, the other two helped spread them around the world.

The first was an Indian prince called Siddhartha Gautama. He is thought to have lived from about 563 to 483 BC, and was born in Lumbini, in modern-day Nepal. The only historical evidence of his life comes from texts written by his followers some 400 years after his death, so some of the details may well have merged into myth over centuries of oral rendition. His mother, Queen Maya, died a few days after his birth, leaving him to be brought up by his father, Suddhodana, a king or tribal chief, who had three palaces built in honour of his newborn son. His father wanted to shield Siddhartha from religious teaching and knowledge of human suffering, thinking that this would allow him to become a strong king.

The Buddha falls into a deep meditative trance, as carved on a wall at the Gangara Temple, Sri Lanka, a land that was converted to Buddhism by Ashoka's children.

23:59:59

A scene from a Hindu poem called the Gita Govinda in which a cowgirl, Radha, is persuaded to meet the god Krishna, the same deity who drove Arjuna's chariot at the Battle of Kurukshetra.

But, at the age of twenty-nine, Siddhartha left his palaces to meet his subjects. His father tried in vain to remove all signs of poverty and suffering, but to no avail. On his first outing Siddhartha saw an old man – until then he knew nothing of the trials of old age. On further visits he met diseased and dying people. Greatly disturbed by what he had seen, Siddhartha fled from the luxuries of his palaces to live as a monk, begging for food in the streets. He then became a hermit and, with the help of two teachers, learned how to meditate and to still his mind.

Next, Siddhartha Gautama and five companions tried to find enlightenment by the total denial of all worldly goods, including food: at one time they ate no more than a single leaf or nut a day. After collapsing in a river and nearly drowning, Siddhartha discovered what came to be known as the 'Middle Way' – a path towards enlightenment (liberation of the *atman*) that could be accomplished without the need for extremes, whether of self-indulgence or self-denial. After receiving the gift of some rice pudding from a village girl, Siddhartha sat under a tree until he found the Truth. After forty-nine days of meditation, aged thirty-five, he at last attained enlightenment, and from then on became known as the Buddha, meaning 'awakened one'.

Tapussa and Bhallika, two merchants, became his first disciples. Legend has it that they were given some hairs from the Buddha's beard, which are now believed to be enshrined in the Shwedagon Temple, in Rangoon, Myanmar (Burma). For the next forty-five years the Buddha journeyed by foot around the plain of the Ganges River, in north-east India and southern Nepal, teaching his doctrine to a wide range of people, from royalty to terrorists and beggars.[4] After making thousands of converts, he died at about the age of eighty, perhaps of food poisoning.

Buddha's teachings were really an extension, or popular interpretation, of many traditional Hindu beliefs. They were of enormous appeal, especially to the poor, for whom there was little hope of social or material improvement. The Buddha explained how by following his Four Noble Truths and the Noble Eightfold Path, these people could rid themselves of inner desires and free their spirits to eternal liberation without the involvement of any priest, king or other intermediary.

Another prince who lived at about the same time as Gautama also renounced his kingdom, and is said to have attained spiritual enlightenment after wandering for twelve and a half years in deep silence and meditation. This man was known as Mahavira, meaning 'great hero', and he became the twenty-fourth and last prophet (*tirthankar*) of the Jain religion.[5]

Jain scriptures were written over a long period of time, but the most popular work was written by an Indian monk called Umaswati more than 1,800 years ago. In his *Tattvartha Sutra*, or Book of Reality, the main aspects of Jainism are set out, identifying its central belief that all life, both human and non-human, is sacred.

For Jains there is no justification for killing another person, however greatly provoked or threatened. They refuse all food obtained by unnecessary cruelty. Jains are vegetarians and avid supporters of animal welfare. In many Indian towns today animal shelters are run by Jain people. Root vegetables are avoided, as harvesting them destroys an entire plant, whereas fruit, such as an apple, is acceptable, as picking it will leave the tree unharmed. Non-violence, religious toleration and respect for nature are cornerstones of the Jain philosophy, which, like Hinduism and Buddhism, is concerned with liberating the individual's soul through enlightenment accomplished through a series of codes of conduct that involve taking five vows: of non-violence to all living things (*ahimsa*); truthfulness (*satya*); non-stealing (*asteya*); chastity (*brahmacharya*); and detachment from material possessions (*aparigraha*).

Mahavira's teachings, like Buddha's, attracted people from all walks of life: men and women (regarded as equal in Jain teachings), rich and poor, touchable and untouchable – all had the potential to reach liberation (*moksha*) and eternal bliss.

Neither Buddhism nor Jainism would have had nearly such a big impact on human history were it not for the patronage of certain rulers in the secular world. By about 500 BC, sixteen different kingdoms, known as the Mahajanapadas, divided the Indian subcontinent, from modern-day Afghanistan in the west, to Bangladesh in the east.

Some, like the kingdom of Kuru, centred near the modern Indian capital of Delhi, became great centres of art and philosophy. It was here that the legendary war of Kurukshetra took place.

Most of these kingdoms were consolidated into India's first empire by Chandragupta Maurya (ruled c.320–298 BC). Unlike China, India's itch to centralize was less to do with internal power struggles between kingdoms and more as a response to threats from outside, in the form of Persian and Greek armies that from about 500 BC started harassing their borders, especially those in the north-west. By 303 BC Chandragupta is reputed to have had an army of some 600,000 men, with 30,000 cavalry and 9,000 elephants.[6] But towards the end of his life he gave it all up and became a Jain monk. Eventually, it is said, he starved himself to death in a cave.

While Chandragupta established the Jain religion as the preferred philosophy of the most powerful ruling family in India, it was his grandson, Ashoka the Great (ruled 273–232 BC), who had the biggest impact of all. To begin with he was as ruthless and violent as any imperial monarch, controlling his empire through the threat of force. Indeed, the name Ashoka means 'without sorrow' in Sanskrit. But, shortly after the end of one of the biggest and bloodiest wars of the time, he underwent a profound and complete conversion.

The Kalinga War (c.265–263 BC) ended with the famous Battle of Kalinga, which left more than 100,000 people dead on the battlefield. On the day after the battle, Ashoka walked out across the city where, as far as his eye could see, the only sights were burned-out houses, dead horses and scattered bodies. 'What have I done?' he cried.

From that moment on, Ashoka is said to have devoted his life and his reign to non-violence. He became a devout Buddhist, and over the next twenty years dedicated himself to spreading the message of this powerful religion. Prisoners were freed and given back their land. *Ahimsa*, the Buddhist doctrine of non-violence, was adopted throughout his domains, forbidding the unnecessary slaughter of animals. Hunting for sport was banned, branding animals was outlawed

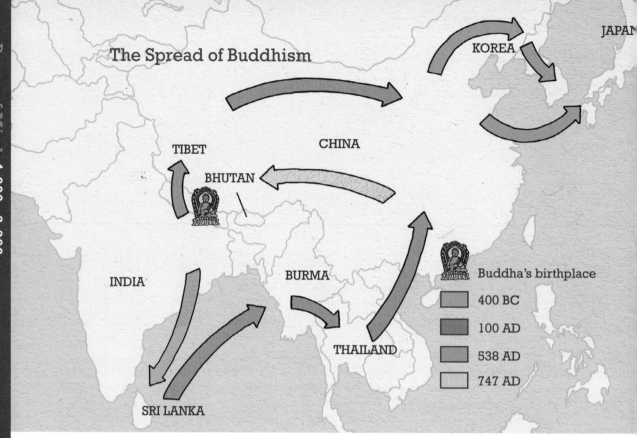

The Spread of Buddhism

KOREA
JAPAN
TIBET
CHINA
BHUTAN
INDIA
BURMA
Buddha's birthplace
400 BC
100 AD
538 AD
747 AD
THAILAND
SRI LANKA

and vegetarianism was encouraged as official policy. Ashoka built rest-houses for travellers and pilgrims, universities so people could become more educated, and hospitals for people and animals alike throughout India.

As many as 84,000 monuments and monasteries (stupas and viharas) were erected for Buddhist followers, many of them built at places associated with the life of the Buddha. His most lasting legacies are probably his Edicts. Dozens of sandstone pillars were erected throughout modern-day Pakistan and northern India. Written in the widely spoken language of the ordinary people, Prakrit, they popularized Ashoka's beliefs in the Buddhist concept of righteousness (*dharma*). Their inscriptions provide details of his conversion after the Battle of Kalinga, as well as his policy of non-violence towards all living things:

'I have made provision for two types of medical treatment. One for humans and one for animals. Wherever medical herbs suitable

for humans and animals are not available, I have had them imported and grown … Along roads I have had wells dug and trees planted for the benefit of humans and animals.'
(Rock Edict No. 2)

They also show how he developed one of the first official policies of religious toleration:

'All religions should reside everywhere, for all of them desire self-control, positive essence, encouraged tolerance and understanding of other religions.'
(Rock Edict No. 7)

The Ashoka pillar at Sarnath, just north of the holy Indian city of Varanasi, was erected to record the Emperor's visit to the city. It still stands there to this day, in its original position. Its depiction of four lions and its Wheel of Righteous Duty – a twenty-four-spoked chariot wheel that represents twenty-four virtues and the endless cycle of time

– were adopted more than 2,000 years later as emblems for the modern Republic of India.

Ashoka sent missionaries to every king and court he could. He wanted everyone to know about the Buddha's message. They travelled as far as Greece, Lebanon, Egypt, Burma and Sri Lanka. The first Egyptian Buddhist colonies in Alexandria date from this time. Ashoka promoted a new notion of kingship, in which a ruler's legitimacy was gained not from the generosity of a divine god, but by advocating Buddhist ideals, establishing monasteries, supporting monks and promoting conflict resolution.

Following Ashoka's reign, Buddhism spread far and wide. His twin children Mahindra and Sanghamitra settled in Sri Lanka, converting its rulers and people to Buddhism. By 100 AD Buddhist monks established a foothold in China, where their teaching fused with a similar philosophy called Taoism, founded by a philosopher called Laozi who lived at the time of the Hundred Schools of Thought (see page 159). His book, called the *Tao Te Ching*, described how violence should be avoided at all costs, and how individuals should rid themselves of strong emotions and desires through stillness and meditation. (It is ironic that in an attempt to find an elixir for immortality in c.850 AD, Taoist monks discovered how to make gunpowder – see page 248.)

Ashoka's influence now looks more powerful beyond India than within it, because by about 1300 AD Buddhism in India had declined into a relatively minor religion, marginalized by a resurgence of Hinduism and the onset of Islam (see page 271). From China, various brands of Buddhism spread to Korea, Vietnam and Thailand. By 538 AD its message had reached the islands of Japan, and by the ninth century Borobudur, in Java, where an immense cluster of Buddhist temples remains to this day. The shrines of Angkor Wat in Cambodia were built 300 years later which, although Hindu in inspiration, include extensive Buddhist sculptures. This religious complex, the apex of a highly religious civilization, is now buried deep in the jungle, spread across 40 square miles.

The brightest modern example of a Buddhist kingship takes us back up into the Himalayas. Today the kingdom of Bhutan, nestled high in the mountains, is home to just over 650,000 people. A Buddhist monk called Padmasambhava is reputed to have brought the Buddha's teachings to Bhutan and Tibet in 747 AD. Jigme Singye Wanngchuck, fourth King of Bhutan (ruled 1972–2006), has stated that Gross National Happiness (GNH) is more important to his people than Gross National Product (GNP) – putting the concerns of social welfare, environmental preservation and cultural protection above economic growth.[7] Uniquely among today's nations, the small, spiritual society of Bhutan is attempting to place material and spiritual well-being alongside the preservation of the natural environment.

Unfortunately, the threats of modern global warming are particularly acute for Bhutan. Huge mountain glaciers high in the Himalayas are nearing the point of meltdown. Sometime soon, no one knows when, a glacial tsunami is expected to wash these people and their society away. As for the Minoans of Crete 3,500 years ago, no amount of preparation and no number of barricades can stop it. The only difference this time is that when it happens, it will come as no great surprise.

East–West Divide

How clashes between wandering nomads and rival civilizations seeded some of the world's most ancient, pernicious and enduring human disputes.

DESPITE ASHOKA'S best efforts, Buddhist kingship did not strike much of a chord in the war-torn region of the Middle East. Between about 900 and 300 BC this part of the world witnessed a surge of conflicts between nomads, settlers and emerging empires to the east and west. In their jostling for supremacy they adopted rival religions, within which lie the origins of several intractable, ongoing human conflicts that persist to this day.

These ancient conflicts seem to have been triggered by the peculiarities of one of the most essential and natural phenomena of all creation – the water cycle (see page 24). Across the wild grassland steppes stretching from eastern Europe past the Black Sea and eventually into north-

eastern China, the world's climate grows increasingly dry and hot. Wet, warm winds from the Atlantic and Mediterranean provide a richer environment for humans and animals at the lower, western end of the Eurasian steppes than further to the east, where, as the altitude rises, the rain runs out, leaving a hinterland of little more than harsh, exposed scrub.

Nomadic people were highly successful provided they had enough lush vegetation and plentiful rain. But when there was an especially hot spell and it became too dry, as happens from time to time, they were drawn irresistibly westwards and southwards. From about 700 BC successful animal domestication meant that the population of the steppes was increasing – so much that when dry spells struck, waves of well-armed nomadic tribes moved westwards in search of more fertile lands, threatening the lives of settled communities living across Europe and the Middle East.

Not far into the Greek historian Herodotus's account of the epic Greek and Persian wars that took place between 500 and 448 BC we hear of the fearsome nomadic Scythians who 'once dwelt in Asia and there warred with the Massagetae, but with ill success; they therefore quit their homes, crossed the Araxes, and entered the land of Cimmeria'.

Scythians and Cimmerians were typical of the many horse-riding tribes who began to push westwards into eastern Europe and south into the Middle East in a search of more fertile, better-watered lands in which to live. Incredible as it sounds, the story of the havoc created in the wake of their migrations more than 2,500 years ago, driven by changes in the climate and rainfall, still reverberates around the world today in the form of some of our deepest cultural, racial and religious disputes.

The first begins with the establishment of a race of people who came to call themselves the Jews. From about 1050 to 700 BC, these people began to settle successfully in what is now modern-day Israel. The description of how they built their nation after the supposed Exodus from Egypt comes from the Old Testament.

Experts believe that these texts probably weren't written down until about 500 BC, leading some to conclude that the history they tell may have blended into myth over generations of oral recitation. However, recent archaeological finds are beginning to verify some of the historic evidence found in the Bible, giving renewed confidence that much of what it contains from the beginning of the Jewish kingdom in c.1050 BC onwards can be regarded as at least one view of what actually took place.[1]

Jewish people claim descent from Abraham, a nomadic herdsman from the Mesopotamian city of Ur who according to the Bible was instructed by God to travel to the land of Canaan, which, in return for being worshipped as the only true God, He gave to Abraham and his 'seed' for ever. No historical evidence exists to verify that Abraham was ever a real person, but the scriptures go on to say that he had two sons by two different women. Ishmael was his first son, born to a servant girl called Hagar. Abraham's wife Sarah consented to his having a child with Hagar because she thought she could not have children herself. Then, after Ishmael was born, to Sarah's considerable amazement (the Bible says she laughed out loud when she discovered she was pregnant) she also had a son, whom they called Isaac. One of Isaac's children, Jacob, had twelve sons, each of whom became leader of one the twelve tribes of Israel, from which Jewish people claim their descent.

After the Jews supposedly fled in search of food to Egypt, where they were enslaved, God repeated the promise He had made to Abraham, this time to Moses on Mount Sinai, in the Egyptian desert. God told Moses that the Jewish tribes were His chosen people, and gave Moses a set of laws to follow which became the foundation of their religion, Judaism. With God's help Moses then delivered the Israelites from slavery in Egypt so they could establish themselves in their Promised Land of 'milk and honey'.

Although Ishmael, Abraham's eldest son, disappears rapidly from the Jewish scriptures, he is to play an essential part later in our story. According to the divine revelation which led to Mohammed's recitation of the Koran beginning in c.610 AD, Ishmael also had twelve sons, from whom all Arabs are descended (see page 232).

So to whom exactly did God give the Promised Land – Arabic or Jewish people, or perhaps, in an effort to help them learn how to share, to them both? The question of who has legitimate claim to this land has become the source of one of the longest, most protracted disputes of all human history. The argument has erupted into sporadic and sometimes lengthy religious and territorial wars that still continue today.

According to the scriptures of the Old Testament, from about 1050 BC to 930 BC the twelve tribes of Israel inhabited the land of Canaan. In c.1008 BC, David, the boy who defeated the giant Philistine Goliath with a stone and a sling, became their king, and established their capital city Jerusalem on the banks of the River Jordan. Here David's son, the wise King Solomon (970–931 BC), built the first Holy Temple. The idea that one group of people was chosen divinely above all others wasn't completely new. The Egyptians believed in the supremacy of their civilization for thousands of years, and the Shang Dynasty in China wasn't slow to justify its rule as being based on the approval of heaven, as testified by their reading of divine oracle bones (see page 158).

But for the Jewish people, theirs was a promise with a difference. It was given to them by the one all-powerful God through two covenants, one with Abraham and the other with Moses. After a series of catastrophic events early on in Jewish history, the documentation of these promises became the key to these people's very survival.

After Solomon's reign, the twelve Jewish tribes squabbled, and determined to split their kingdom into two. In the northern half, called Israel, ten of the tribes settled, leaving the other two further south, in Judaea.[2] In 722 BC, disaster struck the northern tribes. The Kings of Assyria to the east, Shalmaneser V and his successor Sargon II, conquered the northern kingdom's capital, Samaria. The Assyrians later rebuilt the city, but resettled it with Arabs and Syrians.

As many as 40,000 Jewish people were deported to the Assyrian city of Nineveh to work as slaves on irrigation projects. Thousands more fled to Jerusalem in the southern Jewish kingdom, causing the population of the city to increase five times over. From a previously disparate collection of tribes, the Jewish people were quickly consolidated into a more unified community, holding a common belief that they were the chosen people of a single, all-powerful God, Yahweh, who had given them the Promised Land.

Terrorizing the Israelites wasn't Sargon of Assyria's biggest concern. Much uglier menaces lay to the north and east, in the shape of the nomadic Cimmerians and Scythians, both vying for wetter, more fertile land on which to graze their animals. But, in 705 BC, their ferocity got the better of Sargon, and he was slain by them in battle.

His successor, Sennacherib, turned Assyrian firepower back south towards Babylon in modern-day Iraq, the city famous for its laws, science and astronomy (see page 126). Time and time again it was captured. Again and again its people rebelled, before eventually it succumbed and become the Assyrian capital under King Esarhaddon, who then set his sights westwards towards the biggest prize of all: Egypt.

In 671 BC Esarhaddon successfully sacked and plundered that historic land, briefly making the Assyrian Empire, which now stretched from Egypt to India, the biggest the world had ever seen. His legacy wasn't long-lived, however, and by the death of his son, King Ashurbanipal, in 627 BC, the empire, exhausted by war, proved powerless to resist the Scythians and Cimmerians (Ashurbanipal was the king who built the famous library with its 20,000 cuneiform tablets and the royal palace at Nineveh – see page 124).

With the decline of Assyrian power, Babylon became independent again. In 612 BC its armies finished off the Assyrians by sacking their capital at Nineveh, first making sure it had struck as solid an alliance as possible with the troublesome Scythians. Unfortunately for the Jews of Jerusalem, the rise of an independent Babylonian power was not good news at all.

The Babylonian state grew mightier, until its power peaked with the reign of the fearsome King Nebuchadnezzar (ruled 605–562 BC), who transformed the ruined city of Babylon into one of the wonders of the ancient world. He restored temples to the gods, completed the royal palace, built an underground passage and a stone bridge across the Euphrates and an impregnable triple line of walls around the city. His most famous creation, though, was the Hanging Gardens of Babylon, one of the seven wonders of the ancient world. He reputedly built them for his wife, Amyitis, because she was homesick for the mountain springs near her birthplace.

Nebuchadnezzar also built a barrier, called the Mede Wall, between the Tigris and the Euphrates rivers to protect his kingdom from attack by the Scythians and Cimmerians to the north. Now *he*

was ready for the prize of Egypt which everyone coveted most of all. On his way west, of course, stood Jerusalem, which he captured in 597 BC, sending the then King of the Jews, Jehoiachin, as a captive back to Babylon and installing a new King, Zedekiah, to be a loyal vassal in his place.

But a few years later Zedekiah led the people of Jerusalem in a disastrous rebellion. Nebuchadnezzar rode once again to the city. This time he showed no mercy. He plundered it, burned it and razed its temple to the ground. After trying to escape, Zedekiah was captured on the plains of Jericho. His punishment was to witness his wife and children being executed, and then his eyes were gouged out to make sure that the image of his family's death was the last he would ever see. This broken man was taken in chains, along with some 27,000 other captive Jews, to Babylon, where he remained for the rest of his life.

At this point, with both Jewish capitals sacked and burned, the chosen people of God, scattered, broken, homeless and enslaved, looked as if they were about to leave the books of history for ever. But seventy years later they were saved by one of the few people who can legitimately claim to have substantially altered the course of human history. In his lifetime he established a spectacular and enormous empire, bigger even than that of the Assyrians. His name was Cyrus, one of the first rulers in recorded history that posterity has called 'the Great'.

Cyrus had a good start. His father was a Persian King who had fortuitously married the daughter of the ruler of the nearby kingdom of the Medes. As he was their only son, Cyrus could claim both lands, and combine them into a single domain. So when his father died in 559 BC, Cyrus marched against the Medes, taking great delight in defeating their King, his grandfather Astyages, who legend has it tried to have Cyrus killed as a child.[3]

Next, Cyrus marched north and defeated the friend and ally of his grandfather, King Croesus of

Lydia, reputedly after deploying one of the most eccentric and ingenious military tactics of all time. According to Herodotus he placed a herd of camels at the front of his forces; their appalling smell so put off the Lydian horses that they fled, and Croesus was captured. By 542 BC Cyrus's old friend and ally Harpagus added the rest of Asia Minor (Turkey) and Phoenicia (Lebanon) to his territories, forming a substantial Empire. Finally, Cyrus turned his attention south and, after diverting the flow of the Euphrates into a canal, his men waded across the river and entered the city of Babylon. On the night of 12 October 539 BC, they were greeted as liberators. The capture of Babylon added Syria, Phoenicia and Israel to Cyrus's Empire.

Cyrus's Persian Empire was remarkable for its policies of tolerance and respect for other cultures and religions. Local rulers, called satraps, were installed as the governors of conquered lands. After capturing Babylon, Cyrus allowed the Jews to return to Jerusalem, and ordered the building of a new temple to replace the one destroyed by Nebuchadnezzar, 'with the expense met out of the King's household'. He also ordered that 'vessels of gold and silver from the Temple of God which Nebuchadnezzar took … be restored so that everything may be restored to the sanctuary in Jerusalem and be put back in the Temple of God'. The story is told in the Bible in the Book of Ezra, the prophet who led many thousands of Jews back home. As a result, Cyrus is the only non-Jew to be honoured in the Bible as a messiah, a divinely appointed king sent by their one god Yahweh.

We know more than what just the Bible and Herodotus tell us about Cyrus, thanks to the discovery in 1879 of a remarkable inscribed cylinder recovered from under the walls of ancient Babylon. The cylinder confirms the Bible story of the Jews being allowed to return home to Jerusalem. It also talks of Cyrus abolishing all forms of slavery and forced labour, leading many people to describe it as the world's first Charter of Human Rights. Today it is on display in the British Museum, and a replica is kept next to the Security Council chamber in the headquarters of the United Nations in New York. As part of his policies of toleration, and despite his own belief in a single god, Cyrus allowed the people of Babylon to continue worshipping their own many divinities, including their chief god, Marduk.

Cyrus and his successors were Zoroastrians. This ancient religion, still practised today in parts of Iran,[4] was founded by the prophet Zoroaster, who wrote down his divine revelations in a sacred text called the *Avesta*. Zoroastrians believe in a single god, called Ahura Mazda. They see life as a constant struggle between the forces of good and evil. Eventually, however, they believe that good will prevail. Humans must choose for themselves which path to take – truth or falsehood. For Zoroastrians, all humans have the free will to decide. Their religion is marked by a deep respect for truth and integrity, which is at the heart of their faith. As a result it became a hallmark of Persian culture that nothing is considered more sinful that to tell a lie.[5]

Meanwhile, the Jews, almost driven to extinction twice, were now determined to lay claim to their Promised Land once and for all. Back in their shining new Temple in Jerusalem, the stories of the Old Testament – called the *Tanakh* in Hebrew – that had been handed down through generations of oral tradition were now written down by scribes, edited and finally canonized as the official word of God.

Cyrus the Great was a tolerant ruler who established the Persian Empire and, thanks to his generosity to the Jews, was even honoured in the Bible as a messiah.

This sequence of events has led some biblical scholars to believe that the Old Testament stories are best understood as efforts by the Jewish people to prove beyond doubt their historic claim to the Promised Land through a divine covenant with God. Anyone dispossessing them from now on could do so only against a background of prescribed historic illegitimacy, and absolute divine displeasure.

As for Cyrus, he had trouble back up north, where nomads from the steppes were once again threatening the borders of his empire. In 529 BC an offshoot of the Scythians, called the Massagetae, were in full attack at the head of the Tigris River. Their revolt was led by Queen Tomyris, who supposedly had one of her breasts cut off so she could fight more effectively. According to Herodotus, Cyrus's troops began their fight back well, killing both of the Queen's sons and many of her troops. But events swung against them as the battle wore on, and eventually Cyrus was slain. So bitter was the Queen at the death of her sons that she had Cyrus's head chopped off and his skull made into a goblet from which, it is said, she drank fine wine until the day she died.

Despite Cyrus's death, his Persian empire thrived for the next 200 years, thanks in large part to his tolerant policies and respect for local religions and beliefs. His son, Cambyses II, eventually conquered Egypt and defeated the Massagetae, from whom he recovered his father's body, building him a tomb in the old Persian capital of Pasargadae. Cambyses' cousin, Darius I (also called the Great), ruled from 522 to 485 BC, and is credited with reorganizing the empire and building a magnificent new capital called Persepolis, in modern-day Iran, with walls twenty metres high and ten metres thick. Its ruins are spectacular to this day. Darius also dug the first Suez Canal, allowing ships to pass from the Mediterranean to the Red Sea through a channel which, according to Herodotus, was wide enough for two triremes to pass and took four days to navigate.

There is still an inscription on the banks of the Nile near Kabret which reads: 'Saith King Darius: I am a Persian. Setting out from Persia, I conquered Egypt. I ordered this canal dug from the river

called the Nile that flows in Egypt, to the sea that begins in Persia. When the canal had been dug as I ordered, ships went from Egypt through this canal to Persia, even as I intended.'

Darius introduced coinage to the empire, standardized weights and measures, and encouraged commerce by building an enormous 2,500-kilometre road from Sardis, capital of the ancient kingdom of Lydia in western Turkey, to Persepolis. Along the way he had inns built to provide refreshment for the merchants, and stationed garrisons of soldiers to protect them from bandits. Like Cyrus, Darius allowed freedom of worship and maintained a no-slave policy. The workers on his many building projects were all paid.

Such tolerance for different cultures may have lasted were it not for the persistent problem of incursions and invasions by the northern nomadic tribes, who, unbeknown to the Persians,

Darius the Great, King of Persia, was constantly troubled by invading nomads whom he sought to neutralize by invading Europe. As a result, he stirred up a different hornets' nest.

were themselves being pushed westwards by other nomads. Continual harassment by these tribes in the end dragged Persia into what became the biggest and bloodiest conflict yet to take place in human history, one which ultimately led to the collapse of the once proud Persian Empire, and precipitated the rise of the Greeks to the west.

In a bid to put a permanent end to the trouble from the Scythians, Darius took a huge army north in about 512 BC and marched across the Bosporus, the short stretch of sea that divides Europe from Asia, into modern-day Greece. He marched as far as the Danube, so he could attack them from the rear. But unfortunately, thanks to an incorrect understanding of the geography of the region, Darius missed the Scythians altogether, and instead attacked and subjugated the people of Thrace and Macedonia in northern Greece. Having successfully established his presence, he installed his own satraps to oversee the region as a buffer zone protecting the north-west borders of his empire.

This pre-emptive attack backfired. Proud, independently minded Greek cities such as Athens and Eretria encouraged a revolt against Persian rule in western Turkey, led by the Greek Ionians. Darius counter-attacked in 492 BC, but was defeated at the famous Battle of Marathon two years later, sending shock waves through his empire.

Legend has it that after the battle a runner was sent to Athens to tell the city that the Greek army had been victorious, but instructing the city now to prepare itself against a sea-based Persian attack. So exhausted was the messenger by the twenty-six-mile journey that he died on the spot, leaving as his legacy the name of the long-distance race we know today.

Marathon signalled the beginning of Greek independence from mighty Persia. Greek city after Greek city now declared itself free from foreign rule. Instead of having a buffer zone against the nomadic tribes of the western steppes, the Persians had a new foe, led by the increasingly powerful city of Athens. This one had the additional skill of being expert at war at sea.

Darius died soon after Marathon. In 480 BC his son Xerxes prepared to put an end to this new European menace. His army was so large that it took seven days and nights to cross the Bosporus over two pontoon bridges made of more than 300 wooden ships each.[6] At Thermopylae, in the mountains of eastern-central Greece, a local man called Ephialtes told the Persians about a secret pass which would enable them to attack the Greek army from the rear. With a thousand volunteers, King Leonidas of Sparta dug in to resist the Persian onslaught, knowing there was now no way his army could escape certain death. His men held back the Persians just long enough for the rest of the Greek army to escape, giving precious time for Athens to prepare its fleet for war.

Xerxes, notorious for his bad temper, lost many men in the fight against Leonidas, but through force of numbers he still managed to sack Athens and burned it to the ground. However, thanks to the sacrifice made by Leonidas and his men, the citizens of the city had been safely evacuated to a nearby island, from which they watched helplessly as the flames from their homes lit up the night sky. The final act of battle commenced in September, when the Greek and Persian navies locked into combat at Salamis. The Persians' large triremes proved too clumsy against the more manoeuvrable Greek ships, and their navy was destroyed. More than 200 Persian ships sank to the bottom of the sea.

Some historians have even claimed that the Greek victory at Salamis was the most significant battle in all human history, because it led to Greek independence and unification, which helped lay the foundations for modern Western civilization.[7]

Without naval support, Xerxes couldn't supply his huge land-based army, so he was forced to retreat back into Asia over the Bosporus. He left his general Mardonius the task of finishing off the Greeks the following year, but the combined forces of the Greeks finally routed him at the Battles of Plataea and Mycale.

Europe and Asia were now truly at war. Border rivalries between nomads and settlers jostling for access to well-watered lands had mushroomed chaotically into conflicts between rival empires

The Powers of Persia (560–480 BC)

■ Empire of Cyrus (c.559–529 BC)   ■ Conquests of Darius (c.522–485 BC)

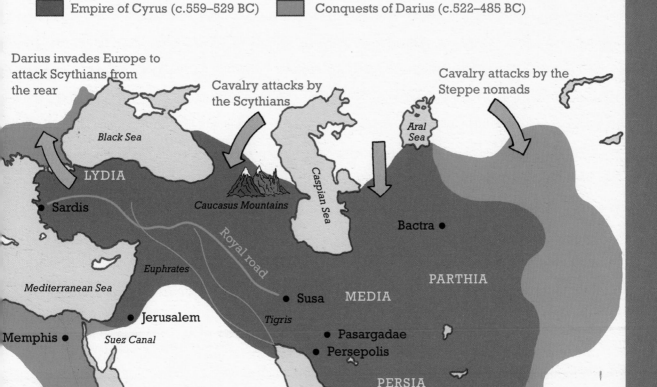

Darius invades Europe to attack Scythians from the rear

Cavalry attacks by the Scythians

Cavalry attacks by the Steppe nomads

Black Sea

LYDIA

• Sardis

Caucasus Mountains

Caspian Sea

Aral Sea

Bactra •

Royal road

PARTHIA

Euphrates

Mediterranean Sea

• Susa   MEDIA

Tigris

• Jerusalem

Memphis •   *Suez Canal*

• Pasargadae
• Persepolis

PERSIA

Thebes •

EGYPT   *Nile*

and city states, which were now establishing their own separate identities, and which justified invasion and conquest by regarding other rival religions and cultures as barbarians and infidels. Ashoka's notions of peaceful kingship and religious toleration were now firmly out of fashion.

But this was just the beginning. Conflicts arising between the Persian cultures of the Middle East and European cultures of the West invariably meant that the Jewish people got caught up in the middle. One of the big surprises in human history is how little some things have changed.

Attacks by increasingly populous northern nomads put pressure on Persian emperors, who in turn invaded Europe to try to make their borders safe.

23:59:59

Olympic Champions

How a range of new experimental lifestyles emerged in a cluster of highly competitive city states that learned to live off the fruits of trade.

NEXT TIME YOU'RE in the kitchen, spare a thought for any extra-virgin olive oil you may have in store before slopping it into the frying pan. For thousands of years oil from olives was the ancient world's premier source of natural fuel, used by households for everything from cooking to lighting lamps. Delicious to eat, easy to store and highly nutritious, olives were a seriously important part of the economy of the ancient human world.

Olive trees grow naturally along the shores of the Mediterranean Sea in and around Greece. Nowadays they are one of the most widely farmed fruits in the world, but already by about 2,500 years ago olives were to the embryonic city states

of ancient Greece what silk was to the increasingly centralized China – a vital source of wealth.

Athens, Thebes, Sparta, Corinth and Argos were just a few of the dozens of small, independent city states which from about 650 BC began a most intriguing series of experiments between nature and human civilizations. The foundations of these innovative societies were built on the olive trade, because little else grew in the dry, craggy, rocky terrain that tumbled down from the mountains of Macedonia to these island outposts on the east Mediterranean Sea.

Olives are amazing. A fully mature olive tree may take up to twenty-five years to grow, but once it starts bearing fruit it is almost completely hassle-free. Poor soil is no problem for an olive tree. Their groves require no ploughing, planting or weeding. And fancy irrigation channels, like those needed for rice, are also quite unnecessary. These fruits are among nature's lowest-maintenance, but highest-yielding crops. To harvest them demands little more effort than shaking the trees at the right moment, when the fruit is sufficiently ripe, leaving gravity to do the rest. Olives have one final but most important secret. After they have been harvested, leave them for a month or so in the sun, and a natural chemical reaction takes place inside. Fermentation by a fungus that grows on the skin of the olive produces a substance called lactic acid that naturally preserves the fruit for months, if not years. No need for a fridge or an ice box to keep them cool, so there was no danger that, once transported to market and shipped off to a foreign land, they would go off. The great thing about olives is that once fermented and immersed in a small quantity of salty water, you can completely forget about having to scribble on a sell-by date.

Without the natural occurrence of this miraculous fruit it is highly unlikely that ancient Greece would have become the crucible of Western civilization that almost all historians throughout the ages have claimed it to be.

Greek people soon learned to exchange their olives for life's other vital foodstuffs. By the time of the Greek–Persian wars in the sixth century BC, more than a hundred city states had successfully established commercial links all over the Mediterranean, where olives were traded for grain in Egypt, raw materials such as iron and copper from Spain and Italy, and cedar wood from Lebanon – vitally important for building ships.

After the fall of the early Minoan and Mycenaean civilizations, various waves of migrating nomadic tribes with their newly fashioned iron weapons settled in Greece and on the western coast of Asia Minor (Turkey). They are known as the Dorians, Ionians, Aeolians and Achaeans. By about 600 BC many of them had mastered the art of farming olives, and had discovered how to live well.

A complete dependence on trade for survival, combined with the gift of a crop that didn't take much effort to grow and harvest, provides one explanation why these people were able to experiment with news ways of life. For one thing, an economy based on trade gave these cities a highly developed, market-driven system for generating wealth. For their citizens, coins and loans became their main currency of exchange rather than grain, rice or slave labour.[1] They also had a lot of something most of the rest of the people in the world had very little of – spare time. Olives meant that for many months of the year they could use their wealth from trade to build a crop of new cities. What's more, the fact that their survival depended on trade meant that Greek cities were constantly exposed to other cultures and civilizations, ideal for adopting and adapting other people's techniques for establishing new, more advantageous relationships with nature. What happened inside some of these melting pots of humanity fundamentally shaped the rest of human and natural history to come.

Signs of a distinctive, new and eccentric pattern of human endeavour first occurred in what became the most famous of all the Greek city states – Athens (named after the Greek goddess of wisdom, Athena). Long before the Persians razed the city to the ground in 480 BC Athens was a laboratory for experiments in human behaviour. In c.594 BC a poet called Solon managed to win a victory for the city by capturing the nearby island of Salamis. He used the considerable power and prestige gained from this triumph to reform the highly oppressive political and legal system that had been established by a ruler called Draco about thirty years before. Under these laws, from which we get our modern word 'draconian', any crime, however small, was punishable by death. Another one of Draco's decrees was that if any man got into debt he immediately lost his freedom, and became enslaved.

So unpopular were these measures that when Solon proposed a new constitution it was greeted with huge support. To start with, Solon devalued the currency. Coins had just started to appear by this time, the idea originating from nearby western Turkey.[2] Solon decreed that one mina would now be worth a hundred drachmas instead of seventy-three – immediately resolving the debt crisis which

Solon, the statesman of Athens, talks to students about his new-fangled ideas on politics that involved giving the populace a role in civic life.

afflicted many of the city's poorer people. Next, he recalled and freed previously exiled or enslaved Athenian citizens. But this was only the beginning. His greatest reforms involved redistributing political power so that it wasn't just the most powerful families who participated in politics and the judicial system. Solon created a system that can now be recognized as the first attempt to create a democratic government – one in which the population at large had a say. Nobles remained the city's magistrates, but Solon introduced juries into almost every social dispute, and so, for the first time, involved ordinary citizens in the deliberations of justice. As Plutarch's *Lives* puts it: [3]

> *'I gave all needed strength to the common people. Yet, I kept the nobles with strong power ...'*

From here it was only a small step to creating two new political bodies – one to represent the nobles, which became known as the Areopagus (or senate), and a second chamber containing one hundred representatives of the four main ancient tribes of Athens. Each member served in the assembly for a year, having been chosen at random through a system of drawing lots. This second chamber was called the Boule.

After introducing a range of other decrees to help regulate civic life, Solon made the city's citizens promise not to change any of them for at least a hundred years. Storing the new laws on special wooden cylinders and placing them safely in the city's treasury, called the Acropolis, he then set off on a ten-year-long journey around the Mediterranean to see what would happen to Athens in his absence.

By the time he got back, the city was in a mess. Mediating between the warring political factions tested Solon's political mettle, and from time to time even he couldn't prevent power being seized by tyrants who occasionally dominated the new political systems. Solon died in 558 BC, but his template for how to involve common people in government without too severely disenfranchising the rich and powerful lived on. Over the next 200 years this early style of democracy helped Athens become, for a time, the most powerful and influential city in the world.

At just about the same time that Solon was experimenting with new forms of civic government, the beginnings of what was soon to become a revolution in scientific and religious thought was emerging just across the narrow stretch of sea separating Europe from Asia – the Bosporus.

Miletus, on the west coast of Turkey, was an ancient, vibrant city, full of trade, wealth and a variety of different cultures. It was also home to a man called Thales (born c.640 BC), who became famous for correctly predicting that a solar eclipse would take place on 28 May 585 BC. In those days people believed that the planets were gods, because of their seemingly chaotic passage across the skies, sometimes appearing to stop in their movement across the heavens and even to move backwards.[4] The only explanation they could come up with was that they were controlled by a pantheon of unpredictable gods. Borrowing from Sumerian and Babylonian traditions (see page 126), each planet was associated with a different god: some were known for causing trouble while others promoted love and peace. Associating events on earth with the movements of the planets became known as 'astrology'.

Thales demonstrated that the movements of the planets could be predicted using a set of astronomical tables originally compiled over hundreds of years by holy men in Babylon and Egypt, who had been studying the movement of the planets and the moon from their temples and ziggurats. The invasions of Darius the Great brought such knowledge, stored on clay tablets, into western Turkey, where Thales lived. When these fell into the hands of someone like Thales, with a keen eye for numbers and mathematics, patterns began to emerge that could then be extrapolated to predict events like a solar eclipse.[5]

In a world where such events were believed to be caused by the arbitrary whims of all-powerful gods, anyone who could make an accurate prediction of an event as dramatic as a solar eclipse made something of a stir. Thales's reputation spread far and wide.

The discovery of a set of rules that governed the movement of the planets in the heavens also caused Thales to wonder what else in nature worked on similarly predictable lines.[6] His lifelong quest for a set of universal laws to explain nature was variously adopted and challenged by other philosophers, many of whom lived in Athens, which, thanks to its victory over the Persians at Salamis in 480 BC, was enjoying several decades of peace (the cultural and philosophical Golden Age of Athens is traditionally taken to be the fifth century BC).

Diogenes (412–323 BC) was a beggar and philosopher who lived in Athens inside a barrel at the bottom of some steps leading to a temple. In his view Thales's search for scientific knowledge was a total waste of time. Instead, mankind should take as its source of inspiration the example set by the dog. After all, dogs don't fuss over what they eat, they're happy to sleep anywhere, they don't lie or dupe people, and best of all, they don't even mind much about where they go to the toilet.

Beneath his cynical and provocative words, Diogenes's philosophy had a serious message. In his view, all forms of society were artificial worlds that took people away from their traditional and rightful relationship with nature. He believed that a virtuous lifestyle was one lived outside the rules of human societies which judge people by their clothes or professions, or by how much money they earn. Diogenes was the first person to claim to be 'a citizen of the whole world' rather than of a specific city or state. He requested that when he died his body be left exposed to the elements so that other worldly citizens, in the form of wild animals, could enjoy feasting off his rotting flesh (see page 140). Diogenes inspired a school of philosophy called the Stoics, who, like Jain monks in the east, believed that personal happiness had nothing at all to do with material circumstances.

Socrates (470–399 BC) was another famous Athenian philosopher. Some people today even credit him with being the founder of modern Western thought. Like Thales he believed in a set of natural universal laws which could be understood through a process of philosophical speculation. Such insights would ultimately lead to great wisdom and personal enlightenment. Like Buddha, Socrates thought a man's soul could be improved over time, but not through mastering stillness of the mind – rather the opposite. For Socrates the path to enlightenment involved the application of problem-solving reason, high-powered discussion and heated debate.

By about 460 BC debate, argument, rhetoric and oratory had become the chief virtues of civic life in Athenian society. For Socrates, these skills were at the core of his philosophical method. Nothing that he actually wrote has survived, but we know a great deal about him and his ideas thanks to his pupil Plato, who also became one of the most influential philosophers of all time.

Some of Plato's works concern the dramatic and bizarre circumstances surrounding the execution of Socrates, who was famously sentenced to death by an Athenian jury for 'corrupting the youth of Athens' – possibly the charges against him were trumped up by those who regarded his ideas as heretical. Despite Socrates having various opportunities to escape, Plato describes how Socrates argued that it was his *duty* to die, having been convicted by a democratic jury; and in 399 BC Socrates performed the customary ritual and drank a fatal dose of hemlock.

Plato's most famous philosophical work, *The Republic*, continued the debate about the best way to rule a human society. Plato was furious at how the democratic assembly in Athens had sentenced his friend and mentor to death, and his work is scornful of all popular forms of government. He argued that it was better to be ruled by a tyrant, since then there was only one person committing bad deeds, than to be a bad democracy, where everyone was responsible for bad decisions. He saved his greatest scorn for the democratic delusion called liberty, which he believed descended quickly into immorality, lawlessness and anarchy:

'In a democratic country you will be told that liberty is its noblest possession … You can see the result: the citizens become so sensitive that they resent the slightest application of control as intolerable tyranny and in their resolve to have no master they end by disregarding even the law – written or unwritten.'[7]

Like Thales, Plato believed that what underpinned the universe was a reality that didn't originate from the traditional ragbag of Greek gods like Zeus, Apollo and Aphrodite, who inflicted their fancies on an unsuspecting world. Instead, like Socrates, Plato believed that the truth could be revealed through philosophical reasoning and contemplation. Therefore, in his description of an ideal society, it was philosopher-kings who ruled, sharing the wisdom of their insights with their subjects.

Plato came up with the ingenious suggestion that if man could successfully control nature by selectively breeding plants and animals, then why not apply the same techniques to humans?

'The brave man is to be selected for marriage more frequently than the rest so that as many children as possible may have such a man for their father.'[8]

This radical idea was put into practice by the southern Greek city of Sparta, which from 431 to 405 BC led a league of city states locked in a vicious war against the growing power of Athens and her empire. The Peloponnesian War was eventually won by Sparta thanks to its famous commander Lysander, who led its forces in an epic sea battle at Aegospotami (404 BC) that annihilated the Athenian fleet (168 out of its 180 ships were sunk). After a short siege Athens surrendered, and for the next thirty years the kings of Sparta ruled most of Greece.

The secret behind the power of this city that overthrew mighty Athens was a prime example of how mankind was beginning to organize itself in experimental new ways. Sparta was a totalitarian military society at its most pure and powerful. Lycurgus (c.700–630 BC), its legendary founder, apparently got his inspiration from the Oracle at Delphi, the famous temple dedicated to Apollo, the Greek god of medicine, healing, light and truth.

Lycurgus decreed that baby boys born too weak to become soldiers should be abandoned and left to die on the desolate slopes of Mount Taygetos. All the others were sent off to a military training camp at the age of seven to learn to become fearless warriors. Boys were initiated into the camp by having to run the gauntlet of older youths who would flog them with whips, sometimes causing the weakest to die on the spot, thus furthering the selective breeding principle. Instead of being fed, the boys were encouraged to steal food. They were punished if they were caught – not for stealing, but for being clumsy enough to have been spotted.

Spartan society was strictly divided between 'hoplites', native Spartans who ruled and trained for military service, and 'helots', slaves who had been captured during campaigns overseas, and were sentenced to grow crops in the fields. At the end of their training the soldiers would be sent into the countryside to murder any helots who happened to be wandering outside after dark. This exercise, called *krypteia*, had the double benefit of giving the young Spartan soldiers a healthy taste for murder, as well as instilling a climate of fear among the helots to stop them from causing trouble.

Spartan society impressed not only some ancient philosophers like Plato, but later ideologies such as the Hitler Youth Movement of the 1930s, where children were taught that their duty to the state was of greater importance than to the individual or the family (see page 367).

Hoplites were expected to put the welfare of the city above their duty to their families. They were trained in the discipline of the phalanx – a highly effective form of warfare, where troops were linked together by their arms into a human wall. Rectangular formations of tightly packed soldiers, at least four lines deep, with shields and spears locked closely together, were trained to march towards the enemy to the beat of pipes and drums, breaking into a run just before battle commenced. A phalanx required total loyalty and depended on every man in the fighting force. Just one drooping shield, like a weak link in a chain, could expose the whole formation to failure. The forces with the strongest, most fearless soldiers, who pushed hardest from the rear, invariably won. This was a form of war that the Spartans revelled in.

The selective breeding programme was carried through to male adulthood. Young soldiers were rewarded for military victory by being given the opportunity to couple with as many as twenty Spartan women, providing a powerful incentive to be brave in war. Failure on the battlefield was not an option. A hoplite returning from battle alive but without his shield was disowned by his family and sentenced to death.

Spartan women were given a higher status in society than anywhere else in ancient Greece. By cultivating beauty, intelligence and strength in their females, the Spartans believed they could succeed in producing a master race. Men and women, both naked, trained for athletics

Complete confidence in each and every man was the secret of the Greek phalanx's military success.

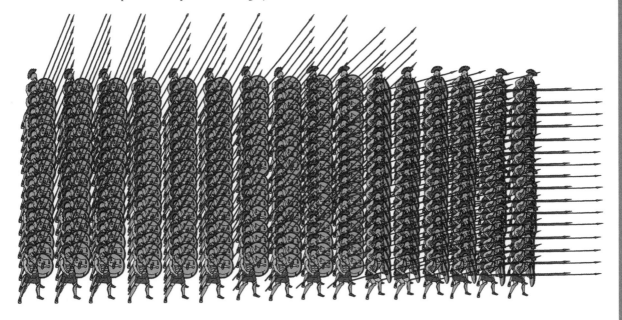

A Spartan hoplite, trained to kill without mercy and selectively bred to cultivate a human master-race.

with poems, statues, and the most prestigious honour of all – a crown made of olive leaves.

The games continued until 393 AD, when, following the Roman Empire's conversion to Christianity, the Emperor Theodosius I outlawed them as a barbarous pagan festival (see page 205). In 1896 they were revived by a Frenchman called Pierre de Coubertin as a way to improve physical fitness and to bring modern nations closer together by having the youth of the world compete in sports.

Like all empires, Sparta's predominance eventually waned. Its experiment in human engineering ultimately failed through a lack of popular support and a dwindling supply of willing and able males to keep its armies big and strong. But long before it reached that point, by about 380 BC, another force was already marshalling itself to the north of ancient Greece – one which, although it put an end to the independence of the city states, proved instrumental in spreading across the world their various templates of how human civilizations could engage with nature, be they the traumas of totalitarianism, the debates of democracy or the systems of science.

Spartan women kept fit through athletics tournaments and even participated in flogging contests.

championships alongside each other, with the women even participating in flogging contests to see who had the most endurance, an ordeal known as *diamastigosis*. This ritual was dedicated to the goddess Artemis and instigated by Lycurgus to replace the previous ritual of human sacrifice in which victims were selected by lot.[9]

With a social structure so completely based around physical fitness, selective breeding and military victory, it is no wonder that an athletics championship should have been established at nearby Olympia to test the absolute limits of human physical prowess in a competitive context. According to Greek legend the first ever Olympic stadium was built by Heracles to honour his father Zeus after completing his twelve labours.[10] By the fifth and sixth centuries BC the games had come to have enormous importance in Greek society. Each city state fielded its best athletes in the hope of winning the ultimate accolade in civic pride. Winners were immortalized

Conquerors of
the World

How humanity's understanding of nature's systems began to express itself in philosophies and laws that were carried by conquest throughout the East and West.

GREEK PHILOSOPHERS were edging towards the radical idea that there were no gods or God who controlled the destiny of life on earth from some detached mountaintop. Rather, it was man himself who, thanks to his own brainpower, could decipher the laws of the universe to become master of all nature.

Supreme amongst such thinkers was Aristotle (384–322 BC), the scope of whose works was truly immense. They covered everything from speculations on the nature of the human soul to the physics of the universe, from city politics and personal ethics to the history of plants and animals, and from public speaking and poetry to music, logic and even the weather.

Like Charles Darwin more than 2,000 years later, Aristotle's primary interest was in the natural world. In his early life he spent several years on the small, fertile Greek island of Lesbos. There he took meticulous observations of its natural ecosystems, using all his senses of sight, hearing, smell, touch and taste. He studied plants and animals, classifying them into groups, examining their behaviours and noting their various similarities and differences. Eventually, in about 335 BC, after several years spent teaching a young prince called Alexander from Macedonia, in northern Greece, Aristotle settled in Athens, where he established his own school of philosophy, which he called the Lyceum.

Aristotle combined what he considered the best of what he had learned from his teacher, Plato, and other Greek philosophers like Thales, with everything he had observed in the natural world. It led him to a single, profound conclusion: underneath all reality there was indeed a fundamental set of universal natural laws that explained everything to do with life on earth and the nature of the universe, from human politics to the weather. To understand these rules of nature was to understand reality. The key was careful observation of the universe and its systems by good use of the human senses, and then by using human reason and intellect to uncover the truth.

Aristotle believed that the universe worked like a giant machine, and that given enough time and attention, its workings could be understood by men. Although not the only Greek philosopher to deserve credit as a forefather of modern science, Aristotle is certainly the best-known, and over time his writings have been easily the most influential:[1]

'Just as puppet-showmen by pulling a single string make the neck and hand and shoulder and eye move with a certain harmony, so too the divine nature, by simple movement of that which is nearest to it, imparts its power to that which next succeeds, and thence further and further until it extends over all things. For one thing, moved by another, itself in due order, moves something else, each acting according to its own constitution, and not all following the same course.'[2]

The question that Aristotle's scientific, rational view of the world provoked was this: in a mechanistic universe governed by rules, what place was there for old-fashioned, whimsical gods? His answer was simple. It was the rules of nature themselves that were the very essence of all that is divine in the universe: 'For God is to us a law, impartial, admitting not to correction or change, and better I think and surer than those which are engraved upon tablets.' (Jewish people, for example, venerated the laws of God as written on tablets of stone passed down to Moses by God – see page 173.)

More than any other person in the whole history of the relationship between *Homo sapiens* and the natural world, it was Aristotle who gave mankind the confidence to explore, discover and learn. But such insights would be useless hidden in the mind of one brilliant man, or stored in a rich patron's library. To fulfil their potential, these ideas needed a force to scatter them far and wide, giving as many human cultures as possible the chance to exert the power of human brains over nature's brawn.

As luck would have it, Aristotle's pupil, the young Prince Alexander of Macedon, was just the right man at just the right time. Quite possibly it was his great teacher's passion for the natural world that fired Alexander, impregnating him with a feverish determination to see everything in the world for himself, conquering whatever empires lay en route.

Alexander's father, King Philip II of Macedon (382–336 BC), had a horse by the name of Bucephalus which no one in his court could tame. Legend has it that, to everyone's amazement, the ten-year-old Alexander spoke soothingly to the animal, and turned its face towards the sun. No longer frightened by the sight of its shadow, the horse became calm. Plutarch, the historian who tells the tale, says that Alexander's father Philip then prophetically said to the boy: 'Oh my son, look thee out a kingdom equal to and worthy of thyself, for Macedonia is too little for thee.'

Thanks to Philip's military success, Alexander was soon given a unique opportunity of journeying on what became the one of the most epic military adventures of all time.

In 338 BC, with the help of Alexander's highly effective cavalry charge, Philip's forces decisively defeated an alliance of Greek city states, including Athens and Thebes, at the Battle of Chaeronea. Philip was now undisputed master of all Greece. His plan was to unite the warring Greek cities by establishing a single Greek army, and then to march on their historic arch-rivals, the Persians, to avenge the wars of the fifth century, especially for the gratuitous sack of Athens in 480 BC (see page 178).

But in October 336 BC, events overran Philip's plan. Soon before his departure, during the wedding

feast of his daughter Cleopatra, Pausanias, one of his bodyguards, assassinated Philip as he was entering the town of Aegae's theatre. No one is sure what Pausanias's motives were, although several accounts describe it as being a lovers' tiff.

Alexander ascended to the throne at the age of just twenty. Within two years Thebes had rebelled in defiance of the new Macedonian King. Alexander cut his royal teeth by razing the recalcitrant city to the ground, leaving just one house standing, the one belonging to his favourite poet, Pindar, who had written a series of hymns to his ancestor Alexander I of Macedon. Alexander then demonstrated his ruthlessness by selling the entire Theban population into slavery. From then on, even Athens, under its belligerent anti-Macedonian ruler Demosthenes, was too fearful not to submit to Alexander's authority.

For the next thirteen years Alexander led an army of 42,000 Greek soldiers across Persia, Egypt and even into India. On his way he famously 'undid' the impossible-to-untie Gordian knot by slashing it with his sword – it was a legend in the city of Gordium in central Turkey that whoever could untie this knot would be the next king of Asia. He routed the Persian Emperor Darius III at the Battle of Issus in 333 BC, capturing his mother, wife and two daughters, along with a great deal of treasure. He then marched down the Mediterranean coast, laying siege to the city of Tyre, which he eventually took after seven months, clearing the way towards Egypt, where, thanks to the decline of Persian power he was welcomed as a liberator and pronounced Pharaoh in 332 BC. Here Alexander founded the most famous of all the cities named after himself, Alexandria, establishing it as the main sea port linking Egypt with Greece, the maritime axis of his new and increasingly powerful Hellenic empire.

Alexander wasn't the type to stay in one place for long. Within just eighteen months he left Egypt, marching back to Persia, where he again faced Darius, this time defeating him at the Battle of Gaugamela, where the Persian King fled from the battlefield and was later murdered by his own troops in the mountains of Media. The way was now open for Alexander to conquer all Persia, first marching on Babylon, then Susa, the ancient Assyrian capital, and finally Persepolis, the magnificent royal home of the Persian kings. After he had rested there for several months, the city was

The Epic Conquests of Alexander the Great

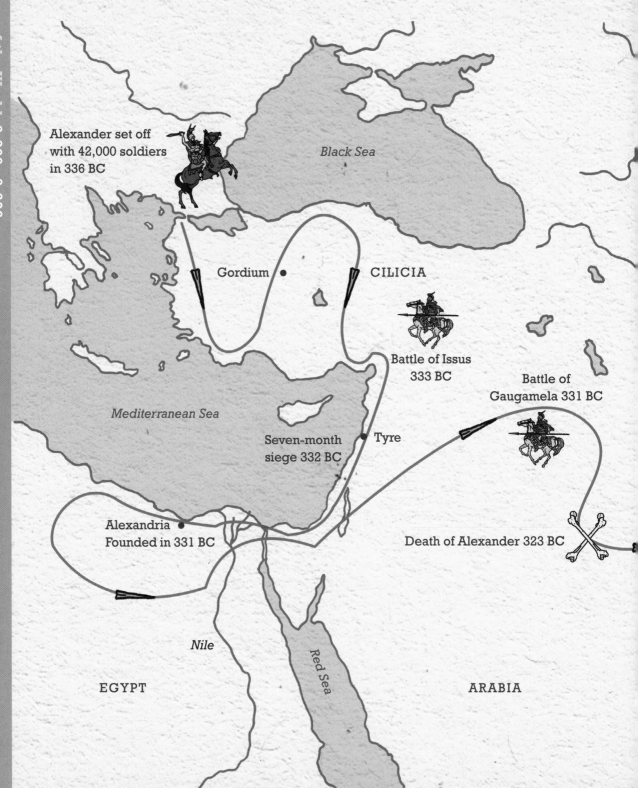

Alexander set off with 42,000 soldiers in 336 BC

Black Sea

Gordium

CILICIA

Battle of Issus 333 BC

Battle of Gaugamela 331 BC

Mediterranean Sea

Seven-month siege 332 BC

Tyre

Death of Alexander 323 BC

Alexandria Founded in 331 BC

Nile

Red Sea

EGYPT

ARABIA

What was it that drove this warrior to want to conquer the world –
a desire to see the end of the earth, or an insatiable appetite for war?
In the end, Alexander's awesome adventures fuelled fusion and
friction between the different cultures of the East and West.

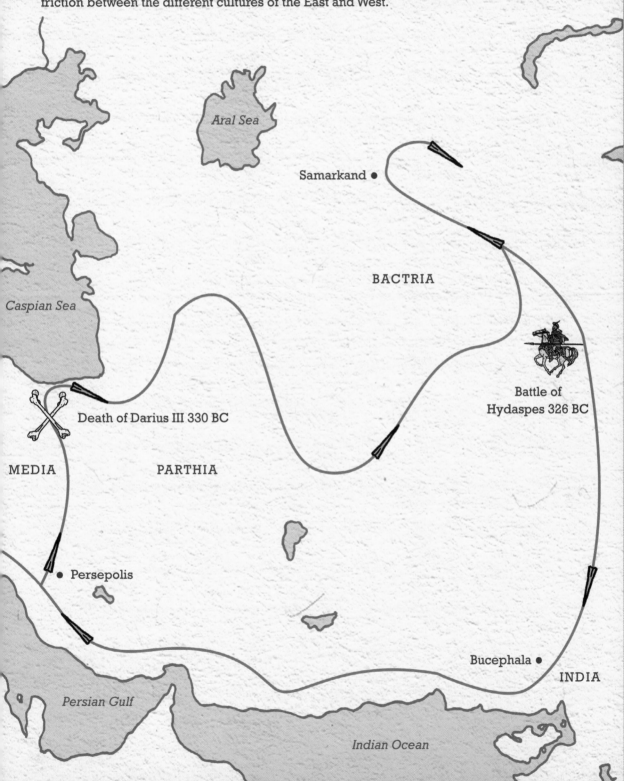

Aral Sea

Samarkand •

BACTRIA

Caspian Sea

Battle of
Hydaspes 326 BC

Death of Darius III 330 BC

MEDIA

PARTHIA

Persepolis •

Bucephala •

INDIA

Persian Gulf

Indian Ocean

burned to the ground – perhaps on purpose, perhaps by accident, no one knows. But great treasures of history were apparently lost, including priceless archives and documents such as an ancient edition of the Zoroastrian *Avesta* 'written on prepared cow-skins and inscribed in gold ink'.

With the death of Darius and the submission of Egypt and Persia, Alexander's military goals had been accomplished. But still the warrior in him could not be controlled. Having sent many of his Greek soldiers back home, he now paid mercenaries to fight for him in a new imperial army, and set off on a three-year campaign to subjugate Scythia and Afghanistan, before reaching the River Indus in northern India. At the Battle of Hydaspes (326 BC) Alexander defeated the Indian King of Porus and founded two cities there, one named after his magnificent horse and lifelong companion Bucephalus, who died in the battle.

Despite Alexander's determination to cross the sacred River Ganges and march into the heart of India, his men had reached their limit. Coenus, Alexander's general, negotiated with him to be allowed to lead them back home to Greece. Alexander, unable to face the prospect of simply retracing his steps, set off home with an elite guard by a different, more southerly route. Along the way he was seriously wounded by an arrow while battling the south Asian tribes of the Malli.

Eventually, Alexander made his way back to Persia across deserts and plains. Cultural hatred between Greeks and Persians was now so intense that Alexander's attempts to gain the respect of his Persian subjects by adopting their customs (for instance by being greeted with a kiss on the hand) led to outbreaks of bitterness and mutiny by his Greek guards. From this time onwards Alexander's great plan was to try to fuse Greek and Persian peoples into a single culture. At one point he even arranged a mass marriage feast, in an attempt to get his senior officers to pair off with Persian noblewomen. To set an example Alexander took a Persian princess, the daughter of Darius III, as his wife.

But bitter enmity between the two peoples remained, and most of these 'arranged' marriages didn't last long. Alexander's frustration finally boiled over in 324 BC when his lifelong friend, battlefield commander and lover, Hephaestion, died of a mysterious disease. Alexander's reaction to the death was brutal. He shaved his head, cropped the manes of his army's horses and cancelled all festivities for an indefinite period. Then, in what seems like an act of mindless revenge, he had his friend's attending doctor crucified for incompetence – a throwback, perhaps, to the laws of Hammurabi (see page 128). For the next six months Alexander mourned Hephaestion's death, and took his body to Babylon to be burned on a magnificent funeral pyre said to have cost 10,000 talents of Persian gold. There, eight months later on the afternoon of 10 June 323 BC in the palace of Nebuchadnezzar II, Alexander himself died, probably of malaria. He was one month short of his thirty-third birthday.

Conspiracy theories have abounded since his death. One story has it that Cassander, the Viceroy of Egypt, brought some poison to Babylon in a mule's hoof, and that Alexander's royal cup-bearer Iollas, Cassander's brother, administered it. This story is unlikely, however, as a full twelve days passed between Alexander falling ill and his death, and such slow-acting poisons were probably not available at that time.

Many historians have devoted their professional lives to the study of this man, yet no one really knows what drove him to try to conquer the world. Was it an innate megalomaniac madness? Was he an ancient Napoleon or Hitler, whose passion was to see how much power and subjugation one man and his army could muster? Perhaps Aristotle's inspirational philosophy fired him with an insatiable desire to find the ends of the earth? Were his later acts of religious inclusion an elaborate attempt to reverse the damage done by decades of war between rival cultures and establish racial unity under a single dynastic domain?

If so, Alexander's premature end cut any such aspirations short. His empire broke up soon after his death into four separate kingdoms, since he left no natural heir or successor. But as a result of his conquests the Greek language became the *lingua franca* across the entire Middle East and Egypt. Thousands of Greek people, some soldiers, others merchants, artisans, scientists and philosophers, moved abroad, taking with them their experimental world views. Of the seven wonders of the ancient world, five were Greek constructions – each one an awesome monument to these people's confidence in man's power over the natural world. The most impressive of all was the incredible hundred-metre-tall Lighthouse of Alexandria, built on the island of Pharos, just off the coast of Egypt. At its apex was a massive polished mirror which reflected sunlight during the day and a fire at night. It was said that ships could see its warnings over fifty kilometres away. The huge construction survived until the beginning of the fourteenth century AD when the unstoppable forces of nature in the form of plate tectonics laid their claim. By 1480 all that remained after a series of earthquakes was a stubby remnant, a relic that was soon absorbed by a new fort built by a medieval Sultan of Egypt.

Archimedes, Eratosthenes and Euclid were three Greek travellers who spent much of their time studying in the new Egyptian royal library of Alexandria. This great centre of learning and knowledge, founded shortly after Alexander's death, accommodated up to 5,000 students, almost certainly making it the ancient world's largest research institution. Perhaps the library's most important influence was in its translation of the Hebrew scriptures of the Old Testament by seventy-two Jewish scholars in c.200 BC. This translation spread prophecies of the coming of the Messiah across the entire Greek-speaking Hellenistic world, without which Christianity is unlikely to have grown into anything more significant than a minor religious sect.

Over the centuries the library at Alexandria suffered several fires and sieges, until, shortly after the Muslim conquest of Egypt in 642 AD, all traces of it disappeared (see page 237). One legend has it that the Caliph Umar, one of Mohammed's closest companions, said to his commander: 'Touching the books you mention, if what is written in them agrees with the Book of God, they are not required;

if it disagrees, they are not desired. Destroy them therefore.' The story goes on to say that the library's scrolls fuelled the city's furnaces for the next six months. It is possible that this story was invented several hundred years later by Crusader Christians, eager to depict Muslims as intolerant vandals. In 2004 a team of Polish and Egyptian archaeologists unearthed what they think are the remains of thirteen of the library's lecture halls.

Meanwhile, Greek confidence, fanned by Alexander's conquests, spread the thirst for knowledge over the entire Hellenistic world throughout the Mediterranean and Middle East. Perhaps it was the reflection of the sun's rays in the Lighthouse of Alexandria that inspired Archimedes (287–212 BC) to come up with his bizarre idea to burn enemy ships by redirecting sunbeams off an array of highly polished bronze shields. Although Archimedes's 'death ray' probably never worked in practice, he has certainly gone down in history as one of antiquity's most important scientists, engineers and mechanics. Amongst other inventions he is credited with designing the world's first pulleys and a system for cranking water uphill, called the Archimedes Screw.

His friend Eratosthenes (276–194 BC) was no less brilliant, being the first human to correctly calculate the circumference of the earth and to measure the distance to the moon. His machine, called an armillary sphere, accurately modelled the movement and positions of the earth, sun and planets. In the Middle Ages Arabic scientists used it as an inspiration for the astrolabe, an instrument later used for trans-oceanic navigation (see page 241). Eratosthenes was also a cartographer. His map of the known world stretched from the British Isles to Sri Lanka, and from the Caspian Sea to Ethiopia. Euclid (c.323–283 BC) was another Greek traveller and frequenter of the Great Library of Alexandria whose supreme mastery of geometry became one of

An armillary sphere was a kind of celestial globe, developed by Eratosthenes to model the movement of the sun, moon and stars as seen from the earth. It was used as a teaching tool.

the cornerstones of modern mathematics. His classic, *The Elements*, written in c.300 BC, is still used as a textbook today.

Persian and Far Eastern philosophies were also absorbed for the first time into the cultures of the western Mediterranean world. Pyrrho, a Greek painter from Elis in the southern Peloponnese, travelled with Alexander's army and for a while settled in the East, studying oriental philosophy under the guidance of Indian religious gurus. It is likely that these 'gymnosophists', as Plutarch calls them, were Buddhist monks or Hindu holy men, and when he returned to Greece, Pyrrho was firmly committed to a life of solitude and peace. He concluded that since nothing can be known for sure, sincere people should hold on to no belief in particular. His legacy was the influential school of thought known as scepticism.[3] So confident was he in his theories that he apparently blindfolded himself to show his friends how little the human senses could determine truth from falsehood. Unfortunately, they were left to mourn his untimely death after he fell head-first off a cliff.

Just as radical was the Hellenistic philosopher Epicurus (341–270 BC), whose ideas heavily influenced the emerging world of the Roman Empire. From about 146 BC, when the Roman army conquered Macedonia and then Greece (see page 197), Roman society began imbibing deep draughts of the Hellenistic culture spread by Alexander's great conquests. Epicureanism taught that pleasure and pain were nature's measures of good and bad. Epicurus was an 'atomist', whose philosophy was a precursor of modern Western notions of atomic theory. The gods do not reward or punish humans, he said, and events in the world are determined by nothing more than the motions and interactions of atoms moving about in free space.

Roman love for all things Greek was particularly focused around the personality and career of Alexander, who after his death became antiquity's greatest role model. His tomb, made of gold and covered with a purple robe, was built in Alexandria by Ptolemy I, who became ruler of the Egyptian and Lebanese portions of Alexander's empire following the Battle of Ipsus, fought between Alexander's former generals, in 301 BC.

The battle determined that Alexander's empire would be divided into four separate kingdoms: Greece, Thrace, Persia and Egypt. His Indian territories had already passed to Chandragupta (see page 169) after a peace treaty with Alexander's satrap Seleucus who had exchanged them for 500 elephants – animals that proved decisive in the Battle of Ipsus.

Roman emperors came to regard Alexander as the epitome of bravery, strength and courage. They regularly travelled to Alexandria to pay their respects. Julius Caesar (ruled 49–44 BC) is reputed to have wept when he saw a statue of Alexander in Spain, and the Roman general Pompey (106–48 BC) is said to have searched high and low all over the Empire for Alexander's famous cloak, which he apparently discovered and wore as a costume of greatness. The mad Emperor Caligula (ruled 37–41 AD) stole Alexander's armour from his tomb and wore it for luck, while the Emperor Caracalla (ruled 211–217 AD) even believed that he was the reincarnation of Alexander himself (reasons for this increasing madness are explained on page 202). Eventually, in 200 AD, the Roman Emperor Septimius Severus closed Alexander's tomb to the public out of concern for its safety, owing to the massive hordes of tourists who came to visit it. Since then its whereabouts have become one of the great mysteries of archaeology.

Unlike Aristotle, Alexander was not an original man. His achievements were not to do with generating new ideas or inventions, but with using his supreme skills of leadership and military prowess to scatter abroad the seeds of a huge range of Greek ideas concerning art, culture, science, politics, invention, power and dominion. With the rise of Roman civilization, for which everything that was Greek was intelligent, cultured and desirable, such seeds fell on fertile ground.

Hurricane Force

How an empire built by violent human copy-cats clung to power far beyond its natural limits despite the birth of Jesus Christ, a man who came to be called Messiah.

ROME'S RISE AND FALL was like a human weather system as destructive as nature's most violent hurricanes. This enormous whirlwind was powered by three essential ingredients, grain, booty and slaves, while its internal momentum, sustained by violence, was hell-bent on keeping a rich ruling class in a lifestyle of luxury. Once the storm was over the landscape of Europe looked quite different – in some places it was changed beyond all recognition, while the legacy of what made the ferocious ancient Roman Empire so long-lasting helped shape the rest of European history to come.

There is no single reason why, from about 600 BC, Rome and its people began to dominate the central region of what is now called Italy. The Romans had no specific military edge, no special invention, no natural advantage, except that, like the Greeks, they had to look beyond themselves to live well, since the land they tilled was not ideal for growing high-yield crops such as wheat. Early on in Rome's history its people turned to easier-to-grow, more productive fruits such as grapes and olives, which they then used as bargaining chips around the Mediterranean in exchange for life's other necessities.

Expansion, growth and violence lie at the heart of the mythological story of how the Romans founded their civilization. In a desperate bid to find themselves wives, the tribe of Rome's twin founders, Romulus and Remus, is supposed to have lured the nearby Sabine people to a luxurious festival. There they abducted their women, married them and started to have children. A few years later the Sabine men fought back. Peace was only re-established thanks to the intervention of the wives, who threatened to kill themselves if the two tribes of men carried on fighting over the lives of their children.

Roman legends give a clue as to what made these people tick, because there is so much similarity between them and the mythological tales from Greece. The Romans were brilliant copy-cats. They excelled in taking everything they found in their outward-looking world view and then adapting and incorporating the best bits to feed their growing empire. Infantry tactics, mythology, art and architecture came from Greece; heavy cavalry and expertise in horses came from Persia; and when the Romans wanted to attack the most powerful maritime city in the region, the Phoenicians' Carthage, they simply captured one of its ships, and within the space of about two months built themselves an entire fleet of more than a hundred similar ships from scratch.

The Roman skill for copying and adapting other people's ideas was equalled only by their brute force and persistence. Several times these qualities helped them escape near annihilation in the early years of the Roman Republic (509–44 BC).

One of the biggest threats came from the Greek general Pyrrhus of Epirus, who arrived in southern Italy in 280 BC to rescue Greek cities from Roman invasion. Pyrrhus's campaigns started well enough, and he defeated the Romans thanks to his superior cavalry and elephants. But the following year, at the Battle of Asculum (279 BC), although victorious, Pyrrhus lost many men. 'One more such victory and I shall be lost,' he declared after the battle – the origin of the saying 'Pyrrhic victory'. Pyrrhus retired to Sicily, returning two years later to find that the Romans had recruited so many new men into their armies that, after a single indecisive battle, he was completely outnumbered. He then fled, giving the Romans the all-clear to take over the remaining Greek cities along the south Italian coast.

Having united central and southern Italy, the Romans now began to look hungrily across the Mediterranean to feed their growing civilization. First in their sights was the island of Sicily, an excellent place for growing crops, but jealously guarded as a province of the Phoenicians, whose capital was now in Carthage on the coast of North Africa. The first war between Rome and Carthage (264–241 BC) ended in a Roman victory thanks to its new fleet. But the Romans didn't just copy the design of Carthaginian ships: they also added their own touch, in the form of an assault bridge called 'the raven' that latched on to enemy vessels. Now Roman soldiers could turn war at sea into hand-to-hand combat as if they were on land, a form of warfare in which they excelled. In 241 BC the defeated Carthaginians signed a peace treaty giving Sicily to Rome and paying a massive financial tribute.

But peace did not last long. A new, more dangerous threat to the growth of Rome came in the form of a famous Carthaginian general called Hannibal, who in 218 BC marched an army of mercenary soldiers and African war elephants all the way up Spain and across the Alps. His surprise attack from the north led to several victories, the most famous being at the Battle of Cannae (216 BC), near Apulia in south-east Italy, where Hannibal's cavalry encircled the amassed ranks of Roman infantry, cutting them to pieces. In the end, though, Roman persistence paid off. Knowing that Hannibal didn't have equipment such as siege engines needed to breach the walls of the city of Rome itself, the Roman forces just waited, shadowing his armies, watching his tactics, but always avoiding war. Meanwhile, another Roman army under the leadership of a young commander, Scipio, defeated the Carthaginian forces in Spain and crossed the short stretch of sea to Africa, where they marched towards Carthage itself. Hannibal had no choice but to return home to try to save his city, but there he was defeated at the Battle of Zama in 202 BC.

Hannibal survived the battle and remained in Carthage to help rebuild the city. He then spent time as a military adviser to kings in Asia Minor before committing suicide in 183 BC to escape capture and torture by Roman soldiers. After its defeat, Carthage was forced to give up all its colonies to Rome, including Spain, and was forbidden to have an army and navy. Once again, it had to pay an enormous ransom.

By now the Romans had become a highly efficient war machine, expanding their frontiers all round the Mediterranean and adapting their tactics to incorporate cavalry and ships. With each conquest they brought home huge hoards of booty in the form of treasure and prisoners of war, whom they turned into slaves. Plunder paid for a fabulously rich lifestyle for Rome's citizens, while imported slaves provided free labour in their homes, on the farms, in the city streets and on the many enormous construction projects that quickly turned Rome into the most advanced artificial world on earth.

By 146 BC a succession of military victories brought Greece into the Empire, followed in 129 BC by Asia Minor (Turkey), whose dying king, Attalus III, had willed his entire kingdom to Rome to avoid disputes between his heirs. From here the Romans had the perfect bridgehead to launch a series of campaigns in the Near East, conquering Armenia, Lebanon, Syria and Judaea by 64 BC under the leadership of the general Pompey, each time adding further riches to their economy in the form of gold, silver and slaves. Roman historian Plutarch says Pompey brought back 20,000 talents of gold and silver from these campaigns, and

imperial tax receipts rose from 50,000 to 85,000 drachmas a year. To the south the Roman general Octavian, who later became the Emperor Caesar Augustus, added the jewel in the crown – the conquest of Egypt in 30 BC.

After success in the Battle of Actium on 2 September 31 BC, Octavian took over Egypt following the suicide of the Egyptian ruler Cleopatra and her Roman lover Mark Antony. Octavian then had Cleopatra's son and heir Caesarion executed, declaring that 'Two Caesars is one too many.'

With its almost limitless supplies of grain from the Nile Valley, the Egyptian bread basket provided the perfect finishing touch, supplying unlimited quantities of food throughout the Roman Empire.

But at the heart of the Roman rise to power lay an intractable problem. What does a civilization that is built on military conquest and financial growth do when it finds that, for various reasons, it cannot expand any more? By 30 BC the city of Rome had a population of somewhere between 500,000 and a million people, making it the largest city in the world. Treasure from military successes

The Italian renaissance artist Jacopo Ripanda (active 1490–1530) painted this fresco depicting Hannibal's march with elephants towards Rome in 218 BC.

and taxes from subjugated kingdoms provided highly luxurious lifestyles for the top tier of Roman society – the nobles, generals, soldiers, politicians and merchants, waited on by an average of four slaves per household. But was such growth sustainable? On the one hand Rome had reached the limits of the lands and riches its armies could conquer, while on the other the number of enslaved and destitute subjects threatened to overwhelm the capacity of its leaders to maintain order and control.

Gaul (modern-day France) had been brought under Roman control after Julius Caesar's nail-biting defeat of Vercingetorix, the heroic Frankish leader who between 52 and 46 BC led a war of revolt in the name of the Franks. The final battle, at Alesia, could have gone either way after Caesar's fortifications were breached. It was only when Caesar personally led his reserves into battle that he won the day, taking Vercingetorix as a captive to Rome, where he was paraded and imprisoned for five years before finally having his head lopped off.

Britain, the first conquest of which began in 55 BC, was finally subjugated after the failed revolt of Boudicca at the Battle of Watling Street in 61 AD. Further expansion towards Scotland proved profitless, and eventually the Romans built a wall to keep out the violent Picts.[1] In the north-west the Romans permanently posted several legions along the natural border of the Rhine/Danube Rivers in an effort to contain Germanic tribes such as the Goths and Visigoths, who, despite numerous attempts, were found to be impossible to bring under control. To the east a new Persian Empire had overthrown the Greek satraps put in place by Alexander. The Parthians were world experts in the use of heavy cavalry, and their *Azatan* knights successfully held the Romans back from the rich lands of the Middle East.

Then there were nature's barriers. To the west, after Spain, there was the edge of the world – the apparently endless Atlantic Ocean. To the south, beyond Carthage and Egypt, there was just dry, barren, lifeless desert.

Boudicca speaks to her British troops before they prepare to battle with the invading forces of Rome.

The story of the later Roman Empire is the tale of how a human civilization, fixed on violence and growth, managed to hold itself together despite reaching expansion's elastic limit. Thanks to a variety of ingenious and often brutal strategies, the ruling Roman oligarchy was able to sustain its luxurious, epicurean standard of living for an impressive 300 years *after* the major phase of expansion ended. They did it thanks to some ingenious tactics in social engineering that preserved their world well beyond its natural limits, and which became model inspirations for later European civilizations. For example, according to some thinkers (e.g. Thomas Malthus and Karl Marx), Western European population levels and economic capitalism had already reached the limits of their natural growth by the middle of the nineteenth century (see page 335).

Imperial Rome's first survival tactic was political. Tyrannical rule was necessary to force through a rapid succession of reforms needed to hold this violent land and sea-locked society together. At least 40 per cent of the capital city's massive population were slaves. They made a doomed attempt at breaking the grip of imperial and dictatorial government when a leader called Spartacus, an escaped Greek gladiator, rallied them to rebellion in 79 BC. Historians of the day, such as Plutarch, say he amassed as many as 120,000 slaves, who fled to join him at Mount Vesuvius, south of Rome. After some initial successes they were eventually routed by the Roman general Marcus Crassus in southern Italy. More than 6,000 were crucified, their crosses set up along the 130-kilometre stretch of road from Capua to Rome. Crassus ordered that their bodies never be removed. There they remained as rotten carcasses for many years, a gruesome memorial of what happened to slaves who disobeyed their masters.

Ruthless suppression was an essential tactic that was employed to preserve the living standards of ancient Rome's rich ruling class. Other strategies were needed, though, to keep the huge population of their capital city under firm control. The three most successful were armies, engineering and entertainment.

One of the most significant reforms made just before the civil war period, in about 100 BC, was a reorganization of Roman armies inspired by the general Gaius Marius. Instead of relying on richer Roman citizens to enlist as soldiers, he began to draw upon the poorest elements of Roman society and train them up as professional servicemen. This made bigger armies for fighting overseas to keep the provinces under control. It also had the important side effect of keeping a potentially hard-to-control segment of Roman society out of trouble and off the streets. But bigger, permanent standing armies had to be paid, and the demands on the imperial treasury (and therefore the need for further expansion) grew substantially as a result of such reforms. Soldiers were now organized in permanent legions under the command of a military governor who was made to swear an oath never to leave his province without permission from Rome. It was this convention that Julius Caesar shattered by crossing the Rubicon River with his legions of soldiers in 49 BC, leaving his province of Gaul and marching to Rome, plunging the Empire into a period of civil war.

A series of civil wars in the period 90 to 27 BC gradually displaced the representative republican government that from c.509 BC had allowed lower-class citizens (plebeians) to have some say in the city's affairs. In its place came a full-blown imperial dictatorship. From the moment in 44 BC when Julius Caesar was declared 'dictator for life', the reality of representative government was over. Julius Caesar's authoritarian rule included reforming the calendar, extending the year from ten months to twelve, and even naming a month – July – after himself.

The brutal manner of Caesar's own departure from politics showed how personalities had taken over from political processes. Losing an election is one way of being removed from office. Being stabbed to death on the steps of parliament by fellow politicians and former colleagues is a great deal more permanent. Offended by Caesar's complete disregard for the Roman Senate, a cabal of more than sixty conspirators were involved in the plot that led to Caesar's assassination, which took place on 15 March 44 BC. Once Caesar was dead, Brutus, one of the chief conspirators, cried out: 'People of Rome, we are once again free!' But the Empire descended into yet more civil wars, followed by more military dictatorship.

Fielding bigger armies was something of a double-edged sword. Whilst it got some of the poorer people out of the city, it cost money, and equipped jealous or ambitious generals with their own armies, giving them the means to provoke serious internal unrest. Better by far was to divert the poor and enslaved into building projects to provide improved amenities for the well-off. With this policy the Roman Empire rapidly became the fountain from which mammoth engineering projects were undertaken all over its conquered lands. Many of their ruins still survive all over Europe, North Africa and the Near East.

The Romans' favourite fancy was to construct artificial water supplies. Aqueducts were the key to the civilized, urban way of life for the nobility. A city without flowing fountains, public baths, laundries and running tapwater was no place to bring up a young Roman family. Wherever they settled, the Romans began engineering projects to take water away from local rivers and mountain sources into city centres. Roman aqueducts were among the most impressive feats of all ancient engineering.

One of the most famous still standing today runs along a fifty-kilometre stretch of rolling landscape in the south of France, from the springs of Uzes to the city of Nîmes. This huge artificial watercourse falls just seventeen metres from its start to its finish, which means precision engineering with a constant gradient of exactly 1/3,000 metres along its entire length. A little too steep and the water would arrive submerged underground – too shallow and the force of gravity would be insufficient to pull the water along its course.

The most spectacular section of this aqueduct is the bridge built to carry the water across the River Gard. Its three levels of arches, stretching some 275 metres across the river valley, were constructed entirely without the use of cement. Huge blocks of stone were lifted into place by treadmill-winches powered by human slaves. The whole project took 1,000 workers three years to complete. Success required persistence. After this aqueduct opened in c.60 AD, the channel was so long and so leaky that it took another twenty-five years to repair before the water ran without interruption. But when it came, the rewards were plentiful. Nîmes was provided with

A Roman aqueduct, built entirely without cement, crosses the River Gard near Nîmes, France, and descends just seventeen metres along a course totalling fifty kilometres.

more than five million gallons of drinking water every day, although poor maintenance after the fall of the Empire meant that its supply had run out completely by about 900 AD.

Exploiting labour from slaves and the poor also gave birth to Europe's first comprehensive road network – essential infrastructure for keeping order and control in an empire that at its biggest, in about 100 AD, covered over five million square kilometres. Slaves, supervised by soldiers, built more than 85,000 kilometres of road, most of them in straight lines, making man's first long cuts of bricks, cement and concrete into Europe's supple surfaces.[2] Everything that got in the way, from forests to farms, was razed to the ground. These people clearly reckoned to be around for a very long time – many of their roads were still in use more than a thousand years after they were built. Some still are.

Spectacular forms of mass-market entertainment were also a key part of the system for keeping the overpopulated Roman capital governable. Riches won after suppressing a Jewish revolt against Roman rule that began in 66 AD financed the cost of building Rome's giant Flavian amphitheatre (the Colosseum), which was constructed under the Emperor Vespasian.[3] When it was opened in 80 AD, the Colosseum could seat more than 50,000 spectators, comparable to many large modern sports stadiums. A new emperor, Titus, celebrated the opening of this temple of entertainment by giving the people of Rome a hundred days of spectacular drama in the form of mock battles, gladiator fights, animal hunts and executions. According to the contemporary historian Dio Cassius, more than 11,000 wild animals were killed in these games. Several of them, such as lions, crocodiles, elephants, giraffes, panthers, leopards, hippos, rhinos and ostriches, were imported from across the Empire.

Like many events staged with an ulterior motive, attendance was free. The Emperor came to the games so his people could admire him in all his glory. He was only too happy to see the most violent of the Roman underclass gratified and in one location, safe under the watchful eye of the imperial guard.

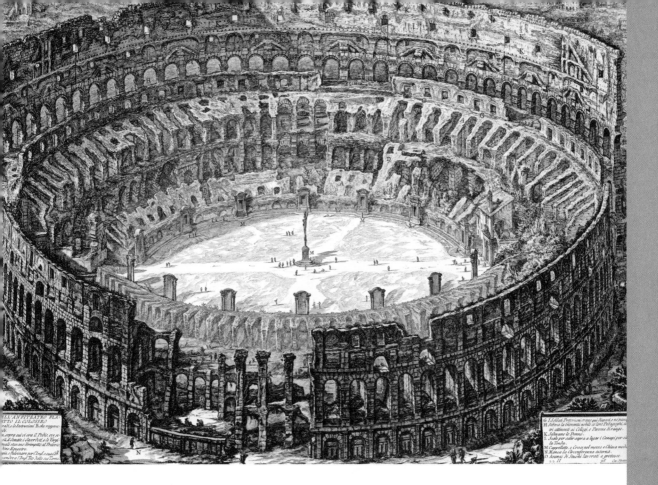

Imperial propaganda and gruesome entertainment as a means of popular control were Roman innovations that substantially reduced the risk of internal unrest in an empire short on expansion but addicted to growth. Amphitheatres like the Colosseum were built in most major towns and cities throughout the Empire, from Bath to Alexandria. It was in one such place of entertainment in Caesarea that 2,500 Jewish rebels were executed following the failed revolt of 66 AD – much to the glee of the spectators.

Another effective way of keeping populations throughout the Empire under tight control was by service in the mines. It was also highly profitable. In fact this tactic solved two problems. So desperate was the Empire for treasure – to pay for its armies, navy, roads and engineering projects – that in the absence of being able to gain more plunder by conquering people like the Jews, attention was focused on trying to mine natural wealth from under the ground.

Silver had been successfully mined, first by the Lydians and then the Greeks, from about 550 BC. Lead was a by-product of extracting silver from ores called *galena*. For the Romans, mining provided a double bonus. While on the one hand it pressed their slaves into silver production to secure more wealth for themselves, it also generated massive quantities of lead, which, with its soft, malleable properties and low melting point proved an ideal construction material for the Romans' many engineering projects. They used lead for making everything, from pots and pans and storage jars to gutters and pipes. In fact, most of the 420 kilometres of water channels in Rome's eleven aqueducts were made of lead.

Exactly how far this exploitation of the earth's natural resources can account for the eventual demise of the Roman Empire is still a matter of debate. New scientific evidence has confirmed that lead poisoning in Roman times was a serious but unknown issue. One study of human bones from

A bird's eye view of the ancient Roman Colosseum, seen here in c.1800. It was originally built using treasure stolen from the Jews after their disastrous revolt against Roman rule in 66 AD.

23:59:59

more than 250 skeletons across twenty separate archaeological sites around Italy has found levels of lead contamination up to ten times greater than those before imperial Roman times. After about 500 AD the levels reduce once again close to normal.[4]

Rome's love affair with lead had its most serious effects among the rich, because their food was cooked in kettles made of lead, which they used for producing a thick grape must, an essential ingredient for their most popular sweetener *defrutum*. According to the historian Cato, this should be prepared in vessels 'made of lead rather than of brass; for, in the boiling, brazen vessels throw off copper-rust and spoil the flavour of the preservative'. Rich Romans added *defrutum* to everything from fine wine to *garum*, their favourite sauce, served with fish.

Lead poisoning causes gout, infertility and memory-loss leading to madness. Were the crises that variously enveloped the Empire from the first century AD exacerbated by the madness of its rulers? There is no shortage of evidence of insanity among Roman Emperors to support the theory, from Nero's strumming of his lyre while watching fire engulf much of Rome in 64 AD to the bizarre orders of Caligula (ruled 37–41 AD), who sent his troops to fight the sea god Neptune, but then changed his mind at the last minute, ordering them to pick seashells on the shores of northern France instead.

In the midst of this hurricane of madness, indulgence, exploitation and violence there was a miraculous moment of calm. In the eye of the imperial storm, almost exactly halfway through Rome's dominance of the Mediterranean world, a son was born to a Jewish carpenter and his wife in a place called Bethlehem, a town situated just south of Jerusalem. His name was Jesus.

Like the Buddha some 500 years before, Jesus was an enlightened charismatic who made a virtue out of poverty and lectured on the benefits of non-violence. His message was simple. Be peaceful. Love your neighbour as yourself. If someone strikes you on one cheek, do not hit back but offer them the other. Do not worship false idols, such as money or material possessions, and, above all, be humble – for one day the meek will inherit the

earth. Jesus is only ever known to have lost his temper once, in the Temple of Jerusalem, where markets had been set up for traders to make a profit. That was shortly before his betrayal by Jewish high priests, who felt threatened by his huge popular following.

Jesus's followers saw him perform miracles, and came to regard him as the earthly incarnation of God as prophesied by Isaiah and others in the Jewish Torah. One of the most deeply held Jewish beliefs was that at the time of the covenants between God, Abraham and Moses, the Israelites were identified as God's chosen people. Yet here was a man whose followers claimed he was King of the Jews, and who offered the prospect of eternal salvation to anyone and everyone who believed in him, regardless of their colour, race or creed.

Jesus was given over to the Roman governor of the province of Judaea, Pontius Pilate, as a heretic, and despite Pilate's misgivings, he condemned him to die on a cross like a common criminal. His body mysteriously disappeared three days after being incarcerated in a tomb, and his disciples began to see visions of him. They wrote about these miraculous events, which they called the Resurrection, and believed it was their divine mission to spread the good news about the son of God coming down to earth and dying on a cross, so that everyone who believed in him might have everlasting life. They set about establishing a Church in his name.

The most prolific evangelist was a Jewish man who never met Jesus. Paul of Tarsus, originally a persecutor of the followers of Christ, wrote about a divine vision he witnessed on his way to Damascus, where God called on him to stop harassing the followers of Jesus. From this moment on Paul travelled far and wide, to Greece, Rome, Cyprus and Crete, spreading the news about Jesus's coming. More than half of the New Testament was written by Paul, the other books being the recollections of Jesus's life composed by some of those who had known him. For many years there was no official New Testament, and no recognisable Church. Disputes about the exact nature of this new religion, first called Christianity in about 130 AD by Ignatius of Antioch, began in earnest.

The early Christian Church developed a huge popular following because it filled a spiritual vacuum inherent in the materialistic, brutal and unequal society of the Roman Empire. Despite this, it did not have a straightforward birth.

An early dispute raged over whether or not someone who wanted to become a Christian first had to convert to becoming a Jew, and be circumcised as was the Jewish custom. At the Council of Jerusalem, in c.50 AD, a meeting of the disciples determined, after heated debate between Peter and Paul, that circumcision and conversion to Judaism were not necessary after all. It was a vital decision, because until that time most followers of Jesus were Jews. Indeed, most of the earliest members of the Church – the disciples, first bishops and clergy – were Jewish.

After this decision, Jewish support for the followers of Jesus waned. Jewish exposure to Christianity was also limited by another devastating massacre following a revolt against Roman rule in Judaea in 135 AD, when as many as 500,000 Jews were slaughtered.[5] Others were scattered far away from Jerusalem and beyond the Roman Empire into Persia. As a punishment for their revolt, the Roman authorities banned Jews from Jerusalem and abolished the kingdom of Israel, incorporating a new province in its place. They called it Palestine.

From this time on, Christianity's main appeal was to non-Jewish poor people, women and slaves. Everyday life in the Roman Empire was proof enough for these people that the pantheon of Greek/Roman gods had nothing much to offer them in terms of spiritual nourishment or hope for the future. The idea that the son of God came to free them and offer them eternal salvation in his Kingdom of Heaven sounded a lot more promising.

Another community attracted towards Jesus's teaching were those keen to establish a new hierarchy to resist the seemingly infinite power of Roman society. Greek thinkers who followed the idea of a universal force of nature first put forward by Socrates, Plato and Aristotle found the concept

of a single universal God who was open to all people rather compelling. The biggest problem for them was how to reconcile this all-pervasive divine force with a carpenter's son from Galilee whose followers claimed he was the incarnation of God.

The problem wasn't finally settled until after Christianity was legalized in the Roman Empire by the Emperor Galerius in 311 AD.[6] In the end, the idea of a Trinity provided the answer. It combined the Jewish God of the Old Testament as the Father, with the person of Jesus Christ as his Son, and the divine platonic or natural force pervading all things as the Holy Spirit. The Father, Son and Holy Spirit make up the Trinity that still marks out Christianity as distinct from other religions. This doctrine was finally ratified and codified into an official creed at the Council of Nicaea in 325 AD, under the auspices of the Roman Emperor Constantine.

Not everyone in the Greek world was convinced. Indeed, some, like Celsus, a Greek philosopher who wrote a blistering attack against Christianity in about 170 AD, found the whole concept bizarre:

Human sacrifice in the name of Jesus Christ, who died to save human souls if only they would turn to him. Painted by the Flemish artist Hieronymus Bosch (1450–1515).

'The idea of an Incarnation of God is absurd; why should the human race think itself so superior to bees, ants and elephants as to be put in this unique relation to its maker? And why should God choose to come to men as a Jew? The Christian idea of a special providence is nonsense, an insult to the deity. Christians are like a council of frogs in a marsh or a synod of worms on a dunghill, croaking and squeaking, "For our sakes was the world created."' [7]

The Emperor Constantine, depicted here on the lid of a stone coffin, is adorned with symbols of Christianity.

Before Christianity was legalized, however, the embryonic religion itself was caught up in the extreme violence of the Roman world. Stephen, James, Peter and Paul, Jesus's most ardent supporters, were all executed and became its earliest martyrs. Nero blamed and persecuted Christians for the Great Fire of Rome in 64 AD. Christian executions then became part of the Empire's popular mass-entertainment system, taking place

in amphitheatres all over the land. But Christ's simple message continued to attract the poor. Displays of bravery and faith by these early martyrs only strengthened the new religion's mass appeal.

The greatest persecutor of all was also its last – the Emperor Diocletian (ruled 284–305 AD). His ambitious reforms saved the Empire from total collapse, giving its ruling class another hundred years of the high life before their final downfall. He did this through the neat conjuring trick of splitting the Empire in two – with a western half ruled from Rome, and an eastern half initially ruled from Split in Croatia.

But Diocletian is best remembered for his vicious suppression of Christianity, which had by now become a major force accelerating the decline and fall of imperial Roman authority. His 'Edict Against Christians' in 303 AD ordered the destruction of all Christian scriptures and places of worship. He then ordered the arrest of Christian bishops and priests. Some 3,500 Christian believers were executed until at last, in 313 AD, a new Emperor, Constantine, saw that the rise of the new religion was now impossible to resist.

Constantine's 'Edict of Milan' is one of the most celebrated moments in the history of Christianity. Although not a Christian yet, Constantine used his Edict to try to heal divisions by making, for the first time in the Empire's history, all religious belief a matter of personal choice, not state dictat.

To what extent Constantine himself became a devout Christian is not clear. He was probably not baptized until late in life, and coins continued to be minted with pagan gods on them for eight years after his supposed conversion at the decisive Battle of Milvian Bridge in 312 AD, which, according to some Christian historians, was the first holy battle fought in the name of Christ. Before the fighting commenced, Constantine is said to have looked up towards the sun and seen a cross of light above it. He then commanded his troops to adorn their shields with a Christian cross, after which he was victorious and marched into Rome as Emperor. [8]

In 330 AD Constantine established a new capital for the eastern portion of the Empire which became known as Constantinople (now Istanbul). Here he promoted Christianity by building churches, and forbade pagan temples. Despite this, most of his imperial staff remained pagans, showing that whatever his own personal beliefs Constantine, like the Persian Cyrus the Great, was a tolerant ruler.

But religious understanding was not to last long. One of Rome's final legacies was to throw out all notions of religious freedom, and instead adopt Christianity as its state creed. This act, sanctioned by Emperor Theodosius I (ruled 379–95 AD), turned the brief light of toleration into a fury of indignation against all non-Christian faiths. Under the influence of Ambrose, Bishop of Milan, Theodosius outlawed all variations of the Christian faith except for the Trinitarian beliefs set down in the Nicene Creed. Bishops who disagreed were expelled, many of them fleeing to the more tolerant Sassanid regime in Persia, which at that time still welcomed people of all faiths. Paganism was outlawed too. The eternal flame in the Temple of Vesta in the Roman Forum was extinguished and the Vestal Virgins disbanded. In their place came the Christian world's first law against witchcraft. Finally, in 393 AD Theodosius abolished the highly cherished Olympic Games, since they were a relic of the pagan past.

All over the Empire people used these decrees as an excuse to legally persecute people of other faiths. One of the most spectacular outbursts was in Alexandria, where a mob, encouraged by Bishop Theophilus, sacked the Great Library and burned many of its most precious books because they were housed in a temple dedicated to Serapis, a pagan god of the underworld. Worst of all was the witch-hunt of Hypatia, head of the Platonist school at Alexandria and one of the most respected female Greek teachers and thinkers of the time, admired by Christians and pagans alike. In 415 AD the Christian Patriarch, Cyril of Alexandria, heaped blame on Hypatia for interfering in a local dispute, and encouraged a Christian uprising against pagans. He aroused a mob who dragged Hypatia from her carriage, after which, according to Socrates Scholasticus's *Ecclesiastical History*:

'They took her to the church called Caesareum where they completely stripped her and then murdered her with tiles. After tearing her body in pieces, they took her mangled limbs to a place called Cinaron, and there burnt them.'

Now it was the Christians' turn to play persecutors.

The Roman Empire ultimately collapsed for many different reasons: invasions by Germanic tribes, the arrival of the Huns from the Mongolian steppes (see page 256), resistance by the early Christians, lead poisoning, plague … Historians usually date its final fall to 476 AD, when the Germanic chieftain Odoacer deposed the last Emperor of the Western Empire, Romulus Augustus, although he was soon overthrown in his turn by another German leader, Theodric the Great, King of the Ostrogoths, who conquered all Italy.

What made Roman civilization remarkable was how long it managed to survive, despite its addiction to a constant pump of economic growth, essential for feeding the insatiable appetites of its rich ruling class. With little territorial expansion to add to its domains after 65 AD, this *ancien régime* demonstrated a genius at the art of staying in power despite ever-lengthening odds. It shamelessly stole the best technologies and ideas of rival empires. It ruthlessly suppressed the poor by enlisting them as soldiers for its armies, or slave labourers for its engineering projects. It learned how to control its huge populations through mass-entertainment programmes and propaganda. It exploited the earth's natural mineral resources when military expansion proved impossible, and it hijacked a minority religious sect to incorporate a new state religion with a fierce intolerance for anything its leaders deemed as heresy.

These were the tactics deployed to postpone the fall of Roman power. They were subject to repeated reincarnation in various forms, initially across the fractious lands of Europe and the arid deserts of the Middle East, but later throughout the whole world. Thanks to the rise and fall of the Roman Empire, the relationship between human civilizations and the natural world lurched into a phase that set the stage for the beginning of the modern world.

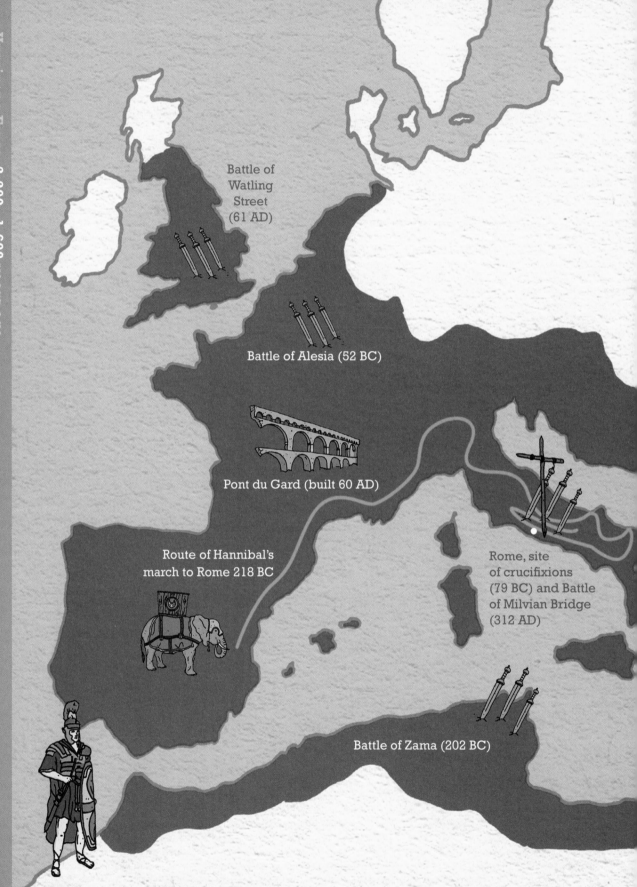

Battle of
Watling
Street
(61 AD)

Battle of Alesia (52 BC)

Pont du Gard (built 60 AD)

Route of Hannibal's
march to Rome 218 BC

Rome, site
of crucifixions
(79 BC) and Battle
of Milvian Bridge
(312 AD)

Battle of Zama (202 BC)

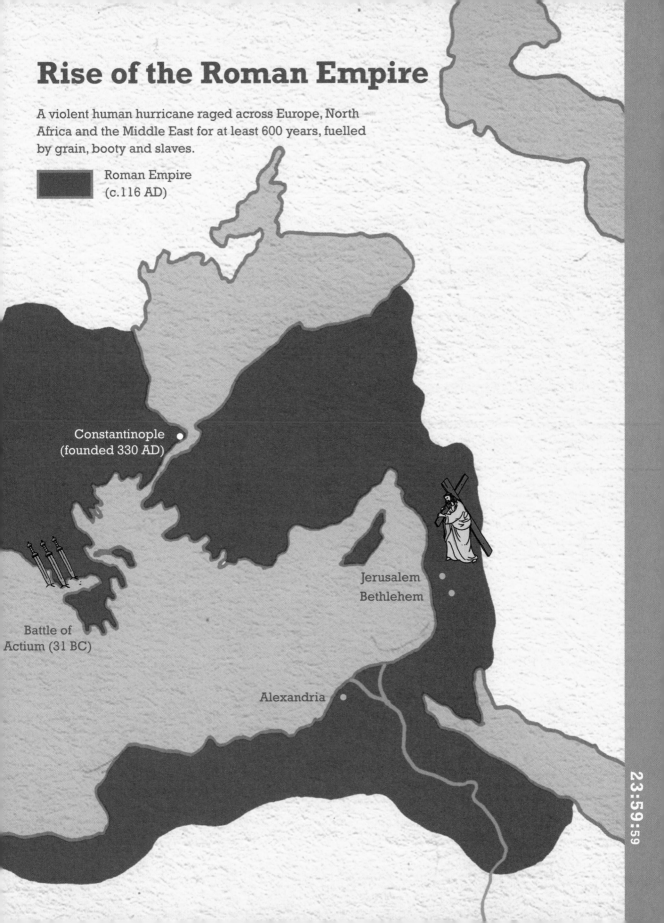

Rise of the Roman Empire

A violent human hurricane raged across Europe, North Africa and the Middle East for at least 600 years, fuelled by grain, booty and slaves.

Roman Empire
(c.116 AD)

Constantinople
(founded 330 AD)

Jerusalem
Bethlehem

Battle of
Actium (31 BC)

Alexandria

23:59:59

Timbuk-taboo

How people living outside the scope of civilizations and wandering herders maintained their veneration for nature, resourcefulness and spiritual well-being.

DESPITE THE MAYHEM and violence of the Mediterranean world and the massive rise of urban human populations around the Mediterranean and in India and China, human beings in much of the rest of the earth were still carrying on much as ever they had. One estimate suggests that of the 200 million or so people living 2,000 years ago, roughly seventy million were still living a Stone Age lifestyle.[1] Most of these people lived outside Europe and Asia. Their beliefs were based on living within nature as the essence of everything that mattered for their well-being and lasting survival.

Living within nature is historically mankind's most robust form of existence. The art was learned over millions of years of cohabiting with animals in the forests of the wild. The most resilient and ancient of all human societies have lived this way. Typically, they are not based on defending against

nature or harnessing its forces to improve material standards of living. Nor do they tend to rely on over-exploiting the natural world. Today, it is only at the very edge of human civilizations that a few of these ancient systems hang on, mostly as fragments, but at the time of the fall of the Roman Empire in about 476 AD, many of them still had the run of the roost.

One of the most ancient natural human habitats is in Australia. We know *Homo sapiens* arrived there more than 40,000 years ago, thanks to the discovery of 'Mungo Man'. This poorly preserved skeleton, found at the bottom of a dry lake in New South Wales on 26 February 1974, is of an old man, about five feet seven inches tall, lying on his back. He had been sprinkled with red ochre, showing the existence even then of elaborate burial traditions.

Exactly what route humans took to Australia is hotly disputed. Land bridges are known to have connected Australia and New Guinea during the last Ice Age, although recent genetic evidence suggests that people came from a broad arc stretching across Africa, India, Japan, eastern Russia and even North America, as well as the closer Polynesian islands and New Guinea.

Since the early sixteenth century AD, when European explorers first encountered the native hunter-gathering inhabitants of previously unknown continents such as the Americas and Australia, these people have come to be known as 'aborigines'. The name comes from the Latin word *ab*, meaning 'from', and *origine*, meaning 'the beginning'. Today, people often think Australian Aboriginals live only in the central part of the continent, the last remnant of the great land mass of Gondwana that split off from Pangaea more than 150 million years ago (see page 72). But before the first Europeans arrived in the eighteenth century, these native peoples lived mostly along the coast, in fertile areas with the greatest rainfall. Population levels would have remained fairly constant for thousands of years before the Europeans came, at somewhere between 350,000 and 750,000 people.[2] With so much land between so few people – nearly fifteen square kilometres per person (compared with about 2.5 people per square kilometre today) – there was little likelihood of conflict over resources.

These people believed that all living things shared a common spirit, a belief which has been called 'dreamtime' since the beginning of the twentieth century.[3] Animals were the ancestral beings of mankind, and their movements even shaped the earth itself. The Aboriginals' creation story, called 'the dreaming', explains how the land was formed. Geological features are attributed to traces left by animal spirits who roamed the ancient land, making it sacred, as described in this dreamtime creation story that comes from the Jawoyn people located in the Katherine Gorge area in today's Northern Territory:

'The whole world was asleep. Everything was quiet, nothing moved, nothing grew. The animals slept under the earth. One day the rainbow snake woke up and crawled to the surface of the earth. She pushed everything

Dreamtime spirits and images of swirling ancestral snakes live on in this colourful Aboriginal painting.

aside that was in her way. She wandered through the whole country and when she was tired she coiled up and slept. So she left her tracks. After she had been everywhere she went back and called the frogs. When they came out their tubby stomachs were full of water. The rainbow snake tickled them and the frogs laughed. The water poured out of their mouths and filled the tracks of the rainbow snake. That's how rivers and lakes were created. Then grass and trees began to grow and the earth filled with life.'

All things, animate and inanimate, share the same dreamtime spirit. There are literally thousands of dreaming stories in Aboriginal folklore, covering all aspects of these people's relationship with nature and other living things.

Although traditional Aboriginal ways of life were left mostly untouched for thousands of years, archaeologists have now discovered that by 500 AD some changes had begun to creep in as a result of contact with other encroaching civilizations. For example, the dingo, the Australian wild dog, was introduced from about 1500 BC by traders from New Guinea. These dogs had a significant impact on Australia's ancient ecosystem, and are thought to have been responsible for driving several species of marsupial carnivores to extinction. The Aboriginals adopted them as companions, domesticating them to help humans hunt. The introduction of eel traps, fish hooks made from shells and the development of smaller, more intricate stone tools, also help account for an overall growth in population, perhaps to as many as a million people by the time of the arrival of the first European settlers in 1788.

It is highly likely that the Australian Aboriginals' deep respect for all life and the sacred earth helped them survive a series of enormous climatic changes throughout their 40,000 years of history.[4] Being adaptable, careful and restrained in the use of natural resources would have been key to survival in the severe periods of drought at the peak of the last Ice Age 20,000 years ago. These were followed by the isolation of Australia from the rest of the world with the subsequent interglacial melt, which flooded the Bass Strait that had connected Australia to Tasmania, and the land bridge known as the Sahul Shelf by which it was attached to New Guinea.

One important survival strategy was to divide small Aboriginal clans into different groups, with each one revering a particular animal or plant as its defining totem. A series of non-contact taboos between men and women of the same clan evolved, ensuring that marriages occurred between groups rather than within a single group, thus avoiding potentially disastrous incestuous matches. Reverence for specific natural resources also meant that in times of scarcity it was less likely for any one species to be unwittingly hunted or gathered to extinction. In this way Aboriginal people maximized their chances of survival in extreme conditions; and through marriage ties between groups, they established a network of obligations between clans to care for others and share precious resources.

Totems and taboos were not limited to native Australians. Thousands of miles to the north-east, amongst the jungle tribes of the Amazon in South America, a system of living with nature evolved that was every bit as ingenious and resilient. The Huaorani were an Amazonian tribe whose responsible use of forest resources has few modern parallels. Their success was founded on an extraordinary expertise in carefully using animals, plants and trees to support a simple forest hunter-gathering way of life.

For these people the animals of the forest had a spiritual as well as a physical existence. They believed that when a person died, his spirit was challenged by an enormous python which guarded the domain of the dead. Victims were returned to the world as animal or insect spirits. Such beliefs gave these people a deep respect for non-human living things, as previous human incarnations. Their diet was based on hunting only certain types of animals, such as monkeys and birds, leaving the rest of the ecosystem balanced, with sufficient predators and prey to avoid the overpopulation or extinction of other species.

Their lifestyle was underpinned by a complex system of hunting and eating taboos – for example, deer were never hunted, because their eyes looked too much like human eyes. The Huaorani believed that they themselves were the descendants of a mating between a jaguar and an eagle, so these creatures were venerated, and never stalked. They believed that Huaorani holy men, shamans, turned into jaguar-like spirits in their trances, allowing them to communicate telepathically with other members of the tribe located many miles away.

Such beliefs caused the Huaorani to hunt carefully, using *curare*, a powerful poison obtained from the plant *Strychnos toxifera*. Blowpipes made from the wood of peach palms meant animals were killed with precision and with as little violence as possible, poisoned darts quickly paralysing the respiratory muscles, leading to rapid death by suffocation.

A belief that animals and plants as well as humans have souls is known as animism. It is striking how common and how widespread such beliefs were before the major monotheistic, people-centred religions of Judaism, Christianity and Islam took root. Animistic beliefs account for a great deal of the considerate and often cautious relationships between man and nature found throughout most indigenous peoples still living in the world today.

In the southern African desert, Bushmen from the Kalahari represent the last throes of a culture that has survived for at least 22,000 years.[5] A small number of them are still trying to live on the land of their forefathers, despite repeated attempts to evict them since the 1990s by the government of Botswana, which wants to turn their hunting grounds into a reserve for tourists.

Their plight, and their way of life, were vividly exposed by Laurens van der Post (1906–96), a South African-born explorer and conservationist who in 1956 made a television series, *Lost World of the Kalahari* (published as a book two years later), about his time spent living with a Kalahari tribe. Largely thanks to his efforts the then colonial government established the Central Kalahari Game Reserve with the aim of securing the tribe's future,

which was under threat from farmers, hunters and industrialists:

> 'His [the Bushman's] world was one without secrets between one form of being and another … He was back in the moment when our European fairy-tale books described as the time when birds, beasts, plants, trees and men shared a common tongue. And the whole world, night and day, resounded like the surf of a coral sea with universal conversation.'[6]

What van der Post describes is animism. The Bushmen's reverence for animals as sacred creatures is best expressed in their extraordinary cave paintings, such as those in the Lapala Wilderness

Strychnine is a powerful poison extracted by Amazonian tribes to make darts for hunting. It kills so quickly that little pain is inflicted on its animal victims.

An image from James Frazer's The Golden Bough in which animals' spirits are conjured out of the ground in response to humans breaking a taboo on felling trees.

American, African, Middle Eastern, Polynesian and even Arctic tribes echoing a similar belief in the one spirit force that touches all nature.

The idea that objects or talismans could impart good or bad luck came from this principle of interconnectedness. Frazer tells the story of how one Polynesian chief refused to blow on a fire with his sacred breath, because it would:

> '... communicate its sanctity to the fire, which would pass it on to the pot on the fire, which would pass it on to the meat in the pot, which would pass it on to the man who ate the meat, which was in the pot, which stood on the fire, which was breathed on by the chief; so that the eater, infected by the chief's breath conveyed through these intermediaries, would surely die.'

Even if passed rather obscurely through breath, fire, pot and meat, this fear of death from touching a chief was a common taboo amongst animist tribes. For example, Frazer recorded that the Cazembes, an African tribe from Angola, regarded their king as so holy that no one could touch him without being killed by his sacred power. Since contact with him was sometimes unavoidable, they devised a means whereby the sinner could escape with his life:

> 'Kneeling down before the king he touches the back of the royal hand with the back of his own, then snaps his fingers; afterwards he lays the palm of his hand on the palm of the king's hand, then snaps his fingers again. This ceremony is repeated four or five times, and averts the imminent danger of death.'

area, now in Botswana, which depict visions of animal gods in the spirit world, including rhinos, elephants and antelopes. Such images are believed to have been painted by holy men in a state of spiritual trance.

Animism includes the belief that all forms of life and other natural materials are inextricably connected by an invisible force or spirit. A vast trove of detailed information about indigenous societies and their animistic beliefs was the subject of *The Golden Bough*, written around the turn of the twentieth century by a Scottish scholar called Sir James Frazer (1854–1941). This massive study of myth and religion caused outrage when it was first published, because it compared the Christian story of Jesus as the Lamb of God, and the timing of Christian festivals such as Christmas, Easter and All Saints' Day, with heathen festivals.

Frazer gathered evidence from hundreds of missionaries and officials throughout the British Empire who were working with, or ruling over, many of the world's native tribes. The book is packed with examples of animistic beliefs – an all-encompassing ideological glue that once stretched across the entire globe, from the Celtic druids of Europe to the Aboriginals in Australia, with Asian,

Taboos were the unwritten laws of cultures that had no need for literacy or writing. Rules against touching royalty helped secure peace and preserve social order by reducing threats to priests, chiefs and kings. Even today, in some cultures, it is still taboo to touch royalty, as a recent Australian Prime Minister, Paul Keating, found in 1993 when he casually put his arm around the waist of Queen Elizabeth II when hosting a British royal visit.

Taboos on killing animals or other people were strictly adhered to by most animistic people, for fear of conjuring up the spirits of the dead, who they thought could come back to haunt the living or shower bad fortune on an assassin's tribe.

On the Polynesian island of Timor, in the South Pacific, it was sometimes deemed necessary for one tribe to wage war against another – perhaps for self-defence, or owing to a dispute over resources. When a tribe's victorious warriors returned home the leader of the expedition was confined to a specially prepared hut, where for two months he would undergo thorough bodily and spiritual purification. During this time he was forbidden to visit his wife or feed himself – food had to be put into his mouth by another person. Sacrifices were offered to appease the souls of their dead enemies, whose heads had been taken as a means of communicating with their spirits. Part of the ceremony consisted of a dance accompanied by a song, recorded by Frazer:

'"Be not angry," they sang, "because your head is here with us; had we been less lucky our heads might now have been exposed in your village. We have offered the sacrifice to appease you. Your spirit may now rest and leave us in peace …"'

It was through elaborate customs and taboos that violence was usually limited to being a last resort, and thanks also to such taboos, the chances of its escalation were minimized by the necessity of an expedition leader undergoing the laborious process of purifying his body and soul.

Tribes living so close to nature ensured that nothing ever went to waste. The Sami are a people who still live in northern Europe, on the fringes of the Arctic in Lapland, Finland, Norway and Sweden. At the end of the last Ice Age they moved northwards from central Europe, pursuing herds of reindeer which dwelt in the forests that had replaced snow and ice.

By 500 AD the Sami had learned how to domesticate reindeer to supply them with just about everything they needed for human survival – from dragging sleighs to providing meat and milk. Clothes and tents were made from their skins, arrowheads and needles from their bones. The Sami's songs, called *joiks*, were dedicated to animals and birds, and their reindeer-hide drumskins were illustrated with maps of the heavens – perhaps for worshipping the gods, or as an aid for polar navigation.[7] It was only after about 1500 AD, when the Sami had to start paying taxes to European states, that overhunting of their reindeer herds became a problem. Eventually most of them moved to the coast, where they could make a living from fishing instead. It is estimated that about 8,000 remaining Sami, 10 per cent of the total population, still live the nomadic life of reindeer herders.

Wasting nothing was just as important for people on the other side of the world, living on small islands surrounded by endless miles of deep blue sea. Polynesia is a group of more than a thousand islands. Formed from the tips of now extinct volcanoes and the tops of mountains constructed by coral, these isles stretch out across the vast central and southern Pacific Ocean. The islands, lying within a triangle bordered by Hawaii to the north, New Zealand to the south-west and Easter Island to the south-east, were first inhabited by humans from about 1000 BC. Archaeologists have discovered a distinctive style of early Polynesian pottery, called Lapita, which is often decorated with delicate tooth-shaped indentations.

Intricate tooth-shaped geometrical designs are typical of the Lapita pottery style and decorate this 3,000-year-old terracotta fragment of a human face, found on the Solomon Islands.

23:59:59

The eastern islands, including New Caledonia, Fiji, Samoa and Tonga, were the first to be peopled. Austronesians came from South-East Asia (Taiwan and southern China), bringing with them survival kits in the form of domesticated pigs, chickens and dogs as well as root and tree crops such as taro, yams, coconuts and bananas. Fish supplemented their diets, while just about everything else they needed for survival, including tools, had to be improvised out of stones, rocks, plants and shells on the islands themselves.

It has long been a mystery to historians and scientists how these people could possibly have then spread further westwards to Hawaii, New Zealand and ultimately Easter Island, probably the remotest habitable place on earth.[8] Prevailing winds blow in the opposite direction, making lucky shipwrecks unlikely – and anyway, the chances of being shipwrecked at all on such tiny islands so far apart in such an enormous ocean are infinitesimally small. The mystery has created reams of research into the ancient arts of Polynesian navigation. It is now thought that the islanders were assisted by a powerful partnership with nature that helped them discover the secret of how to navigate across several thousands of miles of ocean in nothing more than two-man dugout canoes.

First came a piece of new technology. The outrigger canoe featured two or more support floats fastened to the main hull. This allowed small boats to be substantially more buoyant in a rough sea. The *proa*, a small craft built on the same lines as the original Polynesian outrigger canoe, was still one of the fastest sailing boats on the seas well into the twentieth century.

Next came the art of looking to nature as a teacher and a guide – something indigenous people excel in. Experts believe long-distance Polynesian navigation followed the seasonal path of birds. Range marks have been discovered on the shores of various islands. They point out towards other distant islands, following the flight paths of birds. The route from Tahiti to Hawaii, which was first inhabited in c.500 AD, follows the paths of the Pacific golden plover and the bristle-thighed curlew, while the flight path of the long-tailed

cuckoo is thought to have guided sailors from the Cook Islands to New Zealand, where people first arrived in c.1000 AD.

Polynesian sailors are also believed to have taken with them shore-sighting birds, such as the frigate bird that refuses to land on water because its feathers become waterlogged, making it impossible for it to fly. The sailors released the bird, and if it failed to return to the boat, they knew they were close to land, and took their course from its path.

Bird watching was also how the Bushmen of the Kalahari desert found highly prized swarms of bees that provided them with copious supplies of sweet honey. Luckily the honey-diviner bird loved honey just as much they did. By following the paths of these birds, the Bushmen were led to the beehives, from which they could harvest their honey – always making sure to share the comb with their helpful honey-diviner friends.[9]

Sharing resources between all living things, animals or people, was central to the lives of animistic people. Some of the most common taboos didn't prohibit things at all – rather they prompted obligations of generosity. The Penan tribe belongs to the Dayak people of Borneo. They are thought to have been part of the Austronesian expansion which took place about 1000 BC, eventually leading to the populating of Polynesia. A distinctive element of their culture is the requirement of always sharing wisely. This is called *molong*, a word meaning 'never take more than necessary'. To *molong* a sago palm is to harvest the trunk with care, ensuring that the tree will sucker up from the roots. *Molong* is climbing a tree to gather fruit rather than cutting it down, or harvesting only the largest fronds of the rattan, leaving the smaller shoots so that they reach their proper size in another year. Whenever the Penan *molong* a fruit tree they mark it with a knife – a sign that means 'please share wisely'. The greatest taboo in Penan society is *see hun* – a failure to share.[10]

The tougher the living conditions, the more generous the human spirit. In Timbuktu, a city in present-day Mali that lies on the southern edge of the scorching Sahara desert, there's an ancient tradition that still survives amongst some camel herders. It demands that any guest be given what he needs – even if it means slaughtering the last goat whose milk feeds the nomads' children, or sharing the last drop of drinking water.[11]

Some cultures venerated trees as much as animals, and for them the forests were the holiest of holies on earth. They were the sacred places of Celtic European pagans long before the onset of Christianity gave them a new, more abstract God to worship. The pagan beliefs of the Anglo-Saxon Celts, who came from southern Scandinavia, the Netherlands and northern Germany, led them to worship their gods in woods, not temples. According to Tacitus (died 117 AD), a historian of the Roman Empire: 'They judge it altogether unsuitable to hold the gods enclosed within walls, or to represent them under any human likeness. They consecrate whole woods and groves, and by the names of the gods they call these recesses.'

Wyrd, from which the modern English word *weird* originates, was an animistic concept of fate common to Anglo-Saxon and Nordic pagan beliefs. It explained the interconnectedness of all things, linking past actions to future events. Yggdrasil is a gigantic mythological ash tree that connects the nine worlds of Nordic cosmology. Its trunk forms the axis of the world. Beneath one of its roots lies the sacred Well of Wyrd, next to which reside the three Norns, who engrave the Wyrd on the bark of the tree and look after it. So revered were trees that it was customary to offer sacrifices (both human and animal) to the gods by hanging them from tree branches.

Evidence of such rituals emerged in 1950 when a well-preserved body, now known as Tollund Man, dating back to the fourth century BC, was discovered in a Danish peat-bog. This unfortunate man was buried under a lump of peat. A rope made of two twisted leather thongs was drawn tight around his neck and throat, and then coiled like a snake over his shoulder and down his back. Copious quantities of a hallucinogenic fungus called ergot were found in his stomach, leading some experts to believe he was strung up in the branches of a nearby tree as part of a ritualistic sacrifice before being buried in the mud.

Celtic tribes migrated outwards from Germany during the time of the Roman Empire and settled in Gaul, Wales and Ireland. Their priests, called druids, often held ceremonies in forest sanctuaries, since for them also trees were considered sacred.

Animism was mankind's natural global system of beliefs. Oral taboos gave human societies that did not dabble with agriculture sufficient strength, resilience and adaptability to survive the harshest of natural disasters. Tribes that believed in what most modern people consider magic or superstition fostered a spirit of resourcefulness, conservation and a hatred of waste that modern societies are only just beginning to appreciate should be, at the very least, second nature.

Is Tollund Man the eerie remains of a sacrificial Celtic ritual, or did he just eat too many hallucinogenic mushrooms?

Amaizing
Americas

*How humans in the New World created their own civilizations,
oblivious to those established in Europe, North Africa and
Asia, but were fatally handicapped by a lack of large animals.*

IMAGINE AN ALIEN scientist looking down on his latest and greatest experiment – planet earth. More than three billion years have passed since he first sowed the seeds of life, wondering what on earth would happen and how they would take root. Now, literally millions of different life forms have emerged to sustain and take advantage of the planet's living systems. So far, so good.

Right at the end of this epic horticultural experiment, just a tenth of a second before midnight when seen on the scale of a twenty-four-hour clock, he notices that in one part of the world a certain species, a biped ape called *Homo sapiens*, has adopted a rather sudden and dramatic change of lifestyle. By mastering the art of mass food production, this species has started to build enormous new nests, in the form of cities and civilizations, and in the process has been clearing vast tracts of natural forests for fields in which to cultivate crops and keep animals. What's more, his experiment with agriculture has led to an explosive growth of population which shows no sign of abating. As a result, a great deal of innovation, aggression and killing has emerged in a vicious competition for resources and power. Perhaps, thinks the alien, it would be a good idea to conduct a control experiment on the other side of the planet, just to see if the same thing happens there …

Some 5,000 years after adverse climate conditions caused the Natufians and others in the

Fertile Crescent to dabble with agriculture, hunter-gathering people in the Americas were just beginning to reap their first annual harvests. They had absolutely no idea that people on the other side of the world were building huge civilizations based on crops such as wheat, barley and rice, and farm animals like pigs, sheep, cows and goats. Thebes, Jerusalem, Jericho and Babylon were completely unknown to people in North, Central and South America.

For thousands of years these people lived by sharing in the state of nature. Native American people faced several challenges that made their attempts at civilization like no others. In south-central Mexico, where the river valleys provided the right soils for cultivation and the climate was conducive to growing annual crops, the only grass capable of domestication was a rather weedy and unappetizing wild bush called *teosinte* which grew along the banks of the Balsas River. In this part of the world there was no wild wheat, barley or rice.

To begin with *teosinte* had just five to ten seeds, each of which was encased in a hard shell designed by nature to survive the most acidic of animal stomachs so they could be successfully spread elsewhere. By choosing those plants with abnormally numerous seeds and those with the softest shells, the patient people of Central America eventually engineered the crop we know as maize or corn. It took them as long as 5,000 years of painstaking artificial selection to convert unappetizing *teosinte* into a nutritious cob suitable for eating and harvesting on an annual basis.[1]

By 1100 BC a few Native Americans in central Mexico had begun to start their own experiment with living in a settled society. Stores, houses and fixed settlements were followed by terraced fields, annual harvests and seasonal cycles. Enough food was produced to allow former hunters and gatherers to become priests, rulers and artisans, freed up to worship, administer and trade.

The enormous and lengthy struggle to come up with an easy-to-cultivate crop reaped huge rewards for these people, as it has for posterity. The labour of these New World agriculturalists (again, it was mostly females who attended to sowing and choosing the best seeds) eventually produced a huge number and variety of carefully bred crops. The astonishing list they domesticated includes: chillies, sunflowers, pumpkins, peanuts, peppers, squashes, beans, courgettes, marrows, aubergines and avocados. Perhaps more significant today are tomatoes, potatoes and cacao beans – chocolate (the word comes from the Aztec *xocolatl*). Between them these crops, all of which originate from Central and South America, account for over half of all food grown throughout the modern world.[2]

Central American natives developed their own highly efficient crop farming method called the 'Three Sisters'. They planted climbing beans alongside tall cornstalks, to give them support. In return, the beans provided a nitrogen boost for the corn. Between each row they sowed squashes, gourds or pumpkins, whose large fruit and hairy leaves made it difficult for predatory animals to walk through the cultivated area, so protecting the other crops from being eaten. So knowledgeable were these people about how to manage their crops that they learned never to consume too much maize without first adding lime, protecting them from a terrible skin disease called pellagra (for the effects of this disease on later European settlers, see page 309).

Early Native American agriculturalists also cultivated non-food crops like cotton, which they used to construct fishing nets and to make clothing. These were also the first people to extract latex from rubber trees, used to manufacture items that played an important role in their religious rituals. None of these crops spread outside Central and South America until Europeans arrived in the early 1500s. Amazingly, many crops now grown in North America, including the potato, arrived there from South America via Europe, in the stores of European settlers who emigrated in the eighteenth century (see page 310).

The first settled people that have any kind of recorded history came from Mexico. They are known as the Olmecs, meaning 'rubber people'. At first, their civilization looked very similar to those of other early settlers in Egypt and Mesopotamia. Indeed, their achievements were almost identical.

They developed pottery in which to store produce, and built temples for studying the stars which were uncannily similar in style to Mesopotamian ziggurats and Egyptian pyramids. The remains of their earliest city, San Lorenzo (built c.1200 BC near modern-day Mexico City, but abandoned in c.900 BC), shows evidence of an elite class who were actively involved in trade. They made exquisite ornaments such as statues, masks and knives from semi-precious jade which was mined from as far away as Guatemala and exchanged for maize.

A change in river courses, possibly caused by an earthquake, is thought to have caused the demise of San Lorenzo. In its wake rose La Venta, a ceremonial city dominated by a giant pyramid, and located further to the east, close to the Gulf Coast, from where the sea could be farmed more effectively for turtle meat.

Like the Egyptians and the Babylonians, Olmec people developed a passion for arithmetic, driven by their desire to know when was the best time to plant and harvest crops. They used base twenty as their standard for counting. It was the considerable achievement of these Olmec people that they made the world's first ever known use of the integer zero, allowing any number to be expressed simply by placing figures in a series of rows, with zero as a place holder.

The sky was their clock and, like the people of Mesopotamia and Egypt, the Olmecs believed that the planets were driven by gods. Everything in their world moved in cycles depending on the sun, moon and planets, especially the bright 'morning star' of Venus. Their annual calendar of 365 days (which had twenty months of eighteen days each, and five special days left over) was the most precise in the world, and the Olmecs are now thought to have been the first people to develop writing in the New World. Recently, road-builders found a stone block in a pile of debris. It shows sixty-two symbols of an ancient script, some of them representing animals, plants, insects and fish, and probably dates back as far as 900 BC.[3]

Giant stone heads were hewn out of volcanic rock. Seventeen have been discovered to date, most near San Lorenzo and La Venta. Nearly four metres high and weighing up to forty tonnes, these were the Olmecs' equivalent of the Egyptian Sphinx. The gods the Olmecs worshipped were based on representations found in nature, and are not so different from the jackal-headed Anubis, Egyptian god of the dead. There was the feathered serpent and the rain spirit, represented in later Central American civilizations as Quetzalcoatl (Aztec) and Chaac (Mayan). Snakes were highly symbolic, because it was believed they represented an umbilical connection between the earth and the spirit worlds. But specific details of the Olmecs' religious beliefs are scant, because none of their creation myths have survived.

Other New World civilizations, in particular the Mayans, were deeply influenced by Olmec art, science, culture and religious beliefs. The first Mayan settlements emerged in about 1000 BC, south of the Olmecs along the Yucatàn Peninsula. Large-scale towns and cities, such as Tikal, Palenque, Copán and Calakmul, rose up between 200 BC and 800 AD, rivalling others in the rest of the world

This colossal, twenty-tonne Olmec stone head was found at La Venta, a ceremonial city built near the Gulf coast.

both in size and sophistication. The story of these civilizations remained completely hidden from the modern world until 1839, when an American traveller and writer, John Lloyd-Stephens, went in search of ancient ruined cities that Mexican locals claimed lay buried deep in the jungle. With his English architect companion Frederick Cather-wood, he discovered a number of ancient Mayan cities, including Copán and Palenque. Thanks to their accounts, historians have since been piecing together evidence of the people who built these cities, and what it was that made them tick.

Unfortunately, the task has been made a lot more complicated than it should have been. Although as many as 10,000 texts have been recovered from stone engravings and buildings, tens of thousands of precious books, written on paper made from the bark of fig trees, have been lost. The zeal of Christian Spanish invaders in the early sixteenth century, who regarded all Native American writing as cabalistic works of the devil, means that only three of the many thousands of original paper texts (called codices) now survive. One priest, Friar Diego de Landa, personally oversaw the destruction of hundreds of books and more than 5,000 precious works of art at a ceremonial bonfire on 12 July 1562. He later wrote about the event and the effect it had on the native people: 'We found a large number of books … and, as they contained nothing in which were not to be seen as superstition and lies of the devil, we burned them all. Which they [the natives] regretted to an amazing degree, and which caused them much affliction.'[4]

One of the most precious surviving docu-mentary sources, the *Popol Vuh*, was fortuitously written down by an unknown Spanish missionary in the 1540s.[5] It sets out native beliefs that had been passed down orally over many generations. It reveals how crops lay at the very root of Mayan beliefs about how the world was created. Three divine creators, in the form of water-dwelling feathered serpents, decided to create humans to keep them company. First they tried to make them out of mud, but that didn't work. Next they used wood, but that also proved unsuccessful. Finally, 'True People' were modelled out of maize, their flesh made of white and yellow corn and their arms and legs of corn meal.

This creation myth reveals why Central American civilizations eventually evolved in very different ways from those on the other side of the world. The livelihood of their people depended completely on crops like maize, for the simple reason that they had no large domestic mammals. Since the arrival of humans at the end of the last Ice Age, nothing much larger than a turkey had survived (see page 108).

There were no pastoral nomadic people, like those on the Eurasian steppes, whose lives were built on tending domesticated herds of animals while constantly moving from place to place. The intense harassment such people gave those settled civilizations of the Middle East, Europe and Asia never occurred in the New World of the Americas. There were no major military imbalances here – no haves versus have-nots – because no one had the ability to travel quickly on horseback commanding the strategic advantages of surprise, height and speed in battle.

Carts were never invented, because there were no large animals suitable for domestication to pull them. No one had any use for the wheel, which although it has been found in Native American toys, was never deployed in real life. No wheels meant no gears, no pulleys or other civil engineering essentials like treadmills, used by the Greeks and Romans to build their massive lighthouses, waterwheels and aqueducts, all of which were designed to make their worlds less vulnerable to the unpredictable forces of nature.

Without the competitive arms race between nomadic and settled people that caused wave upon wave of war and destruction across Europe and Asia, these people never discovered how to smelt iron. Nothing so strong was necessary. Gold and silver were theirs in abundance, and copper too. Their softness and suppleness made them ideal for crafting long-lasting jewellery and other artefacts for religious worship and to glorify nobility.

The effects of these differences between the New World and the Old became more profound as the centuries rolled by, not that these people had any inkling that their world was militarily inferior to others until Spanish explorers arrived in the early sixteenth century. Without horses, chariots, roads and wheels, the relationship between these civilizations and the natural world travelled on a unique trajectory. Perhaps the history of Central America represents what might have happened to the Egyptian civilization were it not for the invasion of the Hyksos in 1674 BC, that dragged those ancient people reluctantly into the wheel world (see page 135).

For many years historians regarded these Native Americans as essentially peaceful people because archaeological evidence suggests that until about 1000 AD most towns and cities in Central America were unfortified. It is now apparent that there was a much darker side to their way of life. Unable to rely on the enormous benefits of animal power, these people were totally dependent on their annual harvests of maize and other crops and traditionally it was the king's duty to convince the gods to bring sufficient rain. Government policy was therefore fixed on finding effective ways of contacting the spirit world to curry their favour.

The Olmecs came up with a unique form of dialogue with the spirit world which also dates the historical origins of competitive sport. It was constructed around a ball game called *Ulama*. Dozens of prehistoric ball courts dating back to 1400 BC have been excavated in ancient cities all over Central America. The oldest yet discovered is at Paso de la Amada. It is approximately eighty metres long and eight metres wide.

Ancient rubber balls have been found perfectly preserved in swampy sacrificial bogs alongside other religious offerings, suggesting the game had a religious purpose. The object was to score by bouncing the ball through one of two vertical stone rings, up to six metres above each end of the court. Two teams of between two and five players would try to accomplish this using their hips, thighs, forearms and heads, but without touching the ball with either their hands or feet. Hip belts, kneepads, head-dresses and protective masks were all part of the players' apparatus, and were often adorned with symbolic figures and pictures of the gods.

Although the game was sometimes played for fun, championships were usually held during religious festivals, when contests between rival kingdoms and states would be fought, quite literally, to the death. The members of the losing side were ritualistically sacrificed to the gods, their bodies buried underneath the court and their skulls sometimes turned into cores around which new rubber balls could be crafted. For the Mayans and their successors the Aztecs, this game symbolized a battle between the lords of the underworld and the peoples of the earth.

The largest ball court that has been discovered to date is located at the Mayan religious city of Chichen Itza. Wall paintings show two teams just after the 'final whistle'. The leader of the winning team is holding the decapitated head of the opposing captain, whose blood flows from his mutilated body in the form of writhing snakes spewing from his neck.

Sacred sporting events were sometimes used to settle scores between rival kingdoms, as a substitute for war. If a king lost, he paid the ultimate penalty to placate the gods. One of the three surviving indigenous Central American texts, the Codex Zouche-Nuttall (written c.1350) records the outcome of a contest involving the splendidly entitled Lord Eight Deer Jaguar Claw, a Toltec king who was a formidable ball player. He won several cities and states thanks to his prowess – until 1115 AD, when at the age of fifty-two his luck ran out and he had his head turned into a football.

This is a Mayan ball-court marker, used to divide up the sports field, and shows a ball-game player wearing protective clothing.

Kings and their priests gambled everything in their efforts to please the gods to ensure there was sufficient rain for their crops. The extent to which they would go was horrifically revealed by a series of excavations that began in 1895. In that year an American archaeologist and diplomat Edward Thompson purchased, for $75, a Mexican plantation that he knew included the ruins of the sacred Mayan city of Chichen Itza. Like Arthur Evans of Knossos fame (see page 142), Thompson spent much of the rest of his life uncovering the secrets of this ancient place lost to the world in the midst of the Mexican jungle. In the pursuit of his archaeological passion he was hunted by the locals and nearly killed by a poisoned mantrap, losing the feeling in one of his legs.

The focus of his attention was the Sacred Cenote, a ninety-metre-long sacrificial water pool that the Mayans believed provided a portal to the spirit world, possibly because the Yucatàn Peninsula is composed of porous limestone, which makes naturally occurring lakes and pools extremely rare.

Between 1904 and 1911 Thompson and his team recovered more than 30,000 objects from the well, first by dredging it and then by donning diving suits and fumbling in the pitch-dark twenty metres below the water's surface. Among the thousands of objects recovered were knives, sticks, bells, plates, jugs, figurines, jewellery and ornaments. Unfortunately, not all of them have survived, as many were stored in Thompson's house on the plantation, which was burned down in 1920 by Mexican revolutionaries.

Thompson found that many of these objects had been smashed into pieces, to kill off any spirits within them, before being thrown into the lake as offerings to the gods. Among his discoveries was a sacrificial knife with a handle carved into the shape of two writhing rattlesnakes, used to gouge out the still-beating hearts of human victims. A gold plate, dating to c.900 AD, shows a Toltec warrior wearing an eagle head-dress sacrificing a Mayan captive. His costume signifies a descending bird of prey. In his left hand he holds the sacrificial knife, while in his right he grasps the freshly extracted heart of his victim. Four assistants can be seen splaying the victim over the sacrificial stone slab. One looks directly outwards towards you – the witness.[6]

Thompson found the bones of more than forty-two victims in the small lake. Half of them are estimated to have been younger than twenty when they were sacrificed, and fourteen were probably under twelve.[7] The Mayans believed that the younger the victim, the more pleased the gods would be, because younger souls were considered purer. By the time of Aztec dominion (c.1248–1521 AD) child sacrifice was especially common in times of drought. If sacrifices were not given to Tlaloc, the Aztec god of water, the rains would not come and the crops would not grow. Tlaloc required the tears of the young to wet the earth to help bring rain. As a result, priests are said to have made children cry before their ritual sacrifice, sometimes by pulling out their nails.[8] The Mayans' desperation for rain lay at the heart of the reason

Spanish Friar Bernardino de Sahagún (1499–1590) wrote about such Aztec practices as their habit for human sacrifice, in a series of illustrated books called the Florentine Codex, which includes this gruesome picture.

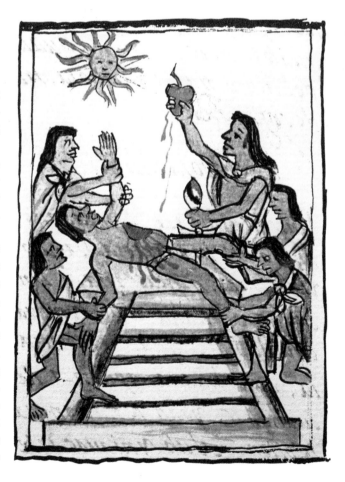

23:59:59

why, by about 900 AD, their civilization fell into decline. Increasingly severe droughts, exacerbated by the effects of deforestation, soil erosion and intensive farming, led to starvation, invasions and violent contests with neighbouring people over scarce natural resources.[9]

In 1428 the Aztecs formed an alliance of three city states – Tenochtitlan, Texcoco and Tlacopan – centred in the valley of Mexico. This was the final native Central American civilization before European invaders arrived in the 1520s. It stretched from coast to coast, except for a small area to the south-east called the kingdom of Tlaxcalteca. It was these enemies of the Aztecs who allied with the Spanish in 1521 to help destroy the Aztec King Moctezuma, bringing the history of independent indigenous American civilizations to a close (see page 285).

Aztec culture closely followed the traditions of previous Central American people, although by now life was getting increasingly desperate because of further droughts and the migrations of other people from the north, affected by climate change.[10] Better access to remaining resources was at the heart of the Aztec triple alliance, which redoubled its efforts to get the gods on side by increasing the levels of human sacrifices to fever pitch. Aztec rulers such as Ahuitzotl (1486–1503) even went to war with neighbouring states specifically to acquire additional prisoners to use as fodder for human sacrifices (conflicts known to history as the Flower Wars).

New buildings were erected in the magnificent Aztec island capital Tenochtitlan, which rose out of an enormous lake connected to the mainland by a giant retractable causeway. This remarkable capital lay on the site of the current Mexico City, although the lake has now been dredged to make way for modern buildings. Little remains of the ancient city, which was destroyed by Spanish invaders in 1521 (see page 286).

The Aztecs believed that if they built a sixty-metre-high Great Pyramid at the heart of the city, Tlaloc, the god of rain and fertility, and Huitzilopochtli, god of war and the sun, would look more kindly on them. Each god had its own temple on top of the giant stepped structure. Consecration rituals for the Great Pyramid's two gods in 1487 were reported to involve several thousand human sacrifices in the hope of persuading the gods to send more rain.[11]

Tales of such acts only reinforced the prejudices of the Spanish invaders, who saw these people and their way of life as barbaric, uncivilized and savage. Sacrificial practices helped them justify their own genocide in the name of the one Christian God, who as far as the Europeans were concerned was the only true portal to the spirit world. Ironically, it was thanks to the power of Christ's own human sacrifice that the gifts of forgiveness, salvation and grace were on offer for all men who believed in Him.

But most European intruders weren't really interested in saving people's souls. Ultimately, it was the lure of enormous quantities of gold and silver that focused their attention. But most of those riches lay further to the south, where another way of life had been quietly gathering momentum, finally blossoming with the rise of the Inca Empire of South America.

New civilizations began to appear further south, along the coastal regions of South America, because living by agriculture in the massive forested interior was simply too difficult. Early Peruvian cultures such as the Nazca (c.300 BC–800 AD) and the Moche (c.100–800 AD) shared many cultural habits with the Central Americans further north. Gold and silver were panned from rivers flowing down from the vast Andes mountains and traded for Mexican maize. Commercial contact led to the exchange of similar systems of religious beliefs.

The Nazca were responsible for what are still regarded as almost superhuman depictions of their animal gods. On a 500-square-kilometre plateau these people carved hundreds of perfectly straight lines and geometric patterns by painstakingly brushing the arid sand and grit to one side. Look down on the landscape from an aeroplane and more than seventy enormous pictures of animals, insects and humans reveal themselves, some of them as much as 270 metres long. Look at them

How on earth could they do this? An enormous Nazca spider geoglyph, as seen from the air.

from the ground, and nothing can be seen but paths in the dust. How on earth could these people have constructed such art (called geoglyphs) without being able to see what they were creating from above? It's one of history's big mysteries. Theories of their purpose have ranged from markers for underground waterways to landing pads for UFOs.

The most likely reason why the Nazcas created these enormous images is that, as for the people from Central America, communicating with the gods was at the heart of everything that mattered. The night sky was their audience chamber, and the arid ground their advertising board. Pictures of monkeys, spiders, hummingbirds and lizards were all venerations to the gods, offered in exchange for the gift of sufficient quantities of water so their crops could grow. Cahuachi is a ceremonial city built by the Nazcas in the middle of the desert that overlooks the lines. More than forty burial mounds topped with ancient clay buildings still stand there. Excavations continue to yield tantalizing glimpses of their lifestyle, including pottery remains, some of which show trophy heads, indicating that these people too depended on gruesome ritual human sacrifice to appease their makers. The remains of more than a hundred ritually prepared heads have been found in various Nazca sites.

The Moche people lived at the same time as the Nazcas. They were farmers, who left an excellent record of their way of life in the vivid art painted on their pots. Scenes of hunting, fishing, war, punishment, sexual acts and elaborate religious celebrations are all clearly illustrated. Pyramids consecrated with the remains of human victims have also been found – yet more attempts to secure divine approval.

The apex of South America's ancient coastal civilizations came with the rise of the Incas. Once just a tribe in the area of Cuzco, they rose to predominance during the twelfth century AD. Strong leadership and a cult that worshipped their supreme rulers as representatives of the gods on earth helped these people build a federal, tribute empire in which other city states and kingdoms submitted to their overlordship in return for protection and assistance during times of trouble.

Unlike Central American people, the Incas benefited greatly from a mammal large enough to be domesticated as the only indigenous means of power – the llama. The presence of this pack animal caused them to build an extensive network of more than 20,000 kilometres of roads and trails, some of them crossing the Andes mountains at heights of up to 5,000 metres.

One of the most famous of these roads leads today from the village of Ollantaytambo to Machu Picchu, a mountain city built in about 1450 as an opulent Inca royal summer residence. This city was completely cut off from the modern world until it was 'rediscovered' in 1911 by the American archaeologist Hiram Bingham, after he was led there by locals. Bingham wrote a number of books about Machu Picchu, and collected as many as 40,000 ancient artefacts, which were sent to Yale University to be stored safely until their return should be requested by the Peruvian government. The Peruvians formally requested their return in 2006, and in 2008 Yale University finally agreed to return approximately 4,000 objects to be housed in a new Inca museum in Cuzco.

In this part of the world, there were no carts or wheels, but human runners, posted at intervals of approximately twenty-five kilometres on each main trail, so that messages could be sent quickly by relay, covering distances of more than 200 kilometres in a day. Instead of paper, parchment or clay, these runners carried pieces of rope which bore messages encoded in an elaborate series of knots that represented numbers and even phonetic sounds. Experts have still not fully deciphered the knotty language of Quipu.

Roads built by Inca emperors converged on their capital city, Cuzco, regarded as the navel of their world. In 1438 the Supreme Inca, Pachacuti (meaning 'world-shaker'), mounted an ambitious expansion programme, and with the help of his son Tupac brought most of Peru, Ecuador and Chile under Inca rule. They divided the empire

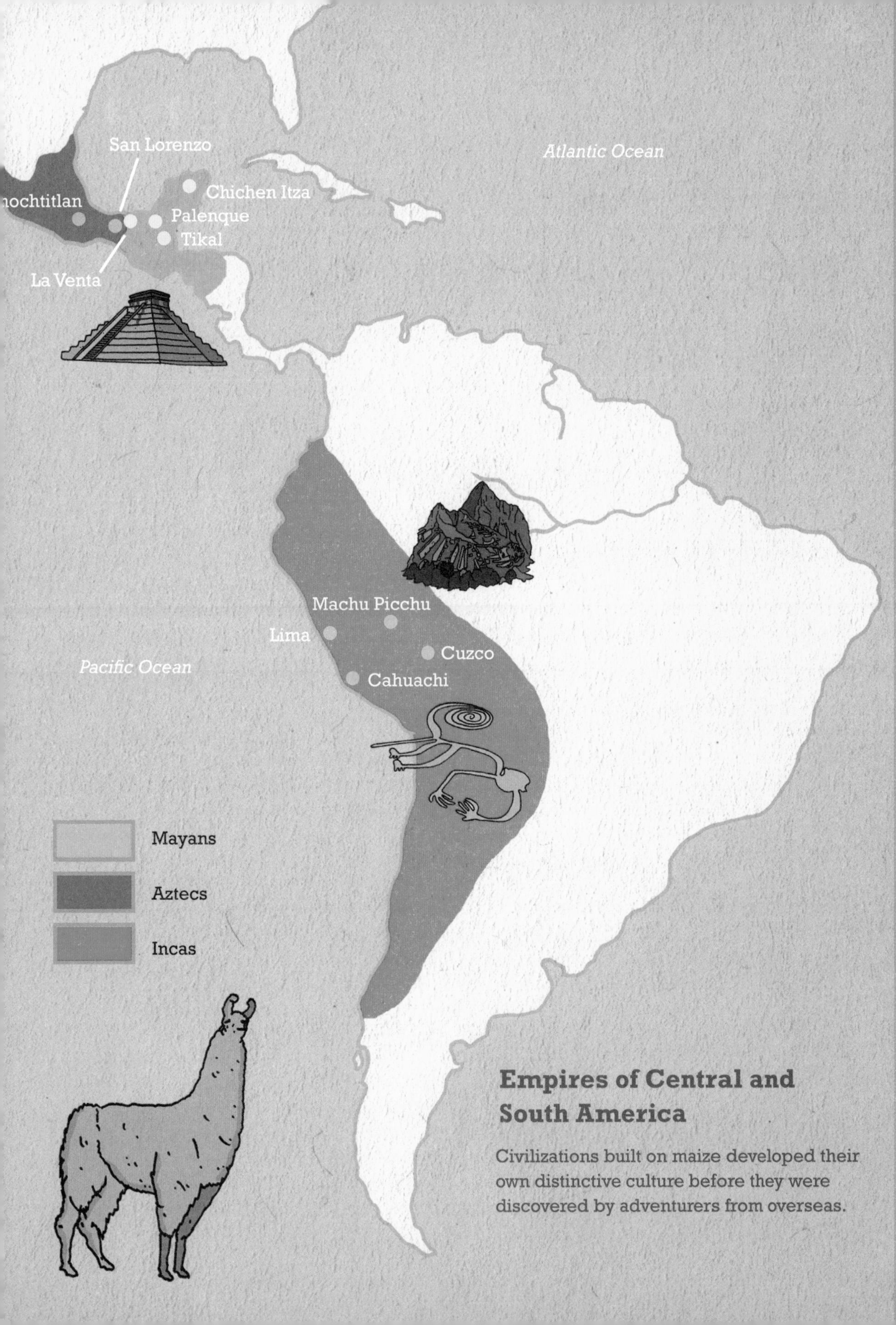

San Lorenzo

Chichen Itza

nochtitlan

Palenque

Tikal

La Venta

Machu Picchu

Lima

Cuzco

Cahuachi

Pacific Ocean

Mayans

Aztecs

Incas

Empires of Central and South America

Civilizations built on maize developed their own distinctive culture before they were discovered by adventurers from overseas.

into four main regions, each with its own governor. Most people accepted Inca rule willingly, since it provided them with a range of powerful paternalistic services to bail them out when times got tough.

No one ever went hungry in the Inca Empire, and there is no evidence of poverty. If roads were damaged or houses fell down, the region's governor would immediately send troops to repair and rebuild them. A national workforce was manned by males between the ages of fifteen and twenty, who were obliged to spend five years serving the state and its people. State storehouses were kept in every major town and city, opening up to the people in times of emergency to provide food and clothing. People were able to pay their taxes in kind by weaving cloth or by donating food supplies. Each village had a record-keeper whose job it was to monitor the

production of goods by the inhabitants, some for use locally, some for dispatch to central stores.

Marriages took place at village festivals. It was the responsibility of the neighbours to build newlyweds a small house in which to live. Married couples enjoyed a year without having to pay tax, to help them get off to a good start. When they had children they were entitled to another two years of tax-free status. Older citizens paid less tax as their productivity waned, and when they could no longer provide for themselves they were given food, help and clothes from the state's central stores.[12]

The Incas' Emperors were fastidious about their bloodline, insisting that their male heirs marry their sisters. In this way the belief in an unbroken line from the gods to the rulers could be preserved. Pretty and talented girls were chosen by state inspectors to be sent to court and join the *acllahuasi*, the House of the Sun Virgins. Maize was chewed by these virgins to help ferment a sacred Incan brew, drunk by thousands at annual religious festivals in an effort to please the gods who, they believed, would then look down on them and be satisfied to see that their grateful people were happy.

Apart from his queen (and sister), every emperor was allowed to choose as many wives as he wished. This meant that by the time of the eleventh Inca Emperor, thousands of children had been fathered by him and his predecessors, forming a unique administrative aristocracy which tightly bound the state to its people.

As with the Pharaoh in ancient Egypt, the Inca Emperor was a god on earth. His currency was gold, which was regarded as the droppings of the gods. Gold was divine – its colour the same as the shining sun. It was easy to craft, and unlike iron, or even silver, which eventually turns to a black oxide, it stays pure and lasts for ever. In the end such treasure is what attracted the sixteenth-century European conquistadors, whose appetite for other people's treasure had no limits.

Neither the religious fundamentalism which reached its peak in the practice of human sacrifice, nor the idea of a benevolent state that looked after all its citizens in times of need, survived the onslaught of the European invaders. Mayan, Olmec, Aztec and Inca beliefs rapidly diffused into

a new mixed culture brought about by the successful invasion of Spanish conquerors, while their cities become buried and forgotten under the dense canopy of the Mexican jungle.

Within a generation of Christopher Columbus's 'discovery' of what he thought was the east coast of Asia in 1492, the awesome power of both the Aztec and Inca empires was devastated by just a handful of Spanish adventurers. How these invaders crossed the world, what it was they were looking for, and why they were so quickly able to conquer the people of these empires are some of the most extraordinary stories in human history. They began in one of the most unlikely, inhospitable places on earth – deep in the dusty deserts of Arabia.

A sun-plate, llama and female idol all hammered out of pure Inca gold. No wonder such treasures caught the attention of European conquistadors.

23:59:59

Part 4
Going Global

(c.570 AD – present day)

DIVINE REVELATIONS that appeared to a man in Mecca, in modern Saudi Arabia, spread like spores on the wind, fusing much of Europe and Asia into a single vast cultural domain. Scholarly inventions pioneered in China were carried by Muslim traders to the hungry but unstable markets of the European West. Glimmers of African gold and rumours of new overseas routes to the East and West helped maritime explorers revive Europe's fortunes. Improvements in ship design scattered Europe's culture to every corner of the globe, fusing the earth's continents into a 'New Pangaea'.

New wealth brought new war. Religious reformation in Europe splintered the Christian Church into a mosaic of separate but related faiths as each nation vied for global supremacy. Refugees and pilgrims fled overseas, swelling the ranks of those looking to get rich quick by settling in non-European lands. African slaves were shipped to the Americas to work in plantations. Crops cultivated in the New World transformed the tastes of people in Europe and beyond.

Other human cultures and the natural world soon felt the effects of this rapid European expansion. Some, like the Russians, joined in the race. Others, like the Ottomans, hit back and seized Europeans as slaves. Persians set up trading stations for selling silk to Europe, unlike rulers of self-sufficient countries in the Far East who tried to shun all cultural and commercial contact with outsiders from the West. Meanwhile, in the Americas, local populations were annihilated by diseases unwittingly brought across the seas by European adventurers and their domestic animals. Native Americans and Aboriginal Australians were rounded up and fenced in on reservations, or had their children stolen from them in a war against traditional ways of life.

By 1800 China's command of technological innovation was overtaken by Europe. High-pressure steam gave mankind its first independent source of power, freeing it from limits set by the earth's natural forces. A conflagration of advances in science and technology accelerated man's conquest of nature. Steamships and railways opened up new possibilities for colonizing previously undisturbed lands in Africa and the Far East. Raw materials shipped back to Europe fed new machines that turned them into mass-produced goods for sale at home and abroad. Growing numbers of wealthy European and American opportunists led the world into a series of global conflicts from which, eventually, an economic system called Capitalism emerged triumphant.

In the last instants of history, an inexorable rise in human population, now approaching seven billion, caused the earth to suffer enormous losses of biodiversity and natural capital (in the form of coal, oil and gas). Consumer waste and industrial pollution led to alarm about the effects of global warming and climate change, prompting three dark questions that today haunt the relationship between planet, life and people, along with their prospects for long-term coexistence.

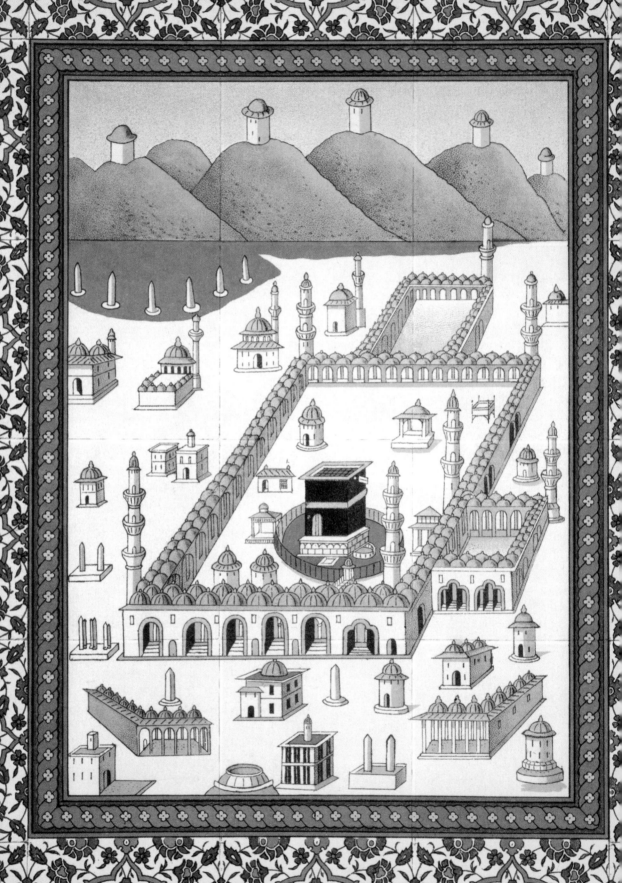

30

What a Revelation!

How a series of visions appeared to Mohammed, a man from Mecca, giving birth to Islam, a new way of life that promised to perfect the errors of mankind.

MOHAMMED was a religious prophet and the founder of Islam, a religion and a way of life that has profoundly affected the course of human and natural history. About 1,400 years ago, this merchant from the city of Mecca was seized by a series of visions in which he saw the Archangel Gabriel reveal the true and final word of Allah, the one almighty God. His family and followers then wrote down these revelations in a series of verses called the Koran. Today, with more than 1.3 billion practising Muslims, Islam is the second most popular religion in the world, after Christianity (there are an estimated 2.1 billion Christians in the world today).

Before the emergence of Islam, the Arabic religion was pantheistic. The Kaaba was a shrine in Mecca, in the middle of the Arabian desert, that contained 360 different gods. Every year nomadic tribes would converge on the Kaaba in the *Hajj*, a pilgrimage. No violence was allowed. It was believed that the Kaaba represented the intersection between heaven and earth. Its cornerstone, a piece of sacred black rock, symbolized that link, having fallen from the gate of heaven as a meteorite. It is still located in the Kaaba, a large cubic building in the al-Masjid al-Haram mosque in Mecca.

Mohammed was born in Mecca in about 570 AD. His family business was the transportation of goods such as salt, gold, ivory and slaves, on the backs of domesticated horses and camels whose wild ancestors had once migrated across from the

The Kaaba, a black building at the centre of the al-Masjid al-Haram mosque in Mecca, is the holiest place in Islam. Muslims all over the world face in its direction during their daily prayers.

The Archangel Gabriel, the same one who visited the Virgin Mary, escorts Mohammed to 'the farthermost mosque' in Jerusalem, from where he ascended to heaven to talk with other prophets.

marriage, at the age of about forty he withdrew from everyday life to take refuge inside a small cave on Mount Hira, near where he lived. There he had the first in a long series of dramatic and vivid visions in which the Archangel Gabriel – the same angel who is said to have visited Abraham and Mary, Jesus's mother – revealed to him the final and absolute word of God.

The angel told Mohammed that there was only one God, not many, and that He was in heaven, not on earth. He said that God had revealed His word many times before through other prophets such as Adam, Abraham, Moses, Jacob, Joseph, Elijah, Jesus and more than fifty others, but that over time, partly by accident but sometimes through deception, humans had corrupted His word and leaped to false assumptions. In so doing they had constructed religions such as Judaism and Christianity, which, although based on the truth of there being only one God, had become misguided and false.

It was a mistake, said the angel, for the Jews to think they were God's only chosen people. The Arabs were also descended from Abraham – not, like the Jews, through his second son Isaac (see page 173), but through his elder son Ishmael. Christians were mistaken when they claimed that Jesus was the son of God, because God is divine and cannot be made flesh. Rather, God spoke through prophets, finishing with Mohammed, who was the last prophet. Nor will there be a 'second coming', when Jesus or any other Messiah (meaning a saviour or liberator of the world) comes to earth in judgement. No: there is only one God, Allah, the God in heaven, and He is the only judge.

Mohammed's visions didn't end at 'perfecting' other faiths' faults. They also provided the foundations for a code of conduct that defines the Islamic way of life. The Five Pillars of Islam are a simple but powerful creed: profess faith in Allah as the one and only true God (and to Mohammed as his messenger); pray to Allah five times a day; give generously to the poor; observe all religious festivals; and, finally, make a pilgrimage to Mecca at least once in a lifetime.

Americas (see page 74). As a youth he gained a reputation for honesty and wisdom. It is said he successfully resolved a heated dispute during the reconstruction of the Kaaba after it had been damaged by flash floods.

The four chief clans of Mecca couldn't decide which of them should have the honour of lifting the sacred cornerstone into place. It was resolved to let the next person who walked into the shrine make the decision – that person was Mohammed. Mirroring the wisdom of King Solomon, he took off his cloak, placed the stone in the middle and instructed the leaders of the clans to lift it into place jointly by taking one corner of the cloak each.

Mohammed was a profoundly unsettled man. Perhaps it was because he never knew his father, Abdullah, who died on a trading trip six months before he was born. Perhaps it was because he lost his mother, Amina, who died of an illness when he was only six. Maybe life as a merchant disillusioned him. Despite what was, by all reports, a happy

From about 613, Mohammed began to preach to the citizens of Mecca, starting with his friends and family. His simple, easy-to-understand faith in one God appealed to many, including the poor and foreigners. As the numbers of his followers increased, the authorities in Mecca began to see him as a threat to their power. Meanwhile, the visions continued. In 620 Mohammed said he had been taken on a journey by the Archangel Gabriel to 'the farthermost mosque', which his followers took to mean Jerusalem, from where he ascended into heaven to speak with the other prophets.

The site from where Mohammed was transported to heaven was later commemorated by the fifth Umayyad Caliph, Abd al-Malik, who built the oldest surviving Islamic building in the world, the Dome of the Rock, constructed between 687 and 691 AD.

By 622 Mohammed had broken with his own tribe in Mecca and accepted an invitation to become leader of a community in Medina, a large agricultural oasis where Islam was already taking hold. His departure from Mecca with caravans of followers demonstrated the power of his message. For the first time Arab people felt able to abandon long-standing loyalties to their clans and families, and instead bound themselves to a new faith whose heartbeat lay in abstract heaven, not on animate earth. Medina had a large Jewish community, and Mohammed believed they would soon see the error of their ways and convert to Islam, further demonstrating that his new, perfected faith was the true way of life for all the world's people, regardless of race or creed.

For ten years Mohammed built up a loyal community of followers, although the Jews stubbornly refused to abandon their own traditions and texts, and remained highly sceptical about the possibility of a non-Jewish prophet. After several battles against the traditional clans of Mecca, Mohammed's followers began to suspect the Jews of Medina of treachery. After the Battle of the Trench in 627, at which Mohammed's loyal cousin Ali killed 'Amr bin 'Abd-e-Wudd, Mecca's most famous leader and war hero, the Muslims of Medina accused one Jewish clan of treason, killed its males and sold their women and children into slavery.

With Muslim claims that Jerusalem was their holy site (thanks to Mohammed's journey with Gabriel from 'the farthermost mosque'), and with Mohammed's hopes to perfect and unite all believers dashed by apparent Jewish obstinacy and treachery, animosity between the Jewish and Islamic faiths got off to a fast start.

The military victory of 627 gave Mohammed and his followers enormous confidence that God was on their side. Anyone who wasn't sure of the new faith, and wanted proof, now had it. In 630 AD, 10,000 followers marched on Mecca from Medina, and the city capitulated without a fight. Mohammed destroyed the many gods in the Kaaba and converted it to a Muslim shrine – the holiest in all Islam. Seeing the power and zeal of Mohammed's followers, other tribes and clans in Arabia soon converted to the simple new faith in which salvation in heaven came from belief in the one true God. Then, just two years after the conquest of Mecca, Mohammed started complaining of head pains and weakness. A few days later he was dead. His body was buried next to the Mosque of the Prophet in Medina.

Islam spread like a raging wildfire. Within a hundred years of Mohammed's death its simple, powerful message had penetrated Egypt, Palestine, Syria and the rest of the Middle East. It then spread to Persia, toppling the Sassanid Empire in 651, extending its reach as far as the Black Sea coast to the north and modern-day Pakistan to the south.

By 711 Muslim warriors had crossed North Africa and moved up into southern Spain, and within five years they had captured the entire Iberian Peninsula as far north as the Pyrenees. By 732 they were engaged in battle near Poitiers, in the heart of France, only to be stopped by a miraculous victory, against all odds, by the Frankish ruler Charles Martel at the Battle of Tours. Thanks to his disciplined, well-trained professional infantry, Martel was able to revive the Greek idea of a tightly packed infantry phalanx (see page 185), outwitting the mobile light cavalry and archers of the Muslim forces.

Some historians, including Edward Gibbon in his famous *Decline and Fall of the Roman Empire*, believe that had Martel not won this battle, Europe might well have become Islamic.

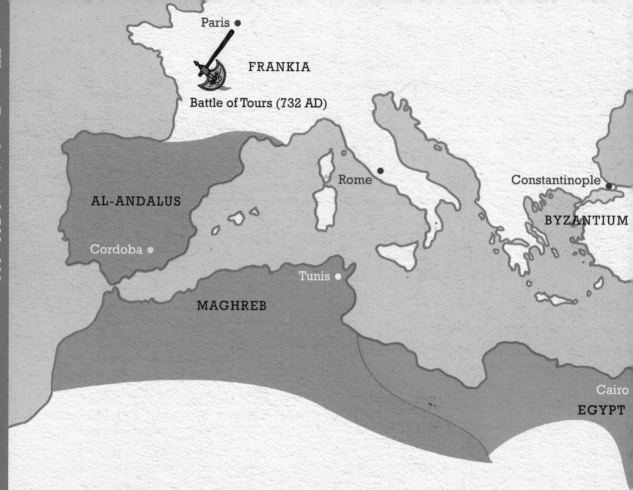

Paris •

FRANKIA

Battle of Tours (732 AD)

AL-ANDALUS

Cordoba •

Rome •

Constantinople •

BYZANTIUM

Tunis •

MAGHREB

Cairo

EGYPT

The Spread of Islam

The power of prayers five times a day whirled like a dervish throughout
the Middle East, North Africa and some parts of Europe.

Islamic world under Mohammed (622–632 AD)

Territory added by the first four Caliphs (632–661 AD)

Territory added by Umayyad Caliphs (661–750 AD)

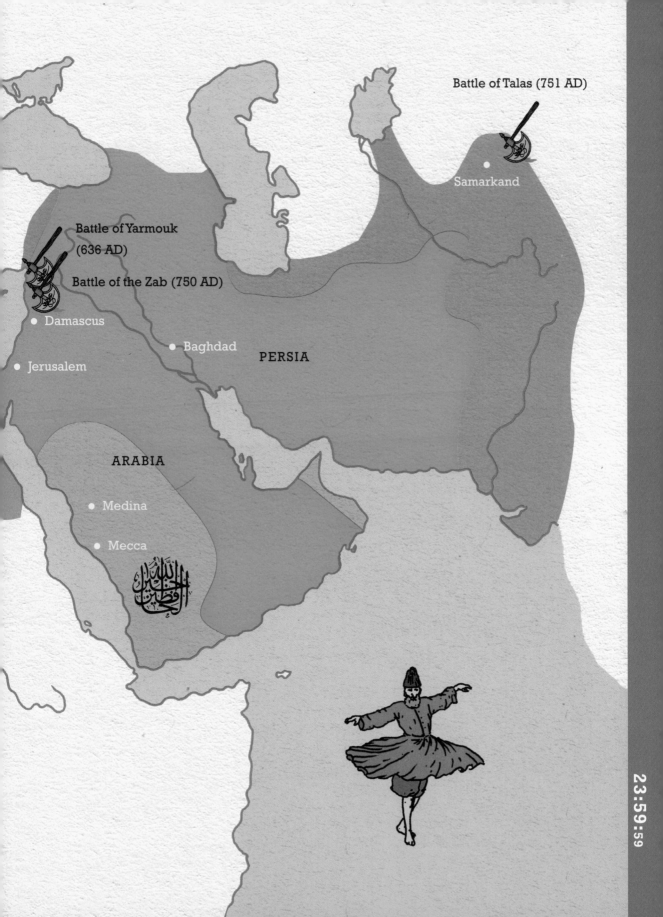

Battle of Talas (751 AD)

Samarkand

Battle of Yarmouk
(636 AD)

Battle of the Zab (750 AD)

● Damascus

● Baghdad

PERSIA

● Jerusalem

ARABIA

● Medina

● Mecca

Meanwhile, at the other end of the Islamic world, in central Asia, a new Islamic dynasty called the Abbasids defeated the Chinese at the Battle of Talas in 751, securing control of the area as far north as the Aral Sea. This Arab victory introduced Islam to inner and central Asia, where it has remained ever since. Within 150 years of Mohammed's death, Islam was the largest and fastest-growing religion in the world.

Islam's furious burst on to the earth's stage was made possible chiefly because of the attractiveness, simplicity and power of its religious message. Anyone could convert to become a Muslim. Few, if any, were forced. Conquered peoples were generally allowed to continue to carry on as Christians or Jews if they wished. Even polytheistic, shamanistic beliefs were tolerated. However, all non-Muslims had to pay a tax (*jizya*). Loading the tax system in favour of conversion was one powerful reason why Islam soon became a religion not just of the ruling class but also of the masses. Still, many rulers actively discouraged conversion for fear of losing important revenues, which suggests that the power of Islam's underlying religious message must account for a large part of its immense popularity.

The Islamic religion is easy to understand. It does not require belief in a Jewish carpenter's son as the miracle-working son of God. The complex Christian concept of the Trinity, which fused God in heaven with the resurrection of a dead man and an all-powerful divine essence called the Holy Spirit, was a barrier Islam did well without. Instead, Islam provided its followers with direct access to God via the Koran, avoiding the need for priests and holy sacraments.

Islam's Five Pillars provided a simple set of instructions that anyone could follow, and through which they could hope for eternal life in heaven. Islamic states could be established based on laws laid down in the Koran (called *Shari'ah*), which included guidance on everything from eating habits to marriage. Civic codes dealt with issues of law and order, and religious festivals such as the month-long fast of Ramadan were observed throughout the Islamic world. Prayer five times a day meant that the thought of Allah, the one true God, was reinforced by the entire community throughout the waking day, from sunrise to sunset. Such a regular, public display of worship caused communities to convert in droves – perhaps it was peer pressure, or even the instincts of a herd. By 1200 it has been estimated that 5.6 million of the seven million inhabitants of the Iberian peninsula were Muslim, almost all of them natives whose families had voluntarily converted to Islam.[1] The number is thought to have fallen to a third by 1600 owing to the policies of forcible conversion to Christianity, the burning of Arab texts and expulsions of Muslims pursed by the Spanish authorities including the Inquisition established by Pope Sixtus IV in 1478.

When Mohammed died, disputes immediately broke out as to who should lead the Muslim community. Since he had no agreed heir, a split emerged that still exists to this day. Sunni Muslims believed that Mohammed passed on his estate, and therefore his authority, to the Muslim community

Precious paper made in newly fashioned mills in Baghdad meant the Koran, like this nineteenth-century scroll, could be copied and distributed around the Islamic world.

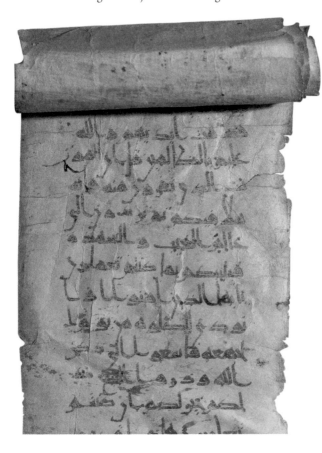

around him. It was from their approval that Mohammed's close friend and ally Abu Bakr legitimized his claim to become the first Islamic Caliph, Mohammed's rightful successor. However, Shia Muslims believed Bakr orchestrated a *coup d'état*, and that Mohammed's cousin and son-in-law Ali, who later became the fourth Caliph, was the Prophet's true heir, owing to his blood relationship. Ever since, Sunni Muslim rulers have claimed their authority from the election or approval of senior Islamic representatives, while Shias believe political and religious legitimacy comes through a direct descent from Mohammed and his family.

Rivalries burst into waves of bitter and violent struggles that exploded in a series of Muslim civil wars called *fitna*, shot through with coups and assassinations. Such rivalries helped spread the new word of God faster and farther. Out of Arabia came men highly charged with political ambitions who scattered in all directions, spreading the word of Allah using whichever version of Mohammed's rightful inheritance suited their own particular claims.

Umar Ibn Al-Khattab, the second Caliph (ruled 634–644 AD), was probably the most successful early Sunni ruler. During his reign Muslim fighters expanded the Islamic Empire by defeating the Persian Sassanids and taking control of Egypt, Palestine, Syria, North Africa and Armenia from the Byzantines. The most decisive battle took place in August 636 at Yarmouk, near Damascus, where after a bitter six-day engagement the heavily outnumbered Muslim forces eventually surrounded and routed the Byzantine imperial army leaving the Emperor, Heraclius, with no choice but to abandon all Syria, including Antioch, and return in defeat to Constantinople.

War between rival Islamic dynasties continued unabated. In 750, at the Battle of the Zab, the Abbasids, claiming direct descent from Mohammed, overthrew the Umayyad regime in Damascus and established a new capital in Baghdad.[2] It was a brutal coup. Only one member of the Umayyad ruling family survived the slaughter at Zab, the twenty-year-old Prince Abd ar-Rahman, who fled

south along the River Euphrates hotly pursued by Abbasid assassins. In desperation he and his brother Yahiya threw themselves into the river in an attempt to cross to the other side. Despite his brother's pleas, Yahiya swam back for fear of drowning. The assassins chopped off his head as a trophy, and left the rest of his body to rot by the riverside.

So overwhelmed was Abd ar-Rahman by these events that he vowed to re-establish his family's supremacy and oust the Abbasid usurpers. Eventually his quest took him to *Al-Andalus* (the Muslim name for Spain), where in 756 he established himself as Emir of Cordoba. For the next 250 years this branch of the Umayyad dynasty built a civilization to rival the court of Baghdad in every way, obstinately refusing to recognize Abbasid supremacy. In 929 Abd ar-Rahman's grandson, Abd ar-Rahman III, declared himself Caliph, by which time Cordoba was probably the richest, most splendid and highly cultivated city in the Islamic world.

Rivalries like these powerfully extended Islam as a religious and political force, and the new word of God spread fast on the backs of horses and camels. Arabs were mostly nomads. Their business was transporting goods and slaves across dusty deserts. As horsemen they were able to make use of the substantial military advantage of cavalry charges, against which settled societies were often defenceless. Following the conquest of Persia, Islamic troops became even more fearsome thanks to the adoption of the Persians' well-bred heavy horses. Stirrups, invented in China, significantly increased a mounted warrior's manoeuvrability, allowing the use of longer and heavier weapons without loss of balance.[3]

Chaos caused by the collapse of the Roman Empire, at least the western half, was another powerful force in favour of Islam. The vacuum of authority in Spain, following invasions by German tribes such as the Vandals and the Visigoths (see page 257), was matched by weak powers in Persia and the Middle East following centuries of warfare between East and West.[4] Exhausted by relentless combat, the magnificent Persian capital at Ctesphon was an early victim of

23:59:59

Muslim expansion after a prolonged siege in 637, surrendering a vast treasury that gave Islamic rulers a much-needed financial boost.

The long-term effects of the voluntary and mostly permanent conversion of millions of people to the new faith of Islam had dramatic consequences for the relationship between nature and mankind. At the heart of the religious philosophy of Mohammed was the complete removal of any concept of God on earth. The only earthly link with divinity was through the immutable, inspired written word of the Koran. Forests, animals, plants, mountains – dreamtime – none was a focus of Islamic veneration. Even the weather, thunder, lightning and other forces of nature were not considered sacred in themselves. Putting a single God in heaven, not on earth, was a trend started by the Jews. Then, barring the temporary exception of Jesus, Christianity reinforced the idea. Now the process was perfected by Islam. God was not to be found in the druidic woods of Europe or in Poseidon's stormy seas, not even in the pyramids of Egypt or on top of the ziggurats of Babylon. God was to be known on earth only through a single set of perfect and immutable rules uttered by Mohammed, and later written down by his followers in their holy book.

Initially the verses of the Koran were either memorized or etched out on anything that came to hand, from stones to pieces of bark. One close follower of the early Caliphs, who was ordered to compile the Koran into a single complete manuscript, protested that 'shifting mountains' would have been easier. In the end he resorted to gathering material from 'parchments, scapula, leaf-stalks of date palms and from the memories of men who knew it by heart'.[5]

The advent of paper-making supercharged the Islamic world's religious lust for the written word. Until the middle of the eighth century the only people who knew the secrets of paper manufacture were from the Far East (see page 244). But following the capture of a few Chinese prisoners by a gang of Arabian knights at the Battle of Talas in 751, this secret's genie was uncorked. The prisoners' knowledge of paper-making helped establish the process outside the Far East for the first time, in the city of Samarkand, now the capital of Uzbekistan. By 794 a paper mill had been set up in the Abbasid capital of Baghdad. From there the art spread to Damascus, Egypt and Morocco, and paper replaced papyrus, silk, wood and parchment.

The first paper book known in Christian Europe was produced using paper made from a mill built by Islamic rulers in Valencia, Spain, in 1151. This book, a religious document called *The Missal of Silos*, is kept in the library of the monastery of Santo Domingo de Silos, near Burgos, Spain.

The mass production of blank books made it easier for calligraphic experts to copy out the poetic verses of the Koran to be distributed all across the Muslim world. From about 900 AD Islamic mystics from Baghdad, called Sufis, began to teach that the Koran could help individuals gain direct experience of divine love.[6] The advent of paper-making encouraged Abbasid Caliphs in Baghdad to commission translations of ancient Greek, Persian and Indian scientific and philosophical texts in Arabic, in the hope of making their language and culture more acceptable to the newly conquered Persian nobility, whose rich past stretched back to the Hellenic age of Alexander the Great. Music, poetry, literature and the concept of courtly love now mingled in the Abbasid world with the wonders of ancient Greek and Roman writers on science, medicine, astronomy and mathematics.

Rulers such as Harun al-Rashid (ruled 786–809 AD) sent diplomats to Constantinople to acquire Greek texts. His son, al-Ma'mun (ruled 813–33 AD), is even said to have made it a condition of peace that the Byzantines hand over a copy of Ptolemy's *Almagest*. Written in about 150 AD, this book of mathematical astronomy explained in precise detail how to predict the position of the sun, moon and planets on any given date, past, present or future. It became an astronomical gospel for Islamic rulers, who used it to determine the dates of future religious festivals, such as Ramadan, that were based on the cycles of the moon.[7]

Thousands of other ancient texts were translated into Arabic or Persian in the Abbasids' 'House of Wisdom', an enormous royal library in Baghdad.

Caliphs lured translators, scholars and philosophers from all over the known world to their courts, and even encouraged debates on how to reconcile the works of rational philosophers like Aristotle to the divine revelations of Mohammed.

Such efforts resulted in the creation of a new school of Muslim thought called Mu'tazili which attempted to provide a rational basis for Islam using Greek philosophy as a foundation. Questions that Mu'tazili attempted to answer included whether or not God created evil or if the words of the Koran should be interpreted allegorically or literally.

One of the earliest philosophers to study in the 'House of Wisdom', al-Kindi (801–73 AD), was reputed to have written over 260 books, including thirty-two on geometry, twenty-two on medicine, twenty-two on philosophy, nine on logic and twelve on physics.

Umayyad Caliphs, never forgetting the brutal slaughter of their dynasty in Damascus in 750, were determined not to be outdone by their Abbasid enemies. Ruling from their rival capital of Cordoba in *Al-Andalus*, they also patronized philosophers, doctors, mathematicians and scientists, but in addition they transplanted an entire Middle Eastern culture, root and branch, to southern and central Spain. Technical experts arrived to help reshape and revitalize the country's earth and soil, bringing with them Arabic knowledge of irrigation, essential for survival in the desert. Thousands of miles of *qanats* – underground aqueducts – were constructed, which from c.900 AD transported water from mountain sources to the fields. Oranges, lemons, apricots, mulberries, bananas, sugarcane and watermelons – crops which had never been grown in Europe before – were brought over from the Middle East and successfully cultivated in Spain. They even brought rice from India, without which today there would be no such thing as Spanish *paella*.[8]

Agricultural riches meant that by 1000 AD the population of Cordoba had grown to more than 100,000. In the words of one Muslim chronicler, al-Maqqari, who is widely suspected of exaggeration, the city had 1,600 mosques, 900 public baths, 213,077 private homes and 80,455 shops!

Islamic merchants brought gold and ivory from across the Sahara, and Cordoba's artisans turned them into coins, jewellery and luxury goods for the Caliph and his entourage. Thanks to the growing population and prosperity of this city, engineers were commissioned to build and later extend a monumental mosque, called the Mezquita. With

Inside the Mezquita at Cordoba, built by Emir Abd ar-Rahman I on the site of an old Visigoth cathedral. Christians turned it back into a cathedral after the Reconquista.

more than a thousand columns made of jasper, onyx, marble and granite, it could accommodate as many as 40,000 faithful for their five-a-day prayers. This mosque, the size of four football pitches, was the second-largest in the world, until the sixteenth century, when Christian conquerors turned it into a Roman Catholic cathedral. When the King of Spain, Charles V, saw the works that had been carried out on the mosque, he said to the Bishop responsible: 'You have built what you or others might have built anywhere, but you have destroyed something unique in the world.'

Islamic courts like those in Baghdad, Cairo and Cordoba acted like throbbing hearts, pumping ideas and inventions around a huge body united by a single faith and the common language of Arabic. They were responsible for fetching and carrying knowledge from as far as China in the East to France in the West, a process that eventually led to the transformation of the fortunes of Europe, helping future explorers develop toolkits for global conquest. Sometimes ideas were exchanged through war, like the transmission of how to make paper, and sometimes through trade.

In about 1200 the Italian merchant Leonardo Fibonacci travelled to Algiers in North Africa to help out in his father's *funduq* – an Arabic trading post. There, for the first time, he saw the incredible power of arithmetic using numbers written on paper replacing the old-fashioned abacus. The idea had come from India via Baghdad, where at the 'House of Wisdom' two scholars, al-Kindi and his colleague al-Khwarizmi (780–850 AD), had written books on newly fashioned paper that illustrated how to replace the convention of using written words to describe numbers (such as *alpha*, *beta* and *gamma)* with symbols ranging from 1 to 9. They showed how these could be arranged to allow fast calculation, and added the symbol 0, to represent no value, so that between these ten digits any number could be written.

After Fibonacci had seen Algerian merchants using the system, he was determined that his Italian counterparts should not miss out on its potential. *Liber Abaci*, his book published in 1202, introduced these new numbers to the Christian West. It showed how Arabic numerals could transform everything from book-keeping to the calculation of interest and money-changing. So successful was this innovation that within 200 years most Italian merchants and bankers had abandoned their old-fashioned ways and converted entirely to arithmetic using ink, pen and paper.

It wasn't just arithmetic that spread to the Christian West via Islamic scholars. Al-Khwarizmi's treatise on *Al-jabr* ('transposition') showed how a process of linear and quadratic equations could determine an unknown value. When al-Khwarizmi's work was translated in Spain, the Arabic word for 'thing' (*shay*) was transcribed as '*xay*' because the letter 'x' was pronounced 'sh' in Spain. Over time this word was shortened to just 'x', which has become the symbol universally used to denote an unknown value. Thanks to the translation of this text into Latin by Christian scholars such as Gerard of Cremona (1114–87) in Toledo, the foundations of modern Western physical sciences were laid. Today algebra is an essential cornerstone in modern science and engineering projects for everything from building particle accelerators to skyscrapers.

Islamic scholars were just as passionate as the ancient Greeks about fathoming out how the laws of nature worked, be they biological, mathematical or astronomical. Modern medicine owes much of its inspiration to Arabic philosophers, who were influenced by ancient Greek writers like Claudius Galen (129–200 AD), a court physician in Rome. A bright young Arabic doctor called Ibn Sina, also known as Avicenna (980–1037 AD), wrote no fewer than 450 books after reading the work of Galen in the royal library at Bukhara, now in Uzbekistan. His *Book of Healing* and *Canon of Medicine* became standard textbooks in European universities for more than 500 years, detailing the symptoms and causes of diseases, treatments using different types of medicines, and the functions of various organs and parts of the body. The fourteen-volume *Canon of Medicine* included the first ever detailed description of how the human eye works, and even detailed how to remove cataracts.

Just as important to the evolution of Western science was a man called ibn al-Haytham (965–1040 AD), who worked out the laws of optics some 600 years before Newton. Al-Haytham was educated in Baghdad, but later travelled to Spain, from where his discoveries made their way via Christian translators to the Latin West.[9] He demonstrated how rays of light bounce off objects by a process of reflection and refraction. Such insights were the key for the future understanding of how lenses work. It was a Latin translation of the work on optics by al-Haytham that early Western scientists used to work out how to build a telescope.

Islamic rulers had a deeply religious reason for wanting to find ways to understand the science of navigation. Mohammed had stipulated that at prayer times all Muslims should face towards Mecca (initially the direction was Jerusalem, but that was changed after the reluctance of the Jews in Medina to embrace Islam), and that mosques

should therefore be built with their prayer shrines (*qibla*) pointing in the right direction. Once Islam spread across the world, knowing in which direction the holy city lay became one of the most significant scientific challenges of all.

The astrolabe was a tool used throughout the Islamic world for precisely this purpose. Ultimately, its value to European explorers was just as important as that of the Chinese-inspired compass. The idea of designing a hand-held instrument that could track the process and elevation of the stars at night to determine one's position on the earth's surface went back to the ancient Greeks – to Eratosthenes (276–194 BC) and Hipparchus (190–120 BC). Using trigonometric tables, Persian scientist Muhammad al-Fazari (died c.777 AD) built the first Islamic astrolabe, although it worked along only a single line of latitude. Later, Andalusian scientist al-Zarqali (1028–87 AD) modified the instrument so it could be used anywhere in the

The Islamic astrolabe was a navigational aid used to find the direction of Mecca for ensuring mosques were built facing the right direction.

world. After his textbooks and astronomical tables were translated into Latin by Gerard of Cremona in the twelfth century, they provided vital navigational resources used by early-fifteenth-century Christian explorers. The first known European astrolabe was built in 1492 in Lisbon by Jewish astronomer Abraham Zacuto (c.1450–c.1510), whose astronomical tables were used by Christopher Columbus as navigational aids for his overseas expeditions (see page 279).

Warfare between the Islamic world and the Christian West conveyed other important innovations from East to West. Well-bred cavalry horses from Persia and stirrups from China were two examples scooped up by Charles Martel after his remarkable victory at Tours, suggesting that the culture of the European medieval knight may have originated in Persia and then diffused via Islamic Spain into eleventh-century France.[10]

When William VIII of Aquitaine captured a group of attractive Saracen slave girls at the Siege of Barbastro in 1064, he so loved their songs that he was moved to write his own love poetry. William later became the founder of the French troubadour movement, a group of itinerant singers and poets who enthralled Crusader Christians with their songs about war, romance and courtly love. If troubadour songs lie at the start of Western Europe's musical tradition, their likeliest origins come from the Muslim world.

The presence of Islamic rulers in Europe, especially in Spain and Sicily, eventually provoked a furious contest between Christian and Muslim civilizations. Initially the dispute centred on control of Jerusalem, holy to both Christianity and Islam, but later hostilities boiled over into a full-blown war, the *Reconquista*, for the return of the Iberian Peninsula to Christian rule.

The idea of a *jihad* was deployed with dramatic effect by Mohammed's early followers in their seventh-century firestorm raids out of the Middle East, their soldiers inspired by the comforting

words of the Koran: 'Consider not those who are killed in the way of Allah as dead. Nay, they are alive with their Lord, and they will be provided for.' By 1095 the Muslim idea of justifying war in the name of religion had matured in the Christian West and included the forgiveness of soldiers' sins and the promise of eternal bliss for the brave, granted by Christ's representative on earth, the Pope in Rome. Once the Islamic concept of Holy War had been adopted by Crusaders in Europe, mankind was truly launched on a new path towards violent global dominion.

Mohammed's fiery revolution thoroughly connected Europe, North Africa, the Middle East and China through war and trade. Like a Whirling Dervish dance, which originated with the Sufi mystics of Persia, the capitals and Caliphs of early Islam sent bold ideas and inventions flinging around the enormous Muslim universe and whatever else it touched in peace or war. When European explorers eventually set out on their great adventures across the seas in the fifteenth century, their fortunes relied on a wide range of Islamic imports including warhorses, mathematics, maps, navigational aids and paper.

31

Paper, Printing & Powder

How Chinese scientific discoveries gradually spread westwards via Islam to Europe, supercharged by a Mongol chief who created the world's largest ever empire.

WHEN HE SAILED across the Atlantic Ocean at the end of the fifteenth century, Christopher Columbus had acquired a very special book. If only a fraction of what was in it turned out to be true, then if his quest was successful Columbus believed he would become a very rich man.

The book was a travel journal, written more than a hundred years before by an Italian who had spent seventeen years living in the court of the Chinese Emperor. His name was Marco Polo. So incredible were the descriptions of China in Polo's journal, called *Il Milione*, that they fired Columbus's determination to find a new way to reach the other side of the world, replacing the arduous and highly dangerous overland route from Europe that could

take up to four years of slogging through deserts and over mountains with a caravan of camels, horses and mules.

Polo's account of life at the Chinese Emperor's court astonished the people of medieval Europe. Until then, hardly any of them knew of the existence of this civilization on the other side of the world. Polo said its palaces were 'so great and fine that no one could imagine finer'. Inside them were many great and splendid halls, 'all painted and embellished with work in beaten gold'. As for the potential for making money from trading spices, Polo described a wonderland dripping with ginger, cinnamon, cloves and 'other spices which never reach our countries'.

aquesta carauana es partida del imperi de sarra panar . alcatayo :

los munts de feber on naxe lo flum Flum Gyhô:

A Catalan map shows Marco Polo travelling in a caravan with his brothers along the arduous overland silk route that could take four years to cross, each way.

Spices such as pepper commanded extremely high prices in medieval Europe (see page 270). In 1511 a kilo of pepper, costing one gram of silver in the Far East, could fetch as much as thirty grams of silver by the time it reached Europe. It was the lure of cheaper pepper that initially attracted the Portuguese to find a new maritime route to the Far East.[1]

Columbus's copy of Polo's book still survives. In it he underlined the points that were most important to him: 'pearls, precious stones, brocades, ivory, pepper, nuts, nutmeg, cloves and an abundance of other spices ...'[2]

Columbus wished to sail westwards to the other side of the world, where he knew he must eventually reach the east coast of China. The only questions he was a little unsure of were how long it would take to reach this magical land; and what, if anything, lay between the south coast of Spain and the New World he was setting out to discover.

Amazingly, the riches of the civilization experienced by Polo and craved by Columbus were actually built on nothing much more than paper.

What we now take for granted as the most commonplace of all natural products was the making of medieval China. For more than 600 years, despite frequent trading links, these people kept the secret of paper-making entirely hidden from Central Asia, the Middle East and Europe. Only when those Chinese prisoners were captured by Arabian knights at the Battle of Talas in 751 (see page 238) did the secret eventually leak out, and even then it was not until hundreds of years later that paper mills became common in Europe.[3]

The Chinese people's expertise in paper-making helped them build the most technologically advanced civilization in the world. Perhaps the Chinese reverence for paper was a reaction against that brutal Great Burning of Books ordered by the paranoid and obsessive absolutist Emperor Qin Shi Huang in 213 BC – the one who built himself the terracotta army to defend him in the afterlife (see page 162). Shortly after his death, a new dynasty called the Hàn came to power. During more than 400 years of almost uninterrupted rule (206 BC–220 AD) its Emperors unified China around a new cultural system. At its core was the beginning of what came to be the world's biggest imperial bureaucracy, begun by Emperor Wu (141–87 BC), in which studying books became central to how people rose through the ranks of the imperial government.

Wu's reforms made it compulsory for court officials to study the classic books of Confucius (see page 160), teaching them that what mattered most was loyalty to the state. In 140 BC Wu ordered the first ever imperial examinations, whereby a hundred official appointments were to be based on performance in an academic test. Most of the candidates were commoners with no connections to nobility. For the first time people from the rural heartlands of China, with no money and from no privileged background, could, if they were clever enough, secure positions in government, positions that brought with them wealth, privilege and influence. Gaining a place bestowed huge honour on the successful candidates' homes, families and villages. As if to prove the point, Wu appointed several scholars from this intake as some of his most trusted advisers.

This simple idea had massive consequences for the effective government of China. By the time of Emperor He (ruled 88–106 AD), materials for writing had become essential tools of government administration. So when a bright court official called Cai Lun came up with a radically new method for making a cheap, versatile writing material, Emperor He took a great deal of notice. So impressed was he by Cai Lun's system of pulping, draining and drying plant fibres into thin matted sheets, usually made from the bark of mulberry trees, that Lun was rewarded with vast wealth and an aristocratic title.

In time the Chinese began to use this soft, cheap marvel of nature for almost everything from wrapping up precious objects to making umbrellas and parasols. Wallpaper, kites, tea-bags, playing cards and lanterns – they all made their first appearance in China.

Even the modern habit of using toilet paper has its origins here. The first evidence of its use comes from a report in 851 of an early Arab traveller to China, who says that the Chinese were 'not careful about cleanliness, and they do not wash themselves with water when they have done their necessities, but they only wipe themselves with paper'. [4]

When paper-making finally reached Europe, via Islam, from about 1200 onwards, the raw material was not the bark of trees, but old rags. Not until the nineteenth century did paper made from wood first start to appear, with the invention of steam-driven paper-making machines that could automate the process of extracting vegetable fibres from wood pulp.

Paper-based examinations increasingly formed the backbone of Chinese government by bureaucracy. Academic achievement turned the Emperor's court into a cauldron of creativity. Zhang Heng, chief astrologer to the Hàn Emperor Ān (ruled 106–125 AD), pioneered the concept of hydraulic engineering by creating a rotating mechanical model of the earth, sun, moon, planets and stars; the water-powered machine used an elaborate system of gears to make the celestial objects move at different speeds. In 31 AD another official, Du Shi, used the idea of waterwheels to operate a series of mechanical bellows that blew air into iron blast furnaces, increasing the temperatures inside to allow the manufacture of steel from cast iron for making better agricultural tools. By this time, iron, steel and salt had become huge monopolies controlled by rich ruling families that employed thousands of poverty-stricken peasants, many of whom were forced to leave their ancestral lands to take up jobs in marshes, factories or mines.

By the time of the Sui Dynasty (581–618 AD), the gap between rich and poor had grown so enormous that China's rulers were forced to intervene to prevent popular revolt turning to civil war. Reforms by Emperors Wen and Yang redistributed land to the poor, increased agricultural productivity by completing the grand canal that connected the two great alluvial valleys of the Yangtze and Yellow rivers, issued a new, more stable coinage, and improved defences against raiders from the north by reconstructing parts of the first Great Wall.

23:59:59

蕩料入簾

prestige – if only their families could produce sons with good brains. Emperor Wu's initial ideas of cultivating a bureaucracy stuffed with scholars were revived by Tang rulers, and competition amongst poor people to succeed in imperial examinations accelerated rapidly.

Now all male citizens were allowed to apply, not just those recommended by existing government officers. Buddhist monks were paid by local communities to pray for and teach the rural poor. The curriculum was extended to include everything from military strategy to civil law, agriculture and geography. Of course, the Confucian ethic of loyalty and obedience to the state ran through everything, providing all candidates, successful or otherwise, with an effective form of cultural alignment.

Success in exams could even earn a place in heaven. Tang rulers claimed descent from Laozi, the founder of Taoism, who lived at about the same time as Confucius, in the third or fourth century BC (see page 171). The Taoist pantheon of heavenly gods was based on a hierarchy that exactly mirrored the bureaucracy of imperial China and its examination system. Human souls were promoted or demoted as deities according to their earthly accomplishments. Appointments were made by the Jade Emperor, the supreme god, who ruled over every aspect of human and animal life.

Failure to make the grade came as the most bitter of disappointments – a more than likely reality when often as few as 5 per cent of candidates passed the tests. Some people took the news very badly indeed.

Huang Chao was one unfortunate poor man who failed his imperial exams. As a result he lost all faith in the justice of Chinese society, and entered the black market for salt-selling. With his resentment bubbling over into fury, he amassed anti-government supporters in the form of disgruntled peasants and merchants who had suffered badly from recent famines and droughts. In 875, Chao's makeshift army rebelled against the Tang government leading to a bitter nine-year civil war. By 880 the rebels had successfully sacked the capital city of Chang'an, causing the entire imperial court to flee. There, Chao briefly established a new

Chinese wealth was boosted by their invention of a new process for making paper from the bark of mulberry trees.

But reforms like these took high taxes and hard labour to achieve, oppressing the poor even further. To sustain itself, China's population had to be large, because it took huge numbers of poverty-stricken peasants to cultivate labour-intensive fields of rice. Civil unrest and dynastic rivalries dominated Chinese politics for more than 350 years following the fall of the Hàn Dynasty in 220 AD. The need to feed ever-growing numbers of soldiers added further to the burden of the poor. The census of 609 AD recorded a population of some fifty million people, most of them living desperate lives.

(Opposite) The Diamond Sutra is the earliest printed text with a known date, 868 AD.

Paper rescued Chinese civilization from perpetual war and civil unrest. It allowed the Sui and their successors the Tang Dynasty (618–907 AD) to expand the examination system and provide poor people with the hope of wealth, privilege and administration. Although the Tang recaptured their city four years later, it was a former commander of Chao's army, Zhu Wen, who eventually brought about the final collapse of the dynasty in 907.

Apart from the exceptional reaction of Huang Chao, Tang rulers successfully deployed the imperial examination system to offer the rank-and-file poor a prospect of state employment. It also meant they had the pick of the best brains in the country. But scaling up the enterprise was a problem. Although mass-produced paper was cheap, thanks to Cai Lun's invention, books were not. Until c.800 every single one had to be hand-written by some patient, skilful and well educated person – usually a Buddhist monk.

The spread of Buddhism from India into China went hand in hand with what some people regard as the second most powerful invention after paper – printing. Books mass-produced by a technique called block printing are known to have originated in China thanks to a remarkable discovery made by Aurel Stein, a Hungarian archaeologist who in the first half of the twentieth century travelled more than 25,000 miles on horseback and foot through inhospitable deserts, plains and mountains because he wanted to find out more about the early culture of Central Asia.

It was worth the effort. His finest discovery was at the Caves of the Thousand Buddhas, located at an oasis on the edge of the Taklamakan Desert in north-west China (see map on page 163). He came across a system of some 492 temples carved out of the rock by pilgrims, monks and travellers along the Silk Road, the ancient overland trade route originally established by the Hàn Emperor Wu in 138 BC (see page 156). The route allowed merchants to travel all the way from the Chinese imperial capital of Chang'an to the east coast of the Mediterranean, although the trek could take up to four years each way, and was often blocked by nomadic raiders from the Eurasian steppes.

Several accounts of Roman embassies to China are recorded. In the *Hou Hanshu*, a history of the Hàn, the Roman Emperor Antonius Pius is said to have sent an overland convoy along the Silk Road which reached the Chinese capital in 166 AD.

Western incense and African ivory were exchanged for Chinese silk.

After a series of successful conquests in the west, Tang rulers reopened the Silk Road in 639 AD, which is about the time many of these Buddhist shrines were built.

When Stein arrived at the caves, in 1907, he heard rumours of an enormous cache of ancient documents that had recently been found hidden behind a temple wall. A shy Taoist priest called Wang Tao-shih had taken it upon himself to become their self-appointed guardian, and had locked up the treasures in a storeroom. When Stein persuaded the priest to show him inside the tiny room, measuring no more than nine feet square, he saw that it was packed floor to ceiling with precious manuscripts, apparently thrown away by monks centuries before, when the advent of mass printing made hand-written scrolls redundant. Here, perfectly preserved by the dry atmosphere of the Central Asian desert, were thousands of untouched ancient texts, some dating from as early as the fifth century AD. After weeks of careful diplomacy Stein managed to purchase as many as 40,000 of these documents, in return for a generous contribution towards the temples' further restoration.

The jewel in the crown of Stein's hoard, now kept in the British Museum, is the world's oldest known complete, dated, printed book. It is a Buddhist text called the *Diamond Sutra* printed with the date of 868 AD – towards the end of the Tang Dynasty. It is made of seven sheets of white paper, pasted together to form a scroll of just over seventeen feet. The manufacturing technique used is called 'woodblock printing', in which sheets of paper were pressed on to wooden blocks intricately carved with words and illustrations. Although making books this way required advanced craftwork skills to carve the wooden blocks, once these were made the number of copies that could be printed was almost limitless.

China wasn't the only state in the Far East to develop the arts of printing and paper-making. Korea had paper by 604, and Japan by 610. It is from about this time that Japan's emerging imperial family, the Yamato, began to model itself on the imperial Chinese court, using paper on which to write down a new generation of creation myths linking its origins to the ancient shamanistic Japanese Shinto, god of the sun.[5] Japanese imperial rulers also introduced paper-based legal reforms, called the Taihō Code (701 AD), that followed Chinese Confucian ideals of obedience to the state. Chinese influence from this time was also evident in a new Japanese imperial capital built at Nara in 710, almost exactly following the grid-like pattern of the Chinese imperial capital of Chang'an.

But China's golden age of innovation in science and technology really took off with the founding of the Song Dynasty by Emperor Taizu, who seized the imperial throne in 960. According to the *Song Shi*, a history of the dynasty written in 1345, he spoke to his military commanders at a victory banquet in words that sound remarkably like a modern manifesto for sustainable government:

> *'The life of man is short. Happiness is to*
> *have the wealth and means to enjoy life,*
> *and then to be able to leave the same*
> *prosperity to one's descendants.'*

One of Taizu's first acts was further to expand the concept of a scholastic imperial bureaucracy. By the end of his reign, in 976 AD, roughly 30,000 candidates per year – all of whom had already passed a pre-qualifying test (called *jinshi*) – were taking the imperial administration's exams. By the end of the eleventh century this had risen to 80,000, and by the end of the Song Dynasty (1279) it reached 400,000, creating an enormous intellectual population.

Printing was pressed into overtime, with more than 500 classic Confucian texts, dictionaries, encyclopaedias and history books carved on to thousands of wood blocks to provide mass-produced study materials for candidates.[6] At least a thousand schools were established throughout China to help prepare students for their civil service exams.

It wasn't just printing books that invigorated cultural and academic life in Song times. Thanks to

an expansion of rice cultivation, by 1102 the Chinese population had more than doubled, to over a hundred million people.[7] Traditional currency, in the form of either silk bolts or copper coins, had become difficult to carry, short in supply and expensive to manufacture. So in the 1120s the Chinese government turned to its most favourite invention. It used woodblock printing to make the world's first known banknotes called *jiaozi*. Within a decade several state-run paper-money-printing factories had been established, employing many thousands of workers.

The idea of printing paper money would have fascinated Europeans, for whom nothing like banknotes appeared for many hundreds of years. The first Western state to adopt banknotes was Sweden in 1661, then America (Massachusetts) in 1690, France in 1720, Russia in 1768, Saxony in 1772, England in 1797 and Prussia in 1806.

Marco Polo was utterly astonished by the Chinese Emperor's ability to manufacture paper money:

> *'He causes every year to be made such a vast*
> *quantity of money which costs him nothing*
> *that it must equal in amount all the treasure*
> *in the world. Pieces of paper are issued with*
> *as much solemnity and authority as if they*
> *were of pure gold or silver ... When all is*
> *duly prepared the chief officer smears the Seal*
> *entrusted to him with vermilion and impresses*
> *it on the paper, so that the form of the Seal*
> *remains on it in red. The money is then*
> *authentic. Anyone forging it would be*
> *punished with death.'*

But in 1127 disaster struck. The ingenious Song government was rudely interrupted by a horde of horse-breeders from Manchuria, the Jurchens, who double-crossed them after defeating a mutual enemy called the Liao in 1125. Their victory over the Song, using innovative siege engines to break the imperial capital's walls, resulted in China being split between a northern region ruled by the Jurchens (who became the Jin Dynasty), and a southern region under the Song, who fled south of the Yangtze River to a new capital at Lin'an.

These events triggered an arms race of epic proportions. Song rulers challenged their best scholarly brains to come up with every conceivable way of creating technological superiority in warfare to guarantee success in the event of further attacks. Their quest led directly to the development of the world's first firearms.

Charcoal, sulphur and mineral ores were first mixed into a primitive form of gunpowder by Taoist monks in the Tang Dynasty, in c.850. These holy men had been charged by their imperial masters to find an elixir for everlasting life, but in the course of their experiments they stumbled across a combination of chemicals which, they warned, could result in disastrous consequences:

'Some have heated together sulphur, realgar and saltpeter with honey; smoke and flames result, so that their hands and faces have been burnt, and even the whole house where they were working burned down.' [8]

Such words must have fired the hearts of those whose preference was for power over others on earth. The first known image of a firearm in actual use dates to c.950. For years the significance of a silk banner found at the Caves of the Thousand Buddhas lay unnoticed in a Paris museum until it was rediscovered in 1978. It shows an army of demons trying to distract the Buddha from attaining Enlightenment. Among the weapons they use is a fire-lance – a long pole held by a demon wearing a head-dress of three serpents. There is a cylinder at one end from which flames spout forth. Beneath him, to the right, is another figure with a serpent entangled in his eyes and his mouth. He is about to throw a small bomb or grenade, from which flames are already pouring out. [9]

Using such clues, brilliant minds at the court of the Southern Song Dynasty (1127–1279) were able to construct an array of powerful new weapons, ranging from catapult bombs and cannon to flame-throwers. Their first use in anger came at

Hellfire: this image of the Buddha under attack from demons with fire-lances and incendiary bombs (c.950 AD) is the earliest known depiction of the use of firearms.

23:59:59

two battles in 1161 on the Yangtze River, the border between the empires of Jin and Song, when Jin forces threatened to fulfil the Southern Song's worst nightmare, and push southwards in an attempt to eliminate them altogether.

According to the *Jin Shi*, a history of the Jin Dynasty, the Jin's admiral at the Battle of Tangdao, Zheng Jia, was utterly astonished by what he saw:

> *'Zheng Jia did not know the sea routes [among the islands] well, nor much about the management of ships, and he did not believe that the enemy [the Song] was near. But all of a sudden they appeared, and finding us quite unready they hurled incendiary gunpowder projectiles on to our ships. So seeing all his ships going up in flames, and having no means of escape, Zheng Jia jumped into the sea and was drowned.'*

For the next hundred years the Southern Song and their weapons ruled supreme on land and sea.

The Song's shift south had other important consequences. Court officials set about commissioning China's first permanent standing navy in 1132, after realizing that for strategic defence the Yangtze River was now effectively their new Great Wall. By 1130 slingshots throwing gunpowder bombs were ordered as standard on all ships. A new generation of paddle-wheel ships, powered by human treadmills, some with as many as eleven paddles on each side, dramatically increased manoeuvrability in the tight conditions of river-based warfare. By 1203 many of these warships had been reinforced with the addition of iron-plated armour.

Naval power gave the Southern Song complete control of the East China Sea, previously dominated by Hindu rulers called the Cholas, a Tamil dynasty that ruled southern India and Sri Lanka between then tenth and thirteenth centuries. Foreign trade was essential for the survival of the Song, since the mountainous terrain of southern China was less suitable for widespread agriculture than the north. With overland trading via the Silk Route firmly in Jin hands, trade by sea was essential for the survival of their civilization.

Such conditions led to the first known portfolio investment schemes, now so common in the modern Capitalist world. Instead of putting all their money into one trading venture, Chinese merchants would collaborate in guilds and spread their money across several expeditions, so that if one were lost, it came as a complete disaster to no one (other than the crew, of course). According to one source, such arrangements proved highly lucrative: 'they invest from ten to a hundred strings of cash, and regularly make profits of several hundred per cent'.[10]

Goods were carried on junks that sailed across the Indian Ocean and into the Persian Gulf. These vessels were waterproofed with tong oil, secured by watertight bulkheads (which meant that if one were damaged the ship would not sink) and equipped with stern-mounted rudders for improved steering. Some of them were powered by a combination of sails and oars, and were able to carry huge cargoes and several hundred men.

Their sailors took with them that most precious of all instruments for seafarers – a compass. The idea of magnetizing a needle by rubbing it against silk and then placing it inside a straw and floating it in water to make a primitive compass had been known about in China since at least the first century AD.[11] In 1044 a Song Dynasty manual on military techniques called *Wujing Zongyao* describes how this knowledge was used to create a south-pointing chariot to help guide troops in gloomy weather and on dark nights (the same book also included a recipe for how to make gunpowder). The biggest breakthrough came a few years later, when court official Shen Kuo, who successfully passed his imperial state examinations in 1061, described how to magnetize the needle of a compass and use it for navigation by stringing it up on a thread of silk.[12] His compass is known to have been used on Chinese ships from as early as 1111.[13]

Astonishing technological breakthroughs were made possible by an exchange of knowledge in an imperial court that cultivated a sophisticated knowledge-base of paper, printing and books. By now Chinese civilization had substantially enhanced man's power over nature by harnessing gunpowder weapons and twenty-four-hour navigation (in all weathers) by magnetic needles.[14] But no dynasty lasts for ever, however ingenious its technology. Nature's cycles were probably the ultimate cause for the collapse of the Song in 1279 AD.

For settled Chinese rulers, danger usually came from the north – which is why the first Great Wall (originally constructed in the fifth century BC, during the Warring States Period, but added to by successive dynasties including the Tang and the Northern Song) was of such strategic importance. Nomadic invaders, like the Jurchens, were brilliant horsemen who bred thousands of animals for trade and war. When environmental conditions got tough in the barren Eurasian steppes, as the ancient Persians and Greeks knew only too well (see pages 172–73), hordes of invaders could be expected to stream across from the east and down from the north, displacing whatever civilizations lay in their way.

This time it wasn't drought that caused massive nomadic migrations, but warmer conditions and increased rainfall between about 1000 and 1200 AD, which meant that more people could be supported by pastoral existence on the plains.[15] But the steppes have strict limits on how many people they can comfortably sustain. In terms of calories produced per acre, plant foods outperform animal foods by a factor of about ten to one. People who live off herds of sheep, goats and horses need much more space than those who raise crops to sustain themselves.[16]

Beginning in about 1200 AD, the climate began to cool. North Atlantic pack ice started advancing southwards towards Iceland, and Greenland's glaciers expanded. (This climate change signalled the start of what is called the Little Ice Age. Cold winters and unpredictable summers continued across Europe until the middle of the nineteenth century.) According to oxygen isotopes captured in Greenland ice cores, the first signs of cooling appear to have been in high latitudes.

Mongol horsemen who burst southwards between 1205 and 1225 would have been early casualties, as bitter Arctic air invaded the heart of Asia with devastating effects. Steppe lands, now populated with more animals and people than ever before thanks to years of increased rainfall, simply couldn't cope. In a cooler climate, their pastoral yield was insufficient. People living in the steppes probably had only one choice if they wished to maintain their numbers and lifestyle. They had to unite and conquer new, more fertile lands. Fortunately for them, they found just the man to help them out. His name was Genghis Khan.

Born in c.1160, this son of a nomadic chief was groomed from an early age to know how to survive the harshest conditions. After his father was poisoned by a neighbouring tribe, the boy Genghis claimed to be the clan's new chief. But, not wanting to be ruled by a boy, his tribe cast him and his family out. After murdering his half-brother with a poisoned arrow, apparently for stealing food, Genghis quickly asserted himself as head of the household. His lucky break came thanks to the generosity of a childhood friend, Toghrul, who had been officially recognized as ruler (Wang Khan) of a cluster of Mongolian tribes by the Jin Dynasty in 1197. After Genghis's wife, Börte, was captured by a rival tribe, Toghrul lent him 20,000 warriors to rescue her, a feat which Genghis successfully achieved.

By 1206 Genghis had united hordes of rival Mongolian tribes (Merkits, Uyghurs, Keraits and Tatars) through a mixture of diplomacy, charisma and leadership skills. Taking climate changes into account, however, his rise to power can also be explained by his being the right man in the right place at the right time. The survival of this population of some 200,000 people depended on a coordinated nomadic expansion.

Genghis was a brilliant military planner and a strict disciplinarian who demanded toughness, dedication and loyalty from all his men. He organized his forces into decimal units of ten, one hundred, one thousand and ten thousand. Soldiers who performed well in battle rose through the ranks. Cowardice was not tolerated. Each unit of ten had a leader who reported up to the next level. If one soldier deserted, his whole unit was executed. If a unit of ten deserted, the whole group of a hundred was slaughtered.

The Mongolian army was extraordinarily self-sufficient. Every soldier's family travelled with him, turning the horde into a huge band of travellers with no need or incentive to head back home (unlike the army of Alexander the Great, for example – see page 192). Its military tactics reflected nomadic instincts of how to hunt animals in packs. Mongol forces spread out in a line, surrounded an entire region and then closed in from all sides, driving everyone in the area together so no one could escape.

Careful planning in military councils (called *kurultai*), and excellent reconnaissance conducted on fast steeds, meant success was swift. The first casualties were the Jin, the enemies of the Southern Song. Genghis declared war in 1211, and advanced with two armies of about 50,000 bowmen each, but found he could not effectively take the Jin cities. After enlisting the help of Chinese engineers and Islamic warriors, who had learned how to build siege engines and giant

Genghis Khan can be seen drawing a long black blade while fighting in the Chinese mountains, in this image from a fourteenth-century manuscript.

catapults (trebuchets) from the Byzantines, Genghis learned how to make them himself, mostly using on-the-spot resources such as nearby trees.[17]

Whenever Genghis faced an enemy city he gave them a simple choice: surrender or die. He was a man of his word. To proud rulers who offered resistance, he showed no mercy. If a ruler agreed to submit, his people were spared, but total loyalty was expected in return. When in 1209 a ruler in Turfan, an oasis city now in the autonomous region of China, submitted to Mongol authority, not only did Genghis spare the lives of his people, the Uighurs, but many of them went on to form the backbone of the Mongols' imperial bureaucracy, introducing writing and literacy to these previously uneducated nomads.

By 1213 the Mongols had advanced as far as the Great Wall, and within two years they had charged into the heart of northern China. In 1215 they successfully besieged and sacked the Jin capital at Yanjing (now Beijing).

Having subdued the Jin, Genghis's anger was next roused by an apparently unprovoked insult from the Islamic ruler of Khwarazm, an empire that stretched from the western edge of China to the Caspian Sea. In 1219 Mongol armies massed on the eastern edge of the Islamic world. Each city was given a choice: surrender or die. By 1223 all Khwarazm was under Mongol control. As ever, resistance was punished without mercy. When Genghis seized the border town of Otrar he not only executed its inhabitants but poured molten silver into the ears and eyes of its haughty governor, one who had refused to surrender.

Genghis then headed north into Russia, where his forces split into two and conquered Georgia and the Crimea. On their way back to Mongolia they defeated a Russian army led by six princes, including the ruler of Kiev. As was customary in Mongolian tradition, the princes were given a bloodless execution: they were crushed to death under the weight of a banqueting platform while the Mongolian generals ate their victory feast.

In 1225 Genghis returned to China, where he again fought and subdued the Jin. Shortly after, he died – no one quite knows how. Some say he fell off a horse. Another legend has it that he was murdered by a beautiful Tangut princess who, as they were about to make love, castrated him with a knife, hidden inside her body, in revenge for the murder of so many of her people.

At the time of Genghis's death the Mongol Empire stretched from the east coast of China to the Caspian Sea. But his children were to take it further still, establishing empires in Russia, Siberia and Central Asia.[18] By 1241 their armies were ready to flood into western Europe following victory over Polish, German and Hungarian forces at the Battle of Mohi, southwest of the Sajó River, in Hungary.

By this time the Mongol Empire was the largest contiguous civilization that the world had, and still has, ever seen. Its inexorable expansion was only halted by the death of Genghis's successor Ögedei Khan in 1242, because tradition said that Mongol leaders must convene a grand council to confirm the appointment of the next Great Khan. So they retreated, at least for a while, to approve their new leader, and Europe was spared.

In the face of the Mongols' ferocious military conquests even the Song Dynasty with its gunpowder couldn't hold out for ever. Möngke Khan, Genghis's grandson and later the fourth Great Khan, invaded Korea and began his first attacks against the Song in a campaign to reunite China. But in a twist of fate, just as his brother Hulagu was on the point of conquering Syria, opening up the immensely rich lands of Egypt, a rock toppled off a cliff and killed Möngke while he was attacking Song forces in Chongqing, western China (some sources say he was killed by cannon fire or some other injury). Once again the Mongol leaders were recalled for a meeting to decide who to approve as Great Khan. This time their western forces would never return to threaten invasion in Egypt or Europe again.

The last Great Khan, another of Genghis's grandsons, Kublai, eventually defeated the Song. By now the secret of gunpowder had spilled into Mongol hands and was being used on both sides.[19] The Song's last stand took place at the Battle of Yamen in March 1279. Their admiral had strung more than a thousand ships together in a defensive line in a bid to seal off the bay of Hong Kong from Mongol forces. But after being caught off guard the Chinese were routed, and it was left to one of the dynasty's loyal court officials to ensure that the last Emperor, a nine-year-old boy called Zhao Bing, under no circumstances fell into the hands of the enemy. When he saw that all was lost the official grabbed the boy, and together they jumped into the sea. Seven days later hundreds of thousands of corpses were seen floating on the surface, including that of the boy Emperor.[20]

After Kublai Khan founded the Yuan Dynasty (1271–1368) he built a new capital called Zhongdu near to the modern city of Beijing and China was united once again. Its resources now ranged from spices, silk and paper money to books, printing, gunpowder, firearms, armoured battleships, the navigational compass and steel.

Visiting Europeans like Marco Polo couldn't fail to be impressed. The inventiveness, scholarship, power, wealth and glory of this civilization was matched only by the contrast it provided with the climate of desperation across medieval Europe, much of which, more than a millennium after the break-up of the Roman Empire, had been reduced to ruin – a wilting wilderness of disease, famine and war.

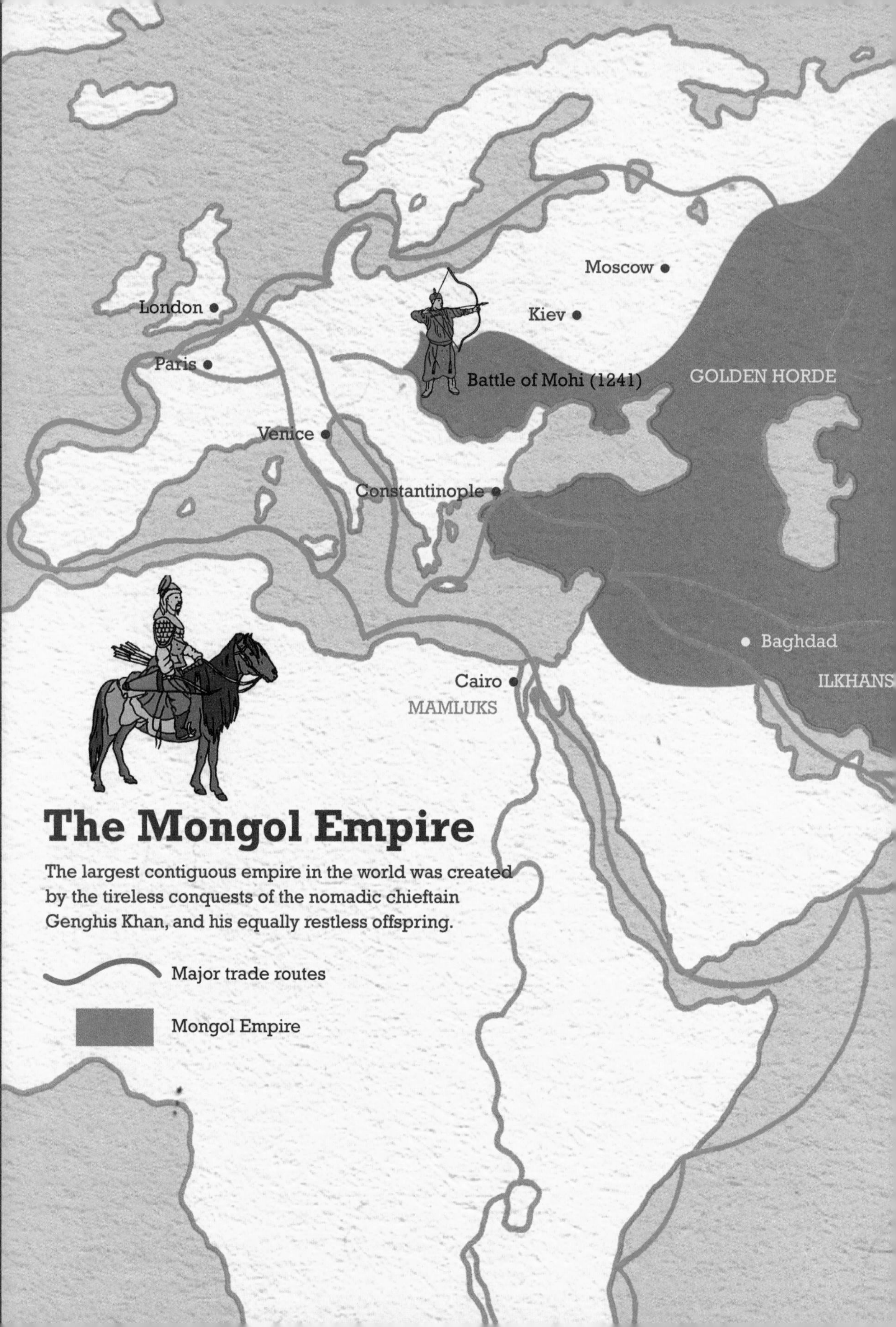

Moscow •

London •

Kiev •

Paris •

Battle of Mohi (1241)

GOLDEN HORDE

Venice •

Constantinople •

• Baghdad

Cairo •

ILKHANS

MAMLUKS

The Mongol Empire

The largest contiguous empire in the world was created
by the tireless conquests of the nomadic chieftain
Genghis Khan, and his equally restless offspring.

Major trade routes

Mongol Empire

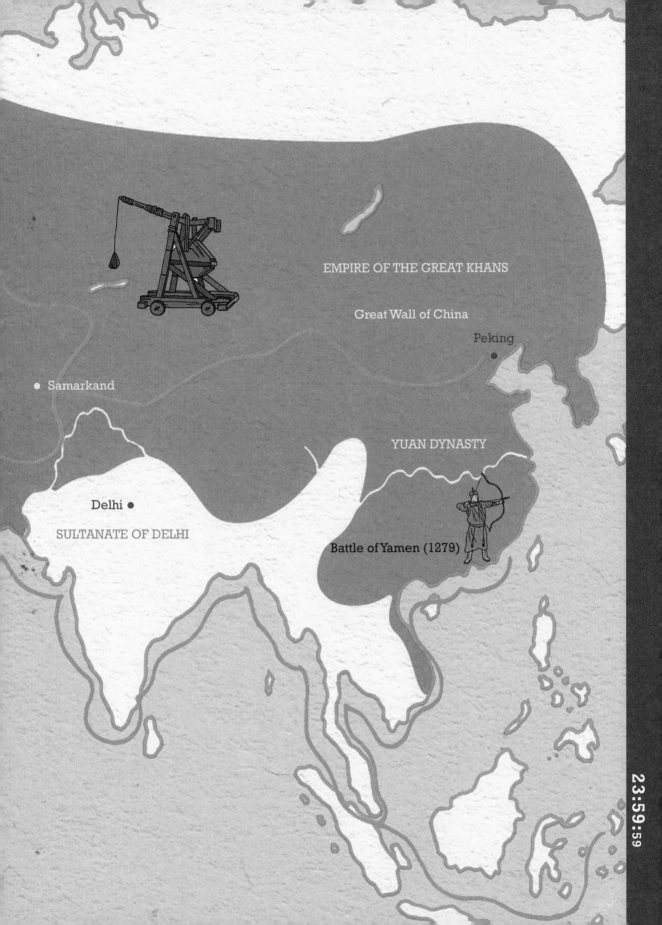

EMPIRE OF THE GREAT KHANS

Great Wall of China

Peking

• Samarkand

YUAN DYNASTY

Delhi •

SULTANATE OF DELHI

Battle of Yamen (1279)

Medieval Misery

How plague, invasions and famine impoverished Christian Europe, which found itself surrounded by Islamic civilizations, impenetrable desert and endless blue seas.

COULD IT BE that the fluttering of a butterfly's wings in China set off a thousand-year-long storm, thousands of miles away, in far-off Europe? Since the start of the twentieth century scientists have been studying the idea, called Chaos Theory, that an apparently tiny change in one part of the world can trigger a dramatic set of events that severely or even permanently disrupts another.

Something like this seems to have happened to medieval Europe, starting at about the time the whirlwind Roman Empire was in its final decline. A military victory by the Chinese Hàn Dynasty in about 100 AD over a northern Mongol tribe called the Xiongnu could well have been the butterfly.

Little evidence survives of what happened to these nomadic horsemen, who were forced westwards by the Hàn forces, until the arrival above the Black Sea of an army of Chinese-looking warriors in the fourth century AD. Archaeologists have found large cauldrons buried under riverbanks, exactly like those used by the Xiongnu tribes of western China. From the accounts of early Byzantine writers, the appearance of this army's famous leader Attila suggests an Eastern origin: 'short of stature, with a broad chest and a large head; his eyes were small, his beard thin and sprinkled with grey; and he had a flat nose and a swarthy complexion showing the evidences of his origin'.[1]

The Huns, as they were called, who charged into Europe during the fourth century AD, were vicious warriors with a powerful technology for fighting. Their small, lightweight, composite bows were legendary for their power, speed and range – ideal for use on horseback. The secret lay in the material they were made of, which generated maximum elastic thrust: the strong but flexible horn of the water buffalo, animals which pulled ploughs in the paddy fields of the Far East.

By 406 AD the butterfly effect was turning into a full-scale rout. Hun archers had already displaced Gothic tribes living north of the Black Sea, who in turn charged westwards and destroyed the Roman army at the Battle of Adrianople (378 AD), which some historians say triggered the final collapse of the western Roman Empire.[2]

On 31 December 406 a terrifying alliance of Germanic and Asian tribes surged across the frozen River Rhine at Mainz, and into western Europe. The Huns reached as far as Orléans in France (in 451) and Ravenna in Italy; meanwhile the Visigoths from eastern Germany went for the jugular, sacking Rome itself in 410 before establishing their own empire across southern France and Spain. Next to come were the Vandals, another east German tribe, who streaked down Spain, crossed into North Africa and along to Carthage, where they built themselves a fleet. They then sailed up the Mediterranean and invaded Sicily, Sardinia, Corsica and Malta before launching their own devastating raid on Rome in 455.

Europe's misery was no temporary setback. A huge effort to restore law and order and to unify the broken territories of the Roman Empire was undertaken by the Byzantine Emperor Justinian I (ruled 527–65 AD), ruling from the new 'Roman' capital in Constantinople. He sent armies and generals all over Europe – to Spain, Italy and North Africa – to reclaim lands seized by the nomadic horsemen, besides having to confront periodic aggression from age-old Persian enemies to the east. Justinian met with partial success, re-forming the legal basis of the Roman Empire, ejecting the Vandals from North Africa and defeating the Ostrogoths, another Germanic tribe, at Ravenna by 552.

It didn't last long. Three years after Justinian's death another wave of raiders came down from the north in the form of the Lombards, originally Baltic people who had settled in Germany near the Danube River. By 561 they had installed themselves as rulers of Italy, a conquest that lasted more than 200 years. Any African and Spanish gains made by Justinian were finally overrun by Mohammed's warriors, who by 732 had conquered all Spain and half of France, only to be stopped from taking the rest of western Europe by the Frankish tribal leader, Charles Martel (see page 253). The Franks were early converts to Catholic Christianity thanks to their founder Clovis I (died 511 AD), who became the Pope's political protector. Clovis is considered by many modern scholars to be the founder of France.

Complete catastrophe for early Christian Europe was only just averted when Constantinople, its new imperial capital, narrowly escaped conquest by the forces of Islam. More than 80,000 Muslim warriors from Damascus descended on the city in 718, supported by some 1,800 ships. Their siege failed thanks only to nature's intervention. The winter of 717–18 was the coldest in living memory, which according to one contemporary chronicler caused

Attila the Hun (406–453 AD) was a fearsome enemy of the late Roman Empire. He invaded as far as Orléans (in Gaul) before eventually being defeated at the Battle of Chalons in 451.

Climate Changes in Medieval Europe

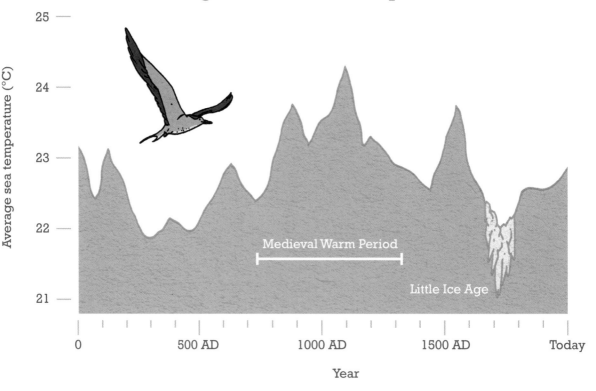

Rising and
falling
Atlantic sea
temperatures
reflect
medieval
Europe's
climactic
see-saw.

appalling hardship for the hungry Muslim forces, who were forced to resort even to cannibalism:

> 'The wind of death grabbed them. The hunger oppressed them so much that they were eating the corpses of the dead, each other's faeces and filths. They were forced to exterminate themselves, so that they could eat. They were looking for small rocks, they were eating them to satisfy their hunger. They ate the rubbish from their ships.'[3]

Famine, disease and the cold saved the great city of Constantinople, preventing Europe from being completely swept away by the tidal wave of Islamic conquest unleashed by the teachings of Mohammed – it was another moment, say some historians, of 'macro-historical' importance.[4]

Over the course of a few hundred years the mighty Roman Empire had been reduced to ruins. By the seventh century the population of Europe was in steep decline, falling from about 27.5 million in 500 AD to just eighteen million 150 years later.[5]

To make matters worse, bubonic plague had laid a major claim on European lives.[6] The disease ravaged the eastern Roman Empire, beginning in 541–42, severely complicating Justinian's efforts to reassert imperial control. An estimated 5,000 people a day died in Constantinople, where the disease claimed the lives of up to 40 per cent of the city's total population. It then spread across much of Europe, with repeated outbreaks over the next 300 years killing, it is thought, as many as twenty-five million people in all. Fields were abandoned, forests regrew and the economy went into steep decline. Outside the occasional kingly court, life in Europe between 350 and 750 AD was extremely nasty, totally brutish and usually rather short.

Scapegoats weren't hard to find. They ranged from Attila the Hun, who sparked off the Germanic invasions, to Jesus Christ, the son of God, who had promised to save mankind but who now seemed to have had second thoughts. One of the most popular movements of the time was a strand of Christianity called Arianism, which subtly relegated Christ's role, suggesting that not even He could be equal to almighty God, arguing that therefore the Catholic idea of the Holy Trinity must be a mistake.[7]

In modern times the blame has been put down to the random and unpredictable forces of nature. A massive volcanic eruption in Krakatoa, Indonesia, recently redated to 535–36 AD, is thought to have caused dramatic swings in the climate after thousands of tonnes of ash were thrown into the earth's upper atmosphere. The sun may have been blotted out for months, causing floods, famine, droughts and storms. Evidence from tree rings and ice cores confirms that temperatures swung wildly during these years.[8]

Bit by bit, hopes of a revival in central authority to help protect all Europe began to re-emerge thanks to the success of the Franks under Charlemagne. Building on his grandfather Charles Martel's victory against the Arabs in 732, Charlemagne successfully extended Frankish rule throughout France and into Italy. By the beginning of the ninth century he had repelled the Lombards and restored power to the Pope in Rome. Charlemagne was crowned Holy Roman Emperor by Pope Leo III on 25 December 800. Leo now reasserted the Pope's claim to be Christ's living apostle on earth, first established by Pope Leo I at the Council of Chalcedon in 451. Perhaps a combination of Papal spiritual supremacy backed up by political power from a newly restored Holy Roman Emperor could re-establish Europe's order and balance, in the way that the Song Dynasty achieved in China.

As if to celebrate Europe's recovery, even the weather perked up. Cold, harsh winters gave way to a 'Medieval Warm Period' (see diagram opposite). Temperatures between 800 and 1300 AD became as mild as they are today. Grapes grew as far north as Britain and the ice sheets retreated, opening up new sea passages to the north-west.

With the previous breakdown of the medieval economy, strong new rulers like Martel and Charlemagne had to use barter to raise troops to re-establish central rule, law and order. The milder climate strengthened their hand with a revival in agricultural wealth. Rulers were able to demand military service in return for granting land which provided a source of income from farming. Knights like those who came from Persia were Europe's new imperial shock troops, but they were expensive to equip. The cost of buying and then maintaining horses and armour meant that only those with good incomes could afford to take part, hence the need for grants of land in exchange for service.

This new political fabric, called feudalism, was given a significant shot in the arm by small butterfly wings still flapping on the other side of the world. High technology for harnessing animal power, in the form of cast-iron stirrups for horses, originated in China sometime in the third century AD. They are thought to have come to Europe via the nomadic Avars of Central Asia, spreading northwards to Baltic tribes near Lithuania and westwards via Islamic warriors pushing towards Constantinople, Spain and France. Once in use by the armies of the Frankish kings of Charlemagne's generation, stirrups had a massive impact on the ability of rulers and their henchmen with horses to exert a new form of social and military control.

The feet of a medieval knight firmly locked into his stirrups, as depicted on this eleventh-century Italian chesspiece.

Stirrups allowed warriors on horses fully to tap into animal power by 'welding horse and rider into a single fighting unit capable of a violence without precedent', immensely increasing their ability to damage the enemy. According to one expert: 'It made possible mounted shock combat, a revolutionary new way of doing battle.'[9] Men with horses and stirrups now became powerful rulers in their own right. Tied into armies through oaths of allegiance and into feudal contracts in exchange for land, order of a sort was restored through the power of a new class of knights-cum-landlords.

But feudalism fuelled friction between local rulers and central kings. In 864 Charlemagne's grandson Charles the Bald (ruled 840–77 AD) issued an edict ordering the destruction of all private structures throughout France that had been erected without his permission. A massive increase in castle-building began soon after his reign so that knights and landlords could protect themselves from Viking invaders equipped with their own stirrups and horses. Castles reinforced the local power that feudalism had fragmented into the hands of individuals and their families, causing the newly resurgent Holy Roman Empire temporarily to stall and shaping the character of European politics for centuries to come.

From as early as the 790s Viking raiders also began to take advantage of improved climatic conditions, bringing with them formidable sailing skills. Viking longboats reached wherever water and oarsmanship took them.

The Vikings were the chameleons of the medieval world – one minute merchants, next pirates, and before long conquerors. For Viking adventurers, trade and raid were different sides of the same coin. By 839 they had sailed deep down into the heart of Europe through its extensive networks of naturally interconnected rivers. They settled along the Danube to Kiev and beyond, establishing themselves as the Rus, a word that is thought to come from an old Norse term, *rods*, meaning 'men who row'. Islamic sources say they then subjugated the Slavic peoples, who provided them with a rich source of slaves who were traded along a network that reached across the Black Sea as far as Islamic Baghdad (the name 'Slav' possibly comes from their being traded as slaves by the Rus).

According to one Persian historian: 'They harry the Slavs, using ships to reach them; they carry them off as slaves and … sell them. They have no fields but simply live on what they get from the Slavs' lands.'[10]

In 845 Viking chief 'Hairy-Pants' Ragnar sailed with a fleet of more than 120 ships and 5,000 men down the River Seine and captured Paris. He departed only after receiving what was then a fantastic sum of 7,000 pounds of silver from the Frankish ruler Charles the Bald. Next stop was across the Channel to England, where things got rather more personal. After invading Northumberland in 865, Ragnar was captured by King Aelle II, who had him thrown into a pit filled with poisonous snakes. So angry were his sons Ivar the Boneless and Ubbe that the following year they crossed the North Sea with a large army, sacked York and captured King Aelle. He was sentenced to death and, according to legend, executed in the most traditional of Viking ceremonies, the Blood Eagle ritual.[11] They cut open his spine with a knife, broke his ribs until they stuck out like the wings of a spread eagle, and then ripped out his lungs. Finally they sprinkled salt into his gaping wounds.

These butchers then moved south, assassinating King Edmond of the Anglo-Saxons along the way. (Originally from north Germany, Anglo-Saxon people had gradually populated parts of Britain since towards the end of the Roman Empire when Germanic people were hired into invading legions to subjugate the British.) The murder of their king provoked what later became the legendary wars for control of all England between Alfred the Great and cousins of the Vikings called the Danes.

The most successful Viking settlement in Europe was around Rouen, where they integrated into Frankish culture, chameleon-like, adopting both feudalism and Christianity. They became known as Normans (originating from the word 'Norseman' or 'Northmen'). Their most famous ruler, William of Normandy, built an army based on feudal ties, horses and stirrups. With the seafaring expertise gained from his people's Viking ancestry, William was confident of victory in his attempted conquest of England, which began with the Battle of Hastings in 1066.

Feudalism thrived on successful conquest. Kings granted newly captured lands to their knights and nobles, who in return provided arms for further conquest and war. As a way of trying to stamp his own authority in England, and to define his overall wealth from the outset, William the Conqueror, as he became known, ordered a comprehensive survey of his new kingdom – the result was the famous Domesday Book of 1086.

For years historians have pored over this unique snapshot of life in medieval England, that details the livings of more than 265,000 people, from farmers, millers and blacksmiths to potters, shepherds and slaves. It shows that 20 per cent of the land was owned by the King and his family, 26 per cent by the Church, and the remaining 54 per cent by about 190 members of the feudal aristocracy, of whom twelve owned 25 per cent between them. Recently, conservationists have been able use it to make an assessment of the impact of medieval agriculture on the natural landscape, since among the questions asked by William's inquisitors was how much woodland, pasture and meadowland was owned by each person. Their analysis is startling.

It seems that a staggering 85 per cent of the English countryside had already been cleared by this time, for use as pastureland for domesticated animals and arable land for growing crops.[12] To support such large-scale production, some 5,624 watermills were in use for grinding grain in 3,000 separate communities. England's remaining forests and woodlands were under tight control, with many reserved especially for royal use and hunting, further indicating that by then substantial deforestation had already taken place.

The arrival of more new technology originating from the Far East had gradually transformed the agricultural fortunes of Europe, and ultimately that was what led to the destruction of huge tracts of natural forests and woodlands. Metal ploughs had been used in China from about 100 BC. The big difference between these ploughs and those used in Europe was that they were built on wheels

– probably originating from a modification of the horse-drawn chariot. Wheeled ploughs meant blades could be a great deal heavier, penetrating thicker soils, without becoming too onerous for animals to pull. A neat touch came with the development of the collar harness, originally adapted by herders in the Asian desert from a wooden ring used for carrying baggage on the backs of camels. Now horses rather than oxen could plough the fields, substantially speeding up the process and increasing medieval Europe's readily available supplies of energy.[13]

By the early ninth century heavy wheeled ploughs pulled by well-harnessed horses could cut through solid clayish earth with vertical and horizontal blades, leading to a huge increase in the number of fields under cultivation. Combined with the discovery that by rotating crops around fields from one year to the next agricultural yields could be further increased, a revolution in farming was complete. It had a dramatic effect on the environment. In 500 AD about 80 per cent of all European land was forest, but by 1300 this had fallen to less than half.[14]

The spread of Christianity, a creed founded on the requirement of Adam to till the ground to

Norman Conquest: Duke William of Normandy spears King Harold II of England at the Battle of Hastings in 1066. He later accounted for his new kingdom in the Domesday Book.

23:59:59

atone for his original sin in the Garden of Eden, accelerated this change. Pagans venerated trees – they were their shrines. Christians did not. Monastic orders such as the Benedictines (established in Italy in 529 AD) and the Cistercians (founded in France in 1098) have been described as the 'shock troops' of clearing and deforestation.[15]

Monasteries established in England, France, Germany, Italy, Spain and Portugal not only spread the word of God but massively increased ecclesiastical wealth by felling trees and turning their land into fields for rent. Between 1098 and 1371 more than 700 Cistercian monasteries were established in Europe; each was a 'nucleus for clearing and farming', following a fashion championed by Charlemagne himself, who had decreed: 'Whenever there are men competent for the task, let them be given forest to cut down in order to improve our possessions.'[16]

Christian rulers also wanted to demonstrate their new agriculture-based wealth in the form of monumental architecture. Forests were cleared to make way for quarries from which were extracted the raw materials needed to build hundreds of magnificent new Christian abbeys and cathedrals. Each one required several million tonnes of stone: Amiens cathedral in France, completed in 1266, was large enough to accommodate 10,000 worshippers, equivalent to the town's entire population. It is still the tallest complete cathedral in France.

Extra food produced extra people. By 1000 AD, the European population had risen to more than thirty-seven million, and then doubled again to seventy-four million by 1340.[17] Such a dramatic increase in wealth, food and people meant that the need to find new lands to conquer and colonize grew ever greater. Viking explorers provided the most spectacular examples, capitalizing on the warmer climate and less icy oceans by finding a new sea passage from Norway to Iceland, which was first settled by Ingólfur Arnarson in 874. This was followed by settlements on Greenland, beginning in 982. According to the Saga of Erik the Red he named it Greenland because he wanted to attract other people to it. The Vikings were even the first

Europeans to reach North America, where in 1006 Erik the Red and his son built a small town.

In 1960 Norwegian archaeologists discovered the remains of a Viking village at L'Anse aux Meadows, on the tip of Newfoundland. This settlement, established nearly 500 years before Columbus's voyage in 1492, marks the first attempt by Europeans to colonize the Americas, although for some reason they withdrew soon afterwards. According to one Viking saga disaster struck after the settlers tried to build good relations with the natives (who they called *skraelings*) by inviting their chiefs into their new village for a convivial drink of milk. Unfortunately, owing to the absence of domesticated milk-producing mammals in America, the chiefs' lactose intolerance caused them to fall sick. Suspecting poisoning, they drove their new neighbours back into the sea.

Pagan plunder westwards was more than matched by the allure of Eastern riches pursued by crusading Christian knights. By 1095, Rome had received repeated appeals for military help from successive Byzantine Emperors to fend off encroaching Muslim forces, following their victory over a Byzantine army at the Battle of Manzikert in 1071. In response, Pope Urban II called for a single European foreign policy to liberate the holy city of Jerusalem, hoping it would help him reassert Papal authority over the Eastern Church of Constantinople into the bargain.

His appeal fell on fertile ground. For the previous one hundred years Christian knights had been fighting the Muslims of Spain, forcing them southwards. Their enthusiasm was fired by Pope Alexander II, who in 1063 had granted the forgiveness of all sins to Christian knights killed in battle against the Muslims in Spain (see page 242). By 1085 European knights had successfully besieged and liberated the strategically important city of Toledo, putting almost half of Spain back in Christian hands. The city's libraries contained many important Arabic translations of ancient Greek texts (see page 240). Now they could be translated into Latin by the Christians, unlocking the secrets of Greek science, geography, mathematics, astronomy and philosophy that had been lost to the medieval

West following the period of anarchy that swept away the last vestiges of ancient Rome.

When Pope Urban extended the promise of earthly forgiveness to all European soldiers, if only they would unite in a common cause to liberate the Holy Land from Muslim infidels, the response was overwhelming. No better example exists in history of how European culture finds the ideal of a united cause so difficult to put into practice, however appealing the prize. Over the next 200 years, what had started off so promisingly turned to disaster.

Success in the First Crusade (1096–99) led to the recapture of Jerusalem and the massacre of its mostly Muslim population. After resurgent Muslim attacks, a second campaign (1147–49) ended up with Christian knights committing Europe's first mass extermination of the Jews. Not considered crimes, these killings were thought of as just revenge for the people guilty of sentencing Christ to crucifixion. In 1187 Saladin, the Muslim Caliph of Egypt, recaptured Jerusalem, giving European leaders a reason to unite once again. In 1189 an unprecedented alliance of the Kings of England (Richard I) and France (Philip II), and the Holy Roman Emperor and King of Germany (Frederick Barbarossa), all supported by Pope Gregory VIII, marched as Christian soldiers to liberate the Holy Land once again.

It was not a success. Frederick drowned while crossing a river in 1190, and Philip fell ill with dysentery and returned to France with his armies in 1191 before they reached Jerusalem. His departure fuelled mistrust between the English and the French, Richard I commenting, 'It is a shame and a disgrace on my lord if he goes away without having finished the business that brought him hither.' As for Richard himself, he concluded a truce with Saladin after realizing that his forces would never be strong enough to retake Jerusalem. Then he was captured on his way home by Leopold V, Duke of Austria, who handed him over as a prisoner to the new Holy Roman Emperor, Henry IV. Richard's freedom was secured only after the English people stumped up 150,000 marks, equivalent to more than twice the country's annual income. He died shortly after his release in 1199, after being struck in the arm by an arrow.

Europe's flailing attempts at unity were finally torn to shreds by the last Crusade, of 1204, when the Christian armies never even got close to Jerusalem. They turned instead on the rich Byzantine city of Constantinople – inhabited by fellow Christians, who they had originally been sent to protect. What followed has often been described as one of the most shameful moments in Christian history. These supposed soldiers of Christ

A Crusader knight kneels to take his oath of allegiance, from the twelfth-century Westminster Psalter.

sacked the city, stole its treasures, raped its women and scattered its citizens.

According to Nicetas Choniates, a contemporary Byzantine chronicler, they used the broken remnants of religious vessels for pans and drinking cups:

> *'All kinds of precious materials admired by the whole world were broken into bits and distributed among the soldiers.'*

Then they sat a prostitute on the bishop's throne:

> *'A certain harlot, a sharer in their guilt, a minister of the furies, a servant of the demons, a worker of incantations and poisonings, insulting Christ, sat in the patriarch's seat, singing an obscene song and dancing frequently.'*

Finally, they attacked the rest of the population:

> *'In the alleys, in the streets, in the temples, complaints, weeping, lamentations, grief, the groaning of men, the shrieks of women, wounds, rape, captivity, the separation of those most closely united. No place remained unassailed, all places everywhere were filled full of all kinds of crime. Oh, immortal God, how great the afflictions of the men, how great the distress!'* [18]

By 1200 the continent of Europe was militarized, antagonized and overpopulated. Efforts at re-creating a stable central authority based on the Pope's Christianity or Charlemagne's revival of imperial authority had failed. Nothing like the paper-based Chinese bureaucracy emerged here. Nothing like the Incas' systems for state aid could knit together its people (see page 226), because thanks to the kings' need for mounted warriors, precious land was bartered into the hands of knights and nobles who wielded ultimate local power from inside bastions of impenetrable stone.

The medieval reality of a highly fragmented but deeply militarized Europe was fully evolved by 1337, when a devastating hundred-year war began,

in which England's kings committed themselves to an epic struggle to reclaim their Norman ancestors' lands in France. English knights, supported by thousands of skilful archers using powerful longbows, pushed deep into French territory, urged on by the prospect of fresh feudal grants of reconquered land.

But by then major warning signs of imminent and dreadful change were already evident. With a population now nearing eighty million, medieval Europe was under severe strain. Despite massive deforestation and the innovations of the horse, plough and crop rotation, its agriculture could supply food only for a finite number of people.[19] A series of three wet, cold summers was all it took to bring the whole of Europe north of the Alps and the Pyrenees to its knees. Grain would not ripen, there was no fodder for livestock, and salt – the only way to preserve meat – was scarce because salt pans could not evaporate in the relentlessly wet weather. Bread prices rose by as much as 320 per cent in some places. No one but the richest landlords could afford to eat; even the King of England, Edward II, was unable to find bread for his entourage while touring the country in August 1315.

Survival was made possible only by slaughtering essential draught animals and eating seeds saved for next year's crops. The consequences for morality and order were horrendous. Children were abandoned to fend for themselves (this is the origin of the Hansel and Gretel story), old people were starved to death for the sake of the rest of their families, and disease soon spread through the malnourished, weakened population. Between 10 and 25 per cent of the populations of many towns and cities perished. In 1276 the average life expectancy was thirty-five; by 1325 it had fallen by a third.[20]

Worse was to come. Having been spared the horrors of plague for some 700 years, medieval Europe was now at its most vulnerable to renewed attack. Traditionally it is thought rats were to blame for the spread of the appalling Black Death that struck Europe between 1347 and 1351, but in fact humans were just as responsible. Once again, butterfly wings in China lay at the source of upheavals in Europe. Plague originated in the province of Hubei in the early 1330s, spreading to a further eight provinces by the 1350s. Mongol traders travelling along the Silk Route are thought to have carried the disease west.

What happened next is a tale of terrorism on a scale never equalled before or since. In 1346 troops loyal to Jani Beg, commander and Khan of the Golden Horde from 1342 to 1357, had to abandon their siege of the Crimean port of Kaffa because plague had struck them. Their response was history's first and most lethal use of biological weapons. Beg's few surviving troops loaded the bodies of their many dead on to catapults and hurled them over Kaffa's walls, infecting the Genoan defenders inside. By the time the Genoese sailed home, most of their men lay dead. Just enough survived, though, to pass on the killer disease, which during the next three years claimed over *forty million* European lives – more than half the continent's total population.

By 1400, after repeated outbreaks, the population of England had fallen from seven million to just two million. Many European cities saw their populations fall by as much as 70 per cent, their cramped and unsanitary conditions providing perfect habitats for disease to spread. Recently, new theories have emerged to try to explain what caused the devastation, casting some doubt on the long-accepted view that plague was spread by rats and fleas. Was it a form of anthrax, or perhaps a highly infectious virus like ebola that caused a gruesome death by internal bleeding?[21] Whatever the causes, the consequences were catastrophic and long-lasting. Between 1348 and 1375 the average life expectancy in Europe fell to just over seventeen years.

Survivors found themselves in a completely different world. Instead of being oppressed, peasants were now so few as to be able to demand higher wages. If they didn't get them, they were in an excellent position to bargain for better deals. The peasants' revolts of 1358 in France (the 'Jacquerie') and 1381 in England (led by Wat Tyler) had their origins in the new dynamics between rich and poor that emerged after the Black Death. Social mobility, political representation and even a change in styles of clothing can all be attributed to the plague that

Black Death devastated European populations. Some experts think the pandemic may nearly have tipped the world into a new Ice Age.

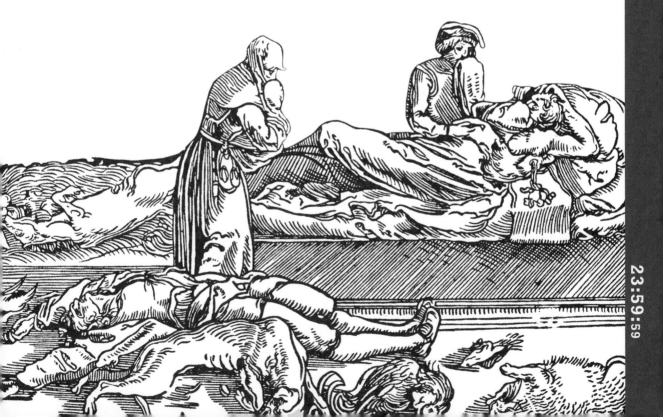

wiped out so many people. The rich panicked, forcing through a series of sumptuary laws, dictating what clothes people could wear, and what breeds of dogs and hunting birds they were allowed to own, to try to stop newly enriched peasants from rising above their social station.

Many of the extraordinary discrepancies between how the English language is pronounced and how it is spelled have been attributed to the new social mobility following the outbreak of plague. Before the Black Death the word *make* would have been pronounced *mak*, and *feet* would have been pronounced *fet*. Vowels became elongated after mass migrations of peasants into towns following the devastation of their rural populations. As a result of this sudden injection of many different regional dialects, a new blended pronunciation developed, accounting for the differences between medieval and modern English. Similar changes happened at about the same time in Germany and Holland.

Scientists are beginning to wonder whether pandemic diseases such as the Black Death may even have caused changes to the climate. Studies of atmospheric carbon dioxide levels in Greenland ice cores show a significant reduction between 1350 and 1500, from 282 to 276 parts per million. There is also a correlating fall in temperatures during this period, which has for a long time been called the 'Little Ice Age'. One possible cause of this cooling is that the massive reductions in human populations as a result of the Black Death reduced levels of atmospheric carbon dioxide because of a sudden decline in agriculture, wood-burning and deforestation.

One study suggests that the effects very nearly tipped the world's climate into another major Ice Age – one that, although long overdue, may now have been indefinitely postponed by the injection of huge quantities of greenhouse gases since the modern industrial age began between 1750 and 1800 (see page 377).[22]

Rapid declines in population also changed the nature of medieval warfare, and caused the fabric of feudalism finally to fray. There simply weren't enough peasants left on the land to supply troops

for epic struggles such as the Hundred Years War. Instead, English kings like Henry V (ruled 1413–22) were forced to pay mercenary soldiers, or even to recruit their own standing armies. Both required money and more taxes on the people, leading to bargains in parliaments to raise funds for waging war in exchange for concessions of power, privilege and prestige.

Europe's rulers desperately needed new ways of fighting that didn't require the deployment of regiments of archers, whose skill with the longbow took years of training and enormous personal strength. Dilemmas like these provided the perfect environment for China's last but ultimately most lethal secret to be played out on the stage of medieval Europe.

Gunpowder had been known in the West since 1267, when a recipe appeared in the English philosopher Roger Bacon's *Opus Majus*, which contains information about everything from microscopes and telescopes to optics, astrology and gunpowder:

'Certain inventions disturb the hearing to such a degree that, if they are set off suddenly at night with sufficient skill, neither city nor army can endure them. No clap of thunder could compare with such noises. Certain of these strike such terror to the sight that the coruscations of the clouds disturb it incomparably less … We have an example of this in that toy of children which is made in many parts of the world, namely an instrument as large as the human thumb. From the force of the salt called saltpetre so horrible a sound is produced at the bursting of so small a thing, namely a small piece of parchment, that we perceive it exceeds the roar of sharp thunder, and the flash exceeds the greatest brilliancy of the lightning accompanying the thunder.'[23]

Gunpowder first came to the attention of English warlords in 1342 at the Siege of Algeciras (in Spain), where Islamic warriors used primitive cannon against Christian invaders with remarkable effect. According to one Spanish historian, 'The besieged did great harm among the Christians with iron bullets they shot.'[24] The English Earls of Derby and Shrewsbury were at the scene, and quite possibly transferred the technology to England. It was tried out for the first time against the French at the Battle of Crécy in 1346.

Within a hundred years gunpowder cannon had transformed the power of Europe's kings, popes and emperors who could afford them. Soon handguns equipped unskilled peasants with the power to kill at a distance. Desperation brought on by famine and disease gave way to a new level of catastrophe ushered in by the arrival of a black powder that could, in minutes, reduce a knight's castle to rubble.

Feudalism and chivalry were now in the process of being replaced by the law of the gun. Meanwhile, cannon so large that they took fifty oxen and a crew of 700 men to fire were being manufactured in eastern Europe and sold to the highest bidder. Urban the Hungarian (not the Pope) was the world's first known arms dealer. Having been told by the Byzantine Emperor Constantine XI that his services were too expensive, he sold his technology to opposing Islamic forces headed by the Turkish Sultan Mehmed II.

Nearly 750 years had passed since the previous attempt by Islamic forces to take the great imperial city of Constantinople, guardian of rich trade routes to the Black Sea. By the spring of 1453 another force of 80,000 Muslim soldiers had gathered just outside its massive walls. The weather wouldn't stop them this time. Enormous eight-metre-long super-guns, crafted out of copper and tin, were slowly rolled across the plains of Asia Minor to a point about a mile outside the city. Giant balls of stone and marble, some plundered from the temples of ancient Greece, relentlessly pounded Constantinople's walls with such force that on impact they burrowed two metres into the ground.[25] On 28 May the Turks finally breached the city walls and their troops flooded inside. Mehmed rewarded his men with three days of looting, as was the custom.

When news of the fall of Constantinople reached Venice on 29 June, and then Rome a week later, it stunned the Christian world. Medieval Europe was on its knees once again. Torn from within by disease, depopulation, gunpowder and war, now its most precious and historic ancient capital had finally fallen into heathen hands.

Smelling success: Sultan Mehmed II successfully conquered Constantinople after hiring the services of a Hungarian arms dealer.

33

Treasure Hunt

How each of the world's settled human societies sought out their own various fortunes through a mixture of trade, toil and theft.

SURVIVAL IS AN INSTINCT as old as life. For tribes living in the wildernesses of America, Australia, Indonesia or Africa, the key to a good life was to exist within nature, preserving resources, moving on from one place to the next, and always ensuring that population levels stayed stable. Settled societies dependent on agriculture and trade had a different challenge. Success for them meant securing wealth – essential for supporting people not involved in making food for themselves, such as kings, priests, prisoners, merchants, artisans and domestic slaves. Without wealth, their worlds fell apart.

By my reckoning there are at least seven ways of getting wealth. You can win it, inherit it, earn it, steal it, extort it, grow it or find it. By 1453, as the Ottoman Turks blasted through the walls of Constantinople, the civilizations at the top of the world's Rich List were either Islamic or Chinese. Christian Europe would have been way down towards the bottom.

Islamic cultures tended to get rich by earning or stealing. Both were natural extensions of their nomadic origins in the Arabian deserts, where carrying goods such as salt, slaves and gold from one oasis, town or settlement to another meant they could take a cut from both seller and buyer. Horses and camels often provided military advantages, so theft was an effective alternative.

By the mid-fourteenth century, wealth earned through trade had turned the Egyptian capital Cairo, founded in 910 by an Islamic dynasty called the Fatimids, into one of the world's greatest cities. These were Shi'ite Muslims whose claim to authority came from direct descent through Mohammed's daughter Fatima.

From here Islamic traders could control the overland trade routes linking China and India to the Mediterranean and Europe, which proved especially lucrative when alternatives such as the Silk Road were blocked by conquest or plague. So tempting were the opportunities to make a good living here that by the fourteenth century the population had grown to a staggering 500,000 – more than fifteen times larger than London at the same period. Just how much commerce was conducted in Cairo was unearthed by Jacob Saphir, a nineteenth-century traveller and rabbi, who found more than 300,000 perfectly preserved documents buried in the storeroom of one of the city's synagogues. Between 950 and 1250 more than 450 different types of trade, carried out by 35,000 merchants are recorded, stretching east as far as India and west to Seville.

Ibn Battuta (1307–77) was a Muslim traveller who spent his life touring around much of the known world, from West Africa to Egypt and from Turkey to India and China. His description of a month-long stay in Cairo in 1326, while on his way to pilgrimage in Mecca, leaves no doubt that enormous wealth was being earned there through control of trade routes – the keys to Islam's increasing riches: 'There is a continuous series of bazaars from the city of Alexandria to Cairo … Cities and villages succeed one another along the banks of the Nile without interruption and have no equal in the inhabited world.'[1]

By then Egypt was ruled by Mamluks, Islamic slave-soldiers from the Turkish steppes originally assembled by Saladin to resist the Western Crusaders. After rising to power through a military coup in 1250, their dynasty acquired great wealth through trade. People, camels and donkeys jostled together in Cairo's central commercial district, where there were literally thousands of shops.

Thirty separate markets specialized in different trades, including butchers, goldsmiths, gem dealers, candlemakers, carpenters, ironsmiths and slave merchants. *Funduqs*, like the one visited by the Italian mathematician Fibonacci in Tunis (see page 240), were commercial stores for highly sought-after goods from the East such as pepper, cloves, cotton, canvas, wool and silk. Inside, rich Islamic merchants haggled with those eager to buy goods for consumption in the Christian West. As a measure of the scale of this trade, one *funduq* built for Syrian merchants in the twelfth century had 360 lodgings above its storerooms, which provided accommodation for as many as 4,000 guests.

Islamic civilizations emerged just about anywhere where new commercial opportunities lay. Traders from Arabia introduced one-humped camels that could survive for more than ten days without water to the Sahara, so that gold mined in West Africa could, for the first time, be transported safely over long distances. They carried the gold either eastwards for sea passage to the Middle East, or northwards through the desert to the Mediterranean coast and Europe.

It wasn't just commodities such as gold that made Muslim merchants rich. Harvesting the tusks of elephants slaughtered in the continent's interior and transported via the port of Mombasa, on the coast of modern-day Kenya, to the markets of the East was highly lucrative, too. Chinese craftsmen prized ivory as the material of choice for carving images of the Buddha and Taoist deities. By the nineteenth century, their skills were diverted into making ivory pipes for smoking opium (for more on China's addiction to opium, see page 361). It has been estimated that before the African ivory trade the continent was home to approximately ten million elephants, a number that has now declined to less than half a million.

Muslim trade in gold and ivory was also supplemented by exporting African slaves to feed Arabia's economy, which like ancient Rome, depended on imported free labour. Even after the banning of the slave trade by the British in the nineteenth century (see page 333), villagers captured in central Africa were herded barefooted in Muslim caravans to Mombasa, where survivors were shipped across the East African seas. As many as 11 to 18 million black African slaves are thought to have been transported eastwards between 650 and 1900 AD, only slightly more than the 12 to 14 million estimated to have been exported westwards by Europeans to the Americas, beginning in the sixteenth century (see page 299).[2]

The Islamic Empire of Mali (1235–1645) grew into one of the most powerful early African states on the back of trade in gold and salt. Its most famous King, Mansa Musa, went on a pilgrimage to Mecca, passing through Cairo in 1324, just two years before Ibn Battuta. It was no ordinary visit. He took with him a caravan of more than a hundred camels fully laden with gold, and on arrival in Cairo went on one of history's most celebrated shopping sprees. According to contemporary historian al-Umari, the merchants of Cairo made incalculable profits out of him and his entourage in buying and selling and giving and taking. 'They exchanged gold until they depressed its value in Egypt and caused its price to fall.'[3] So carried away was Mansa by the allures of shopping in downtown Cairo that, says al-Umari, he eventually ran out of money and had to take out a loan to be able to afford the trip home.

Cartels of medieval Arab merchants with access to precious supplies of raw materials operated much as modern cartels do today with oil. By controlling the flow of valuable commodities, medieval Islam was the OPEC of its day, skimming off massive riches and inflating prices to European consumers. One kilogram of pepper would be worth one gram of silver at Malacca, close to where it was grown in the spice islands of Indonesia. By the time it had been transported to Cairo it was sold for up to fourteen grams of silver to merchants from Venice. They then sold it on to other European merchants for eighteen grams, the price then increasing to between twenty and thirty grams of silver per kilo to consumers in the capitals of Christian Europe. Thus most of the profit was made by Islamic traders, with a cut going to Western merchants in Italy. Ultimately, the high price was paid by medieval Europeans, whose pockets were drained.[4]

Plundering wealth, rather than earning it through trade, was an alternative strategy favoured by some Islamic rulers, especially those who inherited empires from the mayhem caused by the Mongol invasions of Genghis Khan. By the early fourteenth century many of the rulers who

controlled these hordes had converted to Islam, the established religion of their captured people. For example, Uzbeg Khan, ruler of the Golden Horde from 1313 to 1341 and father of Jani Beg, adopted Islam as the state religion. Mahmud Ghazan (died 1304), seventh ruler of the Mongol Empire's Ilkhanate division in Iran, was another Mongolese convert to Islam.

The most notorious thief was Timur, also known as Tamerlane (1336–1405), who launched a ruthless bid to reunite the Mongol Empire, beginning with his home city of Samarkand in Central Asia. After conquering Persia in 1383, he marched in 1398 to northern India, where according to his own memoirs he plundered Delhi, executing more than 100,000 captives, of whom the heads of more than 10,000 were cut off in a single hour:

> 'These infidel Hindus were slain, their women and children, and their property and goods became the spoil of the victors. I proclaimed throughout the camp that every man who had infidel prisoners should put them to death, and whoever neglected to do so should himself be executed and his property given to the informer. All the goods and effects and the treasure became the spoil of my soldiers.'

Like Genghis, Timur was a man of his word. When he left Delhi in January 1399 he took with him ninety elephants loaded with precious stones and jewels. The riches were taken to Samarkand, where he commissioned an almighty mosque in praise of Allah. Historians think this was the enormous Bibi-Khanym Mosque built from jewels looted from India. Its hasty construction caused it to suffer greatly from disrepair and it eventually crumbled into ruins after an earthquake in 1897. However, reconstruction began in 1974 and a new mosque now stands in its place echoing the original design.

Timur's hunger for wealth knew no bounds. In 1402 he sacked Baghdad, massacring 20,000 inhabitants, and then defeated the Ottomans in Turkey, capturing their Emperor Bayezid. He eventually died on campaign in China while attempting to restore the Mongol dynasty of Kublai Khan, which had fallen to an all-too-familiar Chinese tradition, a peasants' revolt, ushering in the Ming Dynasty in 1368. It has been estimated that Timur's orgy of violence resulted in *seventeen million* deaths.[5]

Timur wasn't the first or the last Islamic ruler to covet the jewels of India. The country had been gradually coming under Muslim influence since the early thirteenth century when Mohammed of Ghor, ruler of a province in Afghanistan, was lured by the wealth of its unique diamond deposits. His military campaigns resulted in the creation of the first Sultanate of Delhi (1206–1526), which was later overrun by a descendant of Timur's called Babur, who established the Islamic Mughal Empire

following the Battle of Panipat in 1526 ('Mughal' is derived from 'Mongol'). Until European traders took power in the eighteenth century, India remained a mostly peaceful, tolerant state ruled by Islamic Sultans who lived off its riches, which paid for the construction of some of the world's most fabulous surviving monuments, such as the Taj Mahal.

Central Persia remained locked in civil war and turmoil for more than a hundred years following Timur's conquests. Order was eventually restored after the rise of the Safavids, founded in 1502 by a Shi'ite, Shah Ismail, who conquered what are now Azerbaijan, Iran and most of Iraq. Claiming direct

Turkish Sultan Bayezid I is brought before Timur after being captured at the Battle of Ankara in 1402. Some accounts say Timur kept him chained in a cage as a trophy, or even used him as a footstool.

23:59:59

descent from Mohammed himself, Ismail established a powerful blend of Sufi mysticism based on direct divine contact through poetry and *imams* – religious leaders. For the next 200 years his successors restored Islamic control over the Silk Route. They found that good money could be earned from the exchange of textiles and spices, making the Safavid Persians rich.

⋈⋈⋈⋈

Medieval China's strategies for securing wealth lay in other directions. Its rulers preferred to grow it or extort it. After two brief attempts to conquer Japan by Kublai Khan in 1274 and 1281, both of which were thwarted by appalling storms at sea, China came to rely on increasing agricultural production of goods essential to its settled civilization – rice, paper and silk – and on using its considerable technological expertise to exact protection money from neighbouring states.

As the Mongols were steppe nomads at heart, it was probably not surprising that their neglect of China's agricultural needs eventually led to the downfall of their dynasty, the Yuan. In 1368 a young peasant called Zhu Yuanzhang led a successful rebellion against the Empire with the help of hundreds of thousands of equally fed-up rural supporters. They were furious about abandoned irrigation projects and inflation that was spiralling out of control thanks to the Emperor's over-zealous printing of paper money. When they were ordered to help build flood defences along the Yellow River, they decided enough was enough.

The rebel leader crowned himself Hongwu (ruled 1368–98), and founded the Ming Dynasty. His reforms ensured that wealth was successfully generated through agricultural growth for many years to come. Hongwu confiscated large estates and gave them to the rural poor, encouraging them to build their own self-supporting agricultural communities. New administrative systems were introduced, called the 'Yellow' and 'Fish Scale' records, to ensure peasants were taxed fairly, leaving them enough to live on. Irrigation systems and dykes were restored to improve the quality of the

soil, tax breaks were given to anyone who turned waste land into production, and a new law code, called the *Ta Ming-Lu*, abolished slavery, for so long the cause of immeasurable misery for the poor. (In 1384 an imperial edict required officials to buy back children who had been sold as slaves by their parents in a desperate effort to survive a recent famine.)

While agricultural and rural policies secured the wealth of China and peace from peasant revolts, it was the third Ming Emperor, Yongle (ruled 1403–24), a son of Hongwu, who promoted another traditional source of Chinese wealth, and with it foreign security, through the harvesting of protection money from nearby states. Such payments were extracted by extortion, although the Chinese preferred to called them tributes.

Yongle paid for the construction of the largest and most impressive naval fleet the world had ever known. A Sufi Muslim called Zheng He, who had been castrated at the age of eleven after being captured by rebel Mongol fighters, was appointed admiral. Eunuchs had long been regarded as the most loyal of officials, owing to their inability to build dynastic empires.

Between 1405 and 1433 seven naval expeditions were launched by the Chinese government in a policy designed to impress the rest of its known world with its awesome power, and to exact financial payments and precious gifts in return for peace. The larger the force, it reckoned, the greater the likely return. The first expedition consisted of a staggering 317 ships, holding some 28,000 sailors and troops. Sources describe some of the ships as having as many as nine masts, making them by far the largest marine craft the world had ever seen. Ibn Battuta, who had visited China in 1347, described luxurious four-decked ships with up to twelve sails: 'A cabin has chambers and a lavatory that can be locked by its occupants …'

Stretching to a length of almost 200 metres, these ships were as long as the first aircraft carriers in the early twentieth century. Additional support came in the form of a retinue of vessels such as horse ships, troop carriers and water tankers.

The first three expeditions visited Asia, India and Ceylon. The fourth went to the Persian Gulf and Arabia, while the fifth, sixth and seventh ventured at least as far as the east African coast.[6] Gifts of Chinese silk and porcelain were exchanged with foreign rulers for everything from gold, spices and tropical wood to exotic animals such as zebras and giraffes, which eventually found their way into the Ming imperial zoo. Tributes came from far and wide. By the fourth voyage more than 320 state ambassadors travelled back with Zheng He to pay their rulers' respects and offer gifts to the Chinese Emperor at the Ming court.

Many of the troops taken by Zheng He on his voyages were, like him, Sufi Muslims. As a result, Islam spread further along the coasts of Indonesia, where Zheng established communities of Chinese Sufi Muslims, seeded by his own troops, at important ports in Java, Malaya and the Philippines.

In 1433 Zheng He died, mysteriously lost during his last voyage. What happened next has foxed Western historians for decades. Instead of capitalizing on their enormous naval strength at sea, the next Ming Emperors apparently threw China's foreign policy into reverse. Shipbuilding projects were stopped, yards decommissioned, and the enormous fleets of Zheng He's expeditions were left quietly rotting away in port. What sane ruler would give away such a maritime advantage, allowing Muslim, Japanese and eventually Western powers to become dominant at sea, securing precious trade routes for themselves (see pages 361–63)?

By 1445 the Oirat horde of Mongols were once again threatening the northern borders of China's Empire, causing successive Ming Emperors to divert all their state resources to building a new Great Wall of China further south than the original one built by Qin Shi Huang between 220 and 200 BC. Begun in 1368, the construction of this new line of defence was accelerated substantially in the mid-fifteenth century. The cornerstone of China's foreign policy now became a 6,400-kilometre-long wall guarded by more than one million soldiers. Between two and three million men are thought to have died during the construction of what is the largest human-made structure ever built.

In the context of China's traditional strategies for securing wealth – growing it and extorting it – a sudden policy-switch away from seapower into building protection against the age-old threat of overthrow by northern Mongols looks as obvious as it was consistent. Then, as today, the Chinese made things (porcelain, paper) and grew things (rice, silk). They were not naturally great maritime traders. Their centralized government worked by deploying the massive resources of the state wherever they were most needed.

✕◗✕◗✕◗✕

China's Great Wall mattered little to medieval Christian Europe. Emerging from the Great Famine and Black Death of the fourteenth century, this was the civilized world's least successful continent, with no clear idea how to stimulate its economy. Europe was in a desperate fix, because all the available strategies for creating sufficient wealth to secure its civilizations had ended in abject failure.

The Great Famine meant that *growing* wealth through agriculture had suffered a dramatic setback, and the Black Death had seriously added to Europe's woes by devastating the population. *Stealing* wealth from elsewhere had proved spectacularly unsuccessful following the disaster of the Crusades, whose biggest triumph had been the sack of Europe's most prestigious city, Constantinople, in 1204. *Earning* wealth through trade was an obvious alternative strategy. But, apart from its northern shores, the Mediterranean was now an Islamic lake. As the price of pepper demonstrated, trading on advantageous terms in these conditions was hardly viable.

Yet between about 1300 and 1550, many of Europe's finest artistic and literary achievements were created by poets, philosophers, artists and sculptors such as Dante, Petrarch, Michelangelo and Leonardo da Vinci. During these years, money and wealth on a scale not seen since the height of the Roman Empire flooded into the Italian peninsula. With the rise of wealthy aristocratic patrons such as the families of the Medici, Sforza, Visconti, Este, Borgia and Gonzaga, rich city states studded the landscape across northern Italy, from Venice in the east to Rome in the west, among them Florence, Siena, Pisa, Genoa, Ferrara and Milan.

Here, in the toe of Europe, the reality of Europe's dependence on the success of Mohammed's world was playing itself out. Renaissance Italy was the European terminus of Islam's worldwide trading system. Precious Greek and Roman texts, lost to the Christian West, were translated from Arabic copies captured in cities such as Toledo in Spain and brought to Italy by merchant families enriched by wealth they had earned in cities such as Cairo. Artistic patrons such as the Medici, whose fortunes came from charging other Europeans exorbitant prices for wool and textiles, established banks and double-entry book-keeping, based on knowledge of Arabic numerals brought from Islamic Tunis to Italy by the Pisan trader Fibonacci. Even Dante, the most celebrated Western poet of the period, and perhaps of all time, structured the heaven, purgatory and hell of his epic poem *The Divine Comedy* around the nine-layered Islamic view of the universe first written about by philosophers at the House of Wisdom in ninth-century Baghdad.[7]

The concentration of such enormous wealth in this small part of Europe stimulated new thought, artistic expression and a curiosity about the natural world which was rekindled by the rediscovery of ancient writers like Aristotle and Plato. But this Italian crucible of wealth and patronage, dependent on trade with Islam, could not in itself provide all Europe with a secure strategy for the future welfare of its people. The enrichment of a few dynastic families by virtue of their place at the end of another civilization's trading cartel was hardly a solid foundation for lasting security. Instead, Italian pomp just increased the jealousy of other European nations which, stoked by the failed Crusades and the arrival of gunpowder, became ever more eager to gorge on Italy's treasures for themselves. Between 1494 and 1559 the Great Wars of Italy were fought between France, Spain, the Holy Roman Empire, England, Scotland, Venice, the Papal States and the Ottoman Empire. By the end of the period Spain had become the dominant power in Italy to the detriment of France.

By 1453 the Islamic conquest of Constantinople made Europe's already desperate situation even

Suleiman the Magnificent, closely followed by female attendants, invaded Hungary in 1521 and would probably have captured Vienna were it not for the appalling weather.

worse. Most of the Balkans, including all Serbia and Greece, had fallen into Islamic hands, and their people were converting to Islam (which is why there are still many Muslims in places such as Bosnia today). Economically, Islamic powers tightened their nooses even more, since precious European silver mines in Serbia and Greece, the backbone of the continent's traditional bullion supply, were now firmly in enemy land.

As if Islamic control over Europe's chances of securing wealth through trade wasn't tight enough, in 1517 the Ottoman Sultan Selim I (ruled 1512–20) strengthened it even further by invading Egypt, cutting off historic overland trading links to the East. His successor Suleiman the Magnificent (ruled 1520–66) invaded Hungary in 1521, and after winning the Battle of Mohacs in 1526 Muslim armies massed outside the walls of Vienna. Thanks only to the fortuitous intervention of nature, echoing the salvation of Constantinople more than 800 years before (see page 257), it was only the harshest of winters that saved Vienna from falling, forcing Suleiman to recall his troops.

But if the cold weather provided temporary relief for the inhabitants of Vienna, it proved fatal to another potential European route to riches. The adventurous Vikings, who had established settlements totalling about 5,000 people in Greenland, found the onset of the Little Ice Age too harsh to make survival there viable. Plummeting temperatures led to the collapse of this pilot phase of European maritime exploration. The last ship known to have sailed from Norway to the eastern settlements of Greenland landed in 1406, apparently blown off course while on its way to Iceland. It returned to Norway in 1410. Nothing more is known about what happened to the Viking community that had survived for 450 years as Europe's most remote outpost. It is presumed that they starved to death, having over-exploited the natural resources as a result of the extreme cold.[8]

Europe was surrounded and trapped. To the north lay ice, to the west an ocean too vast to navigate, to the east and south were the lands of the 'infidel' – Muslim rulers who traded only on their own terms and who exercised tight control over Europe's economy. By 1451 the continent was badly in want of a miracle. Unfortunately, at precisely the moment Christ's long-promised return to save His blessed kingdom on earth was needed most, His ambassador on earth, the Pope, had been exposed as a charlatan and a fraud, responsible for probably the biggest forgery the world has ever seen.

For centuries Roman Popes had directly controlled lands in the north of Italy, called the Papal States, which had been secured for them back in the eighth century, in the time of Charlemagne's father Pippin. The justification for giving such earthly power to a supposedly spiritual representative of God on earth came from an ancient document called *The Donation of Constantine*. According to this text, soon after his conversion to Christianity the Emperor Constantine I was stricken with leprosy, but thanks to his new faith he was miraculously cured by the then Pope, Sylvester I (314–77 AD). In gratitude, Constantine entrusted his Western Empire to the Papacy in Rome. For centuries Popes used *The Donation of Constantine* to justify not only direct control of the north Italian lands they acquired through Pippin and later his son Charlemagne but everything from their right to raise their own private armies and to levy taxes all across western Europe, to anointing not just their own chosen bishops, but kings and emperors as well.[9]

Then, in an essay written in 1439–40, a linguist from Florence named Lorenzo Valla proved conclusively that the Pope's precious document was an elaborate fake. Valla demonstrated beyond doubt that the Latin words used in the document came from a later period, probably concocted in the ninth century, at about the time when Pippin delivered the first Papal States to the See of Rome. Desperate to cover up the truth, the Papacy did everything it could to prevent the formal publication of Valla's essay, but in 1517 this explosive evidence of Papal fraud finally emerged in the hands of Christian Protestants, mass-produced by the recently invented printing machine, for all the world to see.

It wasn't just this fraud that undermined the authority of Christ's representative on earth. Popes of the Italian Renaissance were notorious for their appallingly debauched and corrupt ways of life. The most flagrant example was Pope Alexander VI (1492–1503), who among several claims to fame enjoyed a string of mistresses who produced a number of illegitimate children. His love of parties and dancing reached a climax with the infamous 'Banquet of Chestnuts' hosted by his son Cesare at the Vatican, seat of Papal authority in Rome, on 30 October 1501. After dinner, the guests, who included members of the clergy, were invited by fifty specially selected prostitutes to take off their clothes and crawl around naked on the floor picking up chestnuts. Following this bizarre ritual an orgy took place in which, in the words of the master of ceremonies, 'prizes were offered – silken doublets, pairs of shoes, hats and other garments

– for those men who were most successful with the prostitutes'.[10]

With such unholy representatives, it is hardly surprising that hopes for deliverance for Europe were not forthcoming from divine intervention via the Papacy of Rome. Indeed, so corrupt had the papacy become that in 1492 a Dominican friar called Girolamo Savonarola managed to rouse the population of Florence into a frenzy at the abuses of the Catholic Church. After rising to power on a wave of moral indignation, he briefly established a new democratic republic in Florence following the French invasion of Italy in 1494. Determined to stamp out the opulence, orgies and corruption of the city, he conducted a house-to-house purge of all items associated with moral laxity, including mirrors, cosmetics, lewd pictures, pagan books, nude sculptures and gamblers' games. He then had them heaped up and burned on a 'Bonfire of the Vanities' in the city's central square.

Savonarola's attempt to solicit Christ's grace through moral purification was as short-lived as it was ineffective. By 1498 power had shifted back

into the hands of the city's wealthiest families. Pope Alexander VI issued a decree of excommunication, and Savonarola and his leading supporters were charged with heresy, tortured and burned on the same site as their own previous spectacular bonfire. According to one observer, Luca Landucci, their charred remains were several times broken up and burned again so that not the smallest piece of bone was left, in order to prevent supporters recovering pieces and turning them into holy relics.[11]

Medieval Europe was politically fragmented, at war with itself both physically and ideologically, and surrounded on all sides with no obvious means of escape. In such times even the most eccentric ideas of a way forward were bound to attract the interest of at least some rulers. So when one Portuguese dignitary expressed an interest in going on a grand treasure-hunt to find the source of Mansa Musa's West African gold, he was listened to with great attention.

Born into the Portuguese royal family, Prince Henry (known as Henry the Navigator), third son of King John I, spent most of his life trying to find treasure. His interest was first aroused when his father's navy captured the important North African Muslim trading port of Ceuta in 1415. For years Muslim pirates had been raiding the Portuguese coast, stealing villagers to be sold in the lucrative Islamic slave markets of North Africa. Ceuta was the African terminus for gold that was transported by Muslim camel caravans across the Sahara. If only, thought Henry, a way could be found to get to that gold without having to cross the desert, perhaps then new wealth could be accessed without most of it ending up in the pockets of Muslim traders.

Henry's quest led him to establish a village, later called Vila do Infante, or 'Town of the Prince', on the southern coast of Portugal, where explorers could plan maritime expeditions. He soon discovered that light, manoeuvrable ships called caravels worked best for coastal exploration, and that by using a triangular lateen sail, common on Arab boats, it was possible to tack successfully against the wind, allowing his ships to sail not just where the wind blew them.

Henry's goal was to reach the far side of the Sahara desert by hugging the West African coastline and sailing south. With each voyage, he made and updated maps to act as guides for future expeditions. Lagos, a small natural harbour, now a popular tourist destination on the Algarve coast, was his launchpad.

The first major challenge was to navigate south of Cape Bojador, a headland on western Sahara, just south of modern-day Morocco, which was until then the most southerly point known to medieval Europeans. It took ten years and fifteen separate attempts to successfully pass this headland, which because of its fearsome currents and strong winds became known as the place where sea monsters dwelt. Finally, in 1434, one of Henry's captains, Gil Eanes, discovered that by sailing far out to sea, beyond the sight of land, more favourable winds could be picked up which would propel boats further south down the coast.

By 1443 Portuguese sailors had reached the Bay of Arguin, on the Atlantic coast of modern Mauritania, where they later built a powerful fort. By now they had reached south of the Saharan desert. From 1444, dozens of vessels left Henry's port of Lagos each year, bound for south Saharan Africa. In 1452 came the first real results, in the form of slaves and gold delivered directly across the seas by Europeans without reliance on overland Muslim middlemen. Their arrival permitted the minting of Portugal's first ever gold coins, aptly called *cruzados*.

Was this the beginning of a new strategy that could, perhaps, provide an answer to desperate Europe's desire for wealth? Could the enterprise of a Prince from the south-western edge of the continent really offer deliverance from the tyranny of Islam's traders? If gold and slaves could be acquired from just down the coast of West Africa, what other riches lay out there in the unknown world across the deep blue seas?

Here we stand at the threshold of the last 500 years in our world's story, such a tiny fragment of time that it represents just a hundredth of a second to midnight on the twenty-four-hour scale of all earth history. In that tiny speck of time, such terrible tales of disaster laid in wait for the world and its populations of human and non-human beings, mostly as a result of just a few determined European adventurers whose lust for treasure pushed them headlong into mankind's biggest ever global challenge. Soon the new strategic vision that had dawned in Portugal spread like a virus throughout Europe's fledgling states. Whichever of them was to become predominant, one thing was certain: the best prospects for future success lay in becoming mistresses of the sea.

23:59:59

Moules Marinière

How a few maritime explorers accidentally discovered a New World. Their arrival proved fatal to ancient civilizations and provoked a fierce contest between the rival nations of Europe.

POTOSÍ IS THE HIGHEST city in the world. Located in the mountains of Bolivia, at 3,967 metres above sea level, it should be a paradise of clean air and fresh mountain springs. But today, this once-idyllic mountain retreat is one of the most polluted places on earth. Its water has turned to acid, its land is sterile, crops cannot grow and the mountainside is littered with highly toxic waste including cadmium, mercury, chromium and lead.

Towering over the town is the peak of Cerro Rico, known locally as 'the Mountain that Eats Men'. Inside were once the world's richest silver mines, dug out by thousands of African and Native American slaves imported by Spanish explorers who founded the place in 1546. The slaves came

from anywhere that labour could be snatched – locally from the south in Peru, or from as far away as the west coast of Africa or even the islands of the Pacific.[1] Their feet trampled powdered silver ore with deadly mercury into a brew of poisonous metallic slime. Then the mercury was burned off, releasing clouds of lethal vapours that eventually dissolved into the mountain streams, turning the water toxic. Between 1556 and 1783 more than 45,000 tonnes of pure silver were extracted from the mountain using this process, much of it shipped directly to Europe.

No one knows how many men, women and children died from asphyxiation, poisoning and overwork in this Godforsaken place, although the total has been estimated at a staggering eight million.[2] Adult slaves were often made to work inside the mines for more than a week at a time before being allowed back up to the surface. Children were used as mules to transport the silver ore out of the dark, narrow tunnels because conditions were too cramped for animals. Today, the polluted atmosphere still takes its toll. The life expectancy of miners working here, now extracting zinc, is less than forty.[3]

The tragedy of Potosí in far-off South America came about because between 1450 and 1650, Europe's treasure-hunters struck lucky. The arrival of Henry the Navigator's booty in the form of African slaves and gold transformed Portuguese opinion about the best strategy for gaining new wealth. Extra investment in expeditions down the west coast of Africa yielded astonishing results. On one such voyage, in 1488, the explorer Bartholomew Diaz was driven far to the south by a violent thirteen-day storm. Once the seas had calmed he headed east, expecting to find land once again, but by then he had rounded the southern tip of Africa. Confused as to his whereabouts he headed north, travelling up the east coast of present-day South Africa as far as the Great Fish River. On his return journey in May 1488 Diaz was the first European to see the tip of Africa. He realized that here was a previously unknown sea route to the spice-rich lands of India and the Far East, so long monopolized by Muslim traders. Already flush with the profits from African gold and slaves, King John II of Portugal christened this new headland the Cape of Good Hope.

King John assembled the best scholars, mathematicians and mapmakers in his kingdom, and commissioned a series of expeditions in a climate of absolute secrecy. One, led by Duarte Pacheco Pereira, is thought to have revealed an unknown land to the west (Brazil), leading John to conclude that the quickest way to the riches of the East must lie around the Cape of Good Hope.

Much mystery still surrounds how much the court of John II knew about Brazil and the distances between Portugal and India via the eastern and western routes. This is partly because the

information was regarded as top secret, but also because many court records were lost in a devastating earthquake that destroyed much of Lisbon in 1755.

Such secret knowledge is what probably led King John to reject the overtures of a boastful Genoese seaman called Christopher Columbus, who came to his court in 1485 and again in 1488 claiming to be able to reach the East by a westward route across the Atlantic. He demanded to be appointed 'Great Admiral of the Ocean', to become governor of any lands he discovered, and to receive one tenth of all the revenues deriving from his quests.

Columbus's confidence that a route to Marco Polo's China and the spices of India could most easily be found by sailing west across the Atlantic was founded on a basic miscalculation. After studying the works of ancient Phoenician explorers and Arabic mapmakers, he incorrectly concluded that the distance to Asia from the shores of Europe was just 3,600 kilometres (the actual distance from Spain to China, sailing west, is about 24,000 kilometres).[4] Columbus had good reason to want the journey to appear reasonably short, since no patron would support a venture whose sailing time would make it impossible to provide sufficient food and water to sustain its crew.

A mountain that eats men: European lust for South American treasure to finance wars and the high life led to the enslavement and deaths of millions of natives.

In 1492 Columbus's quest for financial support for his mission at last met with success. In that year the Spanish rulers Ferdinand and Isabella, who through their marriage had united the regions of Castile and Aragon, finally ousted the Muslims from Granada, thus turning all Spain into one Catholic kingdom. With Spain no longer able to rely on tributes in gold from the Moors of Granada, and with gangs of armed Christian Crusaders hungry for lands to conquer, the idea of overseas exploration was now enthusiastically received. Most important of all was the Spanish government's desire to find a new route to the riches of the East, not only to usurp the stranglehold of Muslim merchants but to rival Portuguese exploration along the Gold Coast of Africa. Now two of Europe's biggest powers, Portugal and Spain, found themselves aligning their strategic objectives: to fund economic growth by overseas exploration

Columbus meets the locals, who proffer him gifts of friendship. Finding no treasure, Columbus later provoked battles to secure natives as slaves to finance future expeditions.

and thereby break the Ottoman grip on the mostly Islamic Mediterranean sea.

Columbus is widely thought of as the most successful of all the early European overseas explorers and the discoverer of what is now North America. In truth, he was neither. None of his four expeditions discovered gold or silver in any great quantities, and the only mainland he reached was a stretch of Central and South America on his fourth and last voyage.

His goal of reaching the East was never realized, and to his dying day he claimed that his discoveries were part of the Asian mainland, even if he must have realized they could not be so.

The riches promised by Columbus at the outset to his fellow explorers never materialized, causing huge discontent in the early colonies which he established on the islands of Hispaniola and Isabela (these are now part of the Dominican Republic).

Unmet expectations turned his tenure as governor of the islands into farce, and he lost control of his people.

By the end of his second voyage, in 1494, Columbus was desperate to find some source of wealth to placate his expeditions' patrons and financers. On the island of Hispaniola he ordered his men to seize 1,600 natives as prisoners, 550 prize specimens of whom Columbus sent back to Castile to be sold as slaves. Two hundred of them died on the journey, probably from the cold.[5]

Turning natives into slaves to appease Italian paymasters was particularly difficult for early explorers like Columbus because so often indigenous people turned out to be such excellent hosts. Columbus's first descriptions of the Taino tribes of Hispaniola were overflowing with praise:

> 'They are so friendly, generous and accommodating that I assure your highness that there can be no better people, and no better country, in the world. They love their neighbours as themselves and their speech is the sweetest and gentlest in the world, and they always speak with a smile. They go about naked, men and women alike, just as their mothers bore them. But believe me, they have very good morals and the King maintains the most marvellous ceremony with such dignity that it is a pleasure to see it all.'[6]

Yet, thanks to the failure to find significant sources of gold and silver on his first two voyages, Columbus's third expedition could be financed only by the sale of natives as slaves in Castile. And, thanks to a diktat from Queen Isabella herself, slaves could be traded legally only if they had been captured as prisoners of war. It is easy to see how good relations between European explorers and Native Americans soon turned sour, when one side counted on selling the other as slaves to finance its way of life.

Columbus's voyages were significant in at least two respects, however. To begin with, he cracked the code of the Atlantic winds that allowed European sailing ships to cross the ocean by sailing north-west and then returning by heading due east. Second, his discovery of land to the west began an epic rivalry between Spain and Portugal to explore the entire globe until it had been completely charted and mapped. Both countries wished to boost trade by fully exploiting whatever land they could colonize, and by gaining access to precious raw materials such as gold, silver and spices, or lucrative labour in the form of human slaves. By 1494 the Treaty of Tordesillas was signed, which, incredible as it sounds today, divided the world into two halves along a north–south line drawn 370 leagues west of the Cape Verde Islands in the mid-Atlantic. Everything discovered to the east would belong to Portugal, everything to the west to Spain. This was the first clearly articulated globalization strategy entered into by two European nations, explicitly designed to secure the domination of the whole world for their own exclusive benefit. To seal the deal Pope Julius II even sanctioned the arrangement with divine approval by publishing a Papal Bull in 1506.

The first man to confirm that the lands to the west were part of a new continent was Amerigo Vespucci, a slave merchant and jewel dealer from Florence. He found the east coast of South America in at least two expeditions, in 1499 and 1502. Vespucci published several accounts of his voyages in his book *Mundus Novis* (New World), in which, like Columbus, he described the natives as going around naked, conflicting with the then popular belief that life beyond the known world was a fantastic realm of monsters and savages.

One of the most popular late-medieval travel books was Sir John Mandeville's *Travels* (written between 1357 and 1371), which influenced Christopher Columbus and Amerigo Vespucci. Mandeville described a series of monsters that lived beyond the known world, including the giant one-eyed *cyclopes*, the single-legged *sciapodes* and the fearsome *anthropophagi*, whose mouths were positioned within their chests.

Instead of monsters, could there really be people still living in what sounded like the innocence of the Garden of Eden? Had they somehow escaped the curse of original sin? If so, how?

Pictures like this one by Theodore de Bry (engraved in 1562) appeared in sixteenth-century European travel books, showing naked native cannibals feasting off human flesh.

Sadly, the idea of these natives as forgotten, uncorrupted descendants of Adam couldn't be reconciled with the financial imperatives needed to make up for the lack of treasure found on these early expeditions. By 1507 the tales of the early European explorers had been doctored by editors and patrons eager to demonstrate that the lands to the west were a legitimate source of slaves.

Pictures published at the same time as Vespucci's second journal featured a map of the continent of South America, named 'America' after the Italian explorer himself. They also showed cannibalistic savages feasting on the flesh of human legs roasted on a spit. As early as the 1430s, Henry the Navigator had set the precedent when he described his African slaves as 'wild men of the woods'. In their later voyages Vespucci and Columbus reported evidence of cannibalism amongst some of the Native American tribes (the Caribs in particular). It became headline news. The cannibals, Vespucci claimed, bred with captive women 'and after a while, when diabolic fury overtakes them, they slaughtered the mothers and babies and ate them'.[7]

Was this true? Or did Vespucci elaborate his tales for the benefit of his main financial patron and former employer Gianotto Berardi, an Italian ship-fitter and slave-trading investor who, as one of a syndicate of aristocrats and bankers from Genoa, lobbied in favour of Spanish Atlantic exploration? It was this group which first raised Columbus to prominence at the court of Castile.[8]

Transporting and selling slaves, mostly from Africa, had been a profitable enterprise of Muslim merchants for centuries (see page 269). However, European Atlantic explorers very quickly discovered a new purpose for slaves, as an agricultural labour force. In 1419 two Portuguese sea captains in the service of Henry the Navigator discovered the unoccupied Atlantic island of Madeira. It was Henry's idea to attempt to extract wealth from this place by planting sugarcane – then so rare as to be considered a spice – which had been introduced into southern Spain from South-East Asia by the Islamic Caliphs of Cordoba (see page 239).

Growing sugar is especially labour-intensive because the canes must be planted by hand. Henry's plentiful source of cheap African slaves meant that by the 1450s sugar production in Madeira provided a new model for creating wealth: valuable crops grown in the right climate, supported by free labour, proved very lucrative indeed. It wasn't long before every explorer serious about making his venture financially viable adopted the practice. Columbus himself introduced sugarcane to the Caribbean on his second voyage, the same one on which he began to enslave the natives.

By 1500 Madeira had been transformed from a self-supporting community of about 500 settlers into a colony devoted to the growing of sugarcane with a population of about 20,000, the majority of them slaves. Between 1450 and 1500 the Portuguese brought more than 150,000 slaves into their overseas territories.[9]

As much as slave trading appealed to European investors, the idea of saving the souls of these poor wretches was also alluring to the Christian Church. Perhaps through maritime exploration a link could finally be established with the fabled realm of Prester John? This Christian King was thought to have been stranded in the land of the heathen sometime shortly after the life of Jesus. Indeed, European Crusaders had been searching out his lost kingdom for centuries. When knights on the Fifth Crusade in Egypt heard about the invasions of Genghis Khan in 1221, they thought perhaps he was the fabled Prester John.

World maps and epic tales of early explorers became bestsellers for the German publishers who had just developed the first European moveable-type printing press. Traditionally, Johannes Gutenberg is credited with this momentous invention in about 1450, although the first system of moveable type is thought to have been conceived in around 1040 by Bi Sheng, a Chinese inventor who used letters made out of porcelain. Metal moveable type emerged in Korea in about 1230. But it never really caught on in East Asia owing to the difficulties of printing in a language with thousands of different characters. However, when used with an alphabetical system containing just twenty-six symbols, rearrangeable letters turned into printing plates transformed the economics of mass-production printing. Twenty-three editions of Vespucci's *Mundus Novis* appeared between 1504 and 1506, making it the *Harry Potter* of its day.

News of Columbus and Vespucci's adventures travelled fast. Other countries began to sponsor their own expeditions, including Henry VII of England, who sanctioned a trip by Genoese navigator John Cabot, financed by merchants from Bristol. Cabot's aim was to reach the East by sailing westwards finding a passage to the north of modern-day Canada, which he reckoned would be quicker and shorter than Columbus's route further south. He successfully reached the east coast of Canada in 1496, only to be mysteriously lost at sea during a second voyage in the following year. The North-West Passage he was searching for eventually became navigable in the summer of 2007, thanks to recent global warming (see page 377).

Meanwhile, the Portuguese were storming ahead in their search for a sea route to the East, secure in the knowledge that, at least for now, they could legitimately claim whatever they discovered thanks to the agreement with Spain.

Vasco da Gama was the first Portuguese navigator to reach India, arriving at Calicut on 14 May 1498. His voyage proved that the route around Africa was the quickest and easiest maritime passage to the East. Embarrassed at having little or nothing to offer Indian merchants in exchange for silks and spices on his first voyage, he was forced to leave by an angry Muslim crowd who mocked him for having nothing worthwhile to bestow in return. Da Gama set sail again in 1502, accompanied by twenty warships. This time he brought back plenty of silk and gold, despite having precious little to offer in return, thanks to a mixture of piracy and coercion. His expedition provided Portugal with its first exclusive direct trading rights, and terminated previous arrangements with overland and overpaid Muslim merchants. After raiding a Muslim ship returning from Mecca, da Gama locked the 380 men, women and children on board in its hold and then set the vessel on fire. The episode was as symbolic as it was brutal. Indian rulers soon got the message.

Portuguese explorers established settlements along the coasts of the Arabian Gulf, India and Indonesia, even reaching Japan. Goa, Ormuz, Malacca, Kochi, the Maluku Islands and Nagasaki all became key trading posts. Control of the seas was theirs following the successful rout of the Ottoman fleet at the Battle of Diu, off the west coast of India, in February 1509.

During this battle, Portuguese forces defeated not just the Ottoman fleet, but also the Venetians who had sided with the Turks to protect the axis of their wealth that was based on Islamic trade (see pages 273–74). This Venetian–Turkish alliance had been consummated by a peace treaty signed in 1503 that secured new commercial privileges for Venetian merchants trading within Ottoman territories.

Not only could Portuguese merchants now establish their own exclusive Asian trade links, but

Vasco da Gama's caravel took him around the tip of Africa to the spice-rich lands of the Far East where he used a firm hand with incumbent Muslim merchants and rulers.

they could also act as naval shuttles between different eastern ports, earning additional money between seasonal winds.

Just as significant was the Portuguese discovery of Brazil by Pedro Cabral, on 22 April 1500. Whether he sailed there deliberately or was blown off course by a storm is still a matter of debate; however, at least the coastal areas of this land were legitimate Portuguese territory, being just to the east of the line marked out by the Treaty of Tordesillas. Settlers began a trade in the then abundant brazilwood trees from which the country got its name. They were felled for their precious red dye and used to colour rich textiles and velvets. (By the eighteenth century nearly all of these trees had been destroyed.) From the 1530s sugarcane plantations cultivated by indigenous slaves had taken over as the Portuguese colonialists' richest pickings, reinforcing the new trend of crops grown on one side of the world for transportation to the other.

Further north, conquistadors from Spain, fresh from victory over the Muslims of Granada, were eager to increase their fortunes through conquest abroad. Using settlements established by Columbus on Hispaniola and Cuba as their bases, Hernán Cortés and Francisco Pizarro were responsible for Europe's most devastating conquests of all – adventures that resulted in the complete annihilation of the Aztec and Inca Empires of Central and South America.

How did just a few hundred men and horses from Spain manage, within only a few years of landing on the American coast, completely to overwhelm empires that between them comprised an estimated twenty-five million people?[10] For many years historians believed the populations of Aztec and Inca peoples must have been very low, to account for the ease with which the Spanish soldiers took charge. It was also assumed that these were primitive societies living close to the edge of subsistence, or tribal people who could be easily overpowered with primitive guns and swords of steel. But during the late nineteenth and early twentieth centuries, as the sophistication and complexity of these empires began to emerge,

the issue became an area of intense speculation, revision and debate.

Historical records of what actually happened were mostly written by the victorious conquistadors themselves, and many of them are biased against the natives. However, it seems that for Cortés and Pizarro, and their warriors, success came from a lethal mixture of luck, cunning and forces of nature that stacked the odds heavily in their favour.

When Cortés landed on the Yucatàn Peninsula in the spring of 1519 with eleven ships carrying about 110 sailors, 530 soldiers, a doctor, a carpenter, a few women and some slaves, he was directly defying a last-minute order from the Spanish governor of Cuba to abandon his mission. The governor, Diego Velázquez de Cuéllar, tried to revoke Cortés's commission shortly before he left for Mexico, fearing that Cortés might use his arrival there as a launchpad for invasion and conquest. A replacement captain, Luis de Medina, was sent to take over, but Cortés had him murdered before he could reach port.

Cortés then had the good luck to chance upon a Spanish crewman who had been shipwrecked on an earlier expedition that had set off in 1511. Gerónimo de Aguilar had been captured by the native Aztecs and had lived among them as a slave for the previous eight years, learning their language and customs. Cortés also captured a beautiful native woman, called Malintzin, who spoke both Nahuatl – the language of the Aztecs – and Mayan, which was spoken by people living on the Mexican peninsula. Using them both as translators, Cortés could now make himself understood by just about anyone he might encounter on his quest for gold.

Malintzin later became Cortés's mistress and was baptized as Doña Marina. She was also known as La Malinche. She bore Cortés a son, who became one of the first mixed-race people, *mestizos*. Today she is regarded by Mexican people as both the embodiment of treachery and a mother figure.

With the vital asset of smooth communication Cortés was able to negotiate a military alliance with two native civilizations, the Totonacs and Tlaxcalans. Climate change was at least partly responsible for why these people had recently been subjected to vicious assaults by their enemies, the Aztecs. The 'Flower Wars' were brought on by the severe droughts that were sweeping through the region, in response to which the Aztecs sought new victims from neighbouring tribes to be sacrificed to their gods in their desperate bid to secure rain (see page 222).

Cortés marched towards the sacred Aztec city of Cholula, supported by 3,000 native warriors.

Conquistador Hernán Cortés meets his future lover, later baptized as Doña Marina, whose multi-lingual skills proved essential to his invasion's success.

Doña Marina warned him that the city's inhabitants might be planning an ambush, so Cortés sacked it, massacred its citizens and set it on fire. News spread fast. Eager not to suffer the same fate, natives living nearby either joined Cortés or kept well out of his way. Three months later, in November 1519, Cortés and his army finally stood before the glorious capital of the Aztec Empire itself.

Tenochtitlan, a city built in the middle of a lake and accessible by three separate causeways, was a fabulous artificial world of canals, walkways and bridges, the Venice of Central America. Canoes carried goods and people under the causeways and across a vast network of five interconnected lakes. Two terracotta aqueducts, each six kilometres long, took bathing waters from nearby springs to the city centre, since Tenochtitlan's 200,000 people were, like the Romans, devotees of spa baths. Typically

An Italian portrait of Moctezuma, the last king of the Aztecs, whose generous hospitality to visiting Spanish adventurers proved highly misplaced.

they took to the waters twice a day, cleaning themselves with soap made from the root of a plant called *copalxocotl*. Pregnant women and rich citizens even enjoyed the luxury of saunas called *temazcalli*, igloo-like buildings in which therapeutic herbs were thrown on to heated volcanic rocks, giving off soothing vapours to relax and heal those within. Three main streets crossed the city, each leading to one of the causeways that joined it to the mainland. In the centre were about forty-five public buildings, including schools, temples and, most important of all, a sacred ball court.

At the time of Cortés's arrival the Aztec Empire was ruled by a king called Moctezuma, who was renowned for his hospitality. He could afford it. His palace had more than a hundred bedrooms, each with its own en-suite bath. His grounds contained two zoos, a botanical garden and an aquarium, and were looked after by a staff of more than 300. At night-time about a thousand cleaners would take to the city streets. Garbage was collected by boat, and even human waste was recycled, sold to farmers as fertilizer. When Cortés arrived in the winter of 1519 Tenochtitlan was one of the largest cities in the world, and, excepting their traditional thirst for human sacrifice, possibly the most civilized.

Within eighteen months of Cortés's arrival the city, and with it the key to the entire Aztec civilization, had fallen into Spanish hands. After welcoming Cortés and his army as guests – one theory is that Moctezuma thought the Spanish were gods fulfilling an Aztec prophecy – the Aztec ruler soon found himself a captive in his own palace. Cortés demanded an enormous ransom in gold as the price of his freedom. As the weeks passed and treasure flowed into the Spanish coffers, resentment built up inside the city at the continued presence of Cortés's warriors and their native allies, and at the huge ransom demand. Six months into their stay panic struck the Spanish during an Aztec religious festival, when, according to Spanish accounts, they suspected the locals of plotting to kill them. Hundreds, if not thousands, of Aztecs were massacred by a pre-emptive Spanish ambush at the height of the

copolco
zoi micca ỹ
capitan.

evening's festivities. Cortés, who had been away fighting Spanish troops sent to arrest him by the furious governor of Cuba whom he had earlier defied, rushed back to the city to restore order. After hearing his tales of riches and gold, most of the new Spanish warriors switched sides, further swelling his forces.

By now the angry Aztecs were in outright revolt. Stones and spears were hurled at their puppet ruler, Moctezuma, who Cortés paraded on a balcony in an effort quell the riots. A few days later, the King lay dead. Accounts differ as to the exact reason for his demise. Perhaps he was killed by the Spanish, who had come to regard him as a liability. Or maybe one of his humiliated subjects had scored a direct hit.

At the beginning of July 1520, owing to the increasing hostility of the Aztec citizens and suspecting a surprise ambush, the Spanish were once again in fear for their lives. In a desperate attempt to save their skins they secretly slid out of the city by dead of night, only to be spotted as they stole across the causeway, just yards from the safety of the mainland. A fierce battle followed, with hundreds of casualties on both sides: this episode is called *La Noche Triste* – the sad night. Many of the Spanish were drowned in the lake, pulled down by the weight of the gold they were carrying. Cortés and most of his commanders successfully fought their way out, eventually reaching the safe haven of their native Tlaxcalan allies across the lake.

Cunning Cortés promised his new hosts that in return for their loyalty they could take the ultimate prize of the city of Tenochtitlan itself – no tribute payments required – once it had fallen. But by now nature herself was playing a part. Disease, in the form of smallpox, was taking hold of the Aztec people, probably transmitted by an African slave – who may have been a member of the force sent to capture Cortés by the governor of Cuba – killed in the battle to escape Tenochtitlan.

Highly infectious diseases including smallpox, measles, typhoid and influenza originated in large mammals such as cows, pigs and sheep.[11] Following their domestication, the diseases jumped across the species barrier into lethal new human forms. The absence of these mammals in the Americas meant that unlike Europeans, Native American people had built up no immunity against such infections.

The epidemic killed about 40 per cent of the Aztec population within a year. Weakened by disease and besieged by Cortés and his army, who cut Tenochtitlan's supplies of water and food, the city eventually fell. More than 240,000 Aztecs are

Aztec and Spanish forces clash at the battle for Tenochtitlan, although Cortés would never have won without the help of local Aztec enemies and the natural curse of smallpox.

estimated to have died during an eighty-day siege. On 13 August 1521 the Aztecs' new King, Cuauhtémoc, surrendered his Empire and gave himself up as a hostage to the Spanish. He was imprisoned, tortured and later executed on the orders of Cortés himself.

Two hundred thousand natives had fought alongside the Spanish, but despite their crucial assistance, Cortés's promises of reward came to nothing. Instead he took the opportunity of killing off their kings, for fear of further rebellions. Tenochtitlan was looted and then burned to the ground. In its place, the first foundations of modern Mexico City were soon to be laid.

The way was now clear for a massive Spanish colonization and conquest of the whole region. A viceroy of New Spain, as it was called, was appointed in 1524, and over the next sixty years a mixture of Spanish military superiority and smallpox devastated the Native American people.

Spanish soldiers who served in the army that spread throughout Mexico were paid in parcels of newly conquered land. This *encomienda* system, established on Hispaniola in 1498, was now extended to all newly conquered territories. In reality, it meant that the natives had their lands taken away, and became their new masters' slaves. It was not abolished until 1791. In 1546, silver mines like those at Potosí were discovered at Zacatecas, in north-central Mexico, highly prized by another swathe of Spanish settlers eager to get rich quick, who raided local villages as they came to acquire slaves for use as labour.

Meanwhile, not far to the south, Cortés's second cousin, Francisco Pizarro, was ready to try his own luck. Having received imperial approval for a conquest of Peru in 1529, and with natural forces in the form of highly infectious European diseases sweeping through the native population, his odds of success were improving fast. Pizarro landed near Ecuador with a small force in the spring of 1532. His mission, as the Spanish King's new viceroy, was to snatch South America's wealth from the mighty Incan Empire of the Sun.

By now smallpox had seriously weakened the Inca Empire, killing the Emperor Huayna Capac and many of his court in 1526. Their deaths plunged the Empire into a civil war between his sons Atahualpa and Huascar, so that by the time Pizarro arrived the Incas were already in severe disarray.

It helps explain how with just 106 foot-soldiers, sixty-two cavalry and three cannon, Pizarro defeated an Incan army estimated at some 80,000 strong. Faced with these apparently impossible odds, he borrowed tactics learned from his cousin Cortés, and on 16 November 1532 lured Atahualpa into a trap. Through an interpreter (another lucky find), arrangements were made for Spanish delegates to meet the Inca Emperor in the central square of a hilltop town called Cajamarca, where gifts and tokens of friendship would be exchanged.

Atahualpa left his massive army encamped outside the city walls. As he approached the town square he was accompanied by a small household retinue of nobles and concubines. He left his weapons behind as a mark of friendship and trust.

Atahualpa, last sovereign Inca Emperor, was executed by Spanish invaders in 1533 after being caught in a carefully contrived trap.

ATAHVALLPA ͂INGA XIIII
ESPVRIO Y VSVRPADOR D. YMPERIO
VLTIMO YMPERA.R

Our knowledge of what happened next comes from accounts written by the Spanish. They claim that the Emperor was handed a Bible and ordered to renounce his pagan religion and accept the word of Jesus Christ. When Atahualpa threw the Bible to the ground in confusion, Pizarro's men appeared on all sides of the square, then charged on horseback to the accompaniment of cannon fire. The surprise attack provoked panic. Never before had the Emperor and his retinue seen horses or heard gunfire. Many of the unarmed Incas were slaughtered, and the Emperor himself was captured and imprisoned in a small stone room.

With no one to lead them, the Inca army fled in panic. Over the next few months Atahualpa's 'ransom room' was piled high with treasure brought on the understanding of his eventual release. But when he was charged with twelve trumped-up crimes, including revolting against the Spanish, practising idolatry and murdering his brother Huascar, his fate was sealed.

After eight months in captivity Atahualpa was dragged out of his chamber, tied to a stake and made to sit on a stool in front of the assembled conquistadors. Thanks to an enforced baptism he was spared being burned alive as a heretic. Then, according to the diary of one conquistador, the Inca Emperor wept, pleading with Pizarro to look after his two sons and daughter. The remaining Inca lords and women who were still in his attendance began to wail, prostrating themselves on the ground. A rope was tied around Atahualpa's neck, and with one hard wrench of each end, he was garrotted. All night his body remained in the square, seated on the stool and tied to the stake. His head was slumped to one side, and his arms and legs were covered in blood.[12]

As in Mexico, tribes once loyal to the leaderless Incas now switched sides and helped the Spanish forces conquer the Inca capital, Cuzco. Pizarro installed a puppet ruler, but he soon died of smallpox, which was still sweeping through the Native American populations.

A new Spanish city was built at Cuzco, and Pizarro founded Lima as the capital of Spain's South American empire, because it was within easy reach of the sea. For the next ten years, however, fighting among the conquistadors themselves plagued the Spanish conquest, leading ultimately to the death of Pizarro. On the night of 26 June 1541, a gang of twelve assassins slaughtered him on behalf of his arch-rival Diego de Almagro, who felt he had been cheated out of his rightful share of the treasure from Atahualpa's ransom room. They stabbed him several times, after which Pizarro, conqueror of Peru, is said to have collapsed to the floor and painted a cross in his own blood, while crying out for the mercy of Jesus Christ.

It wasn't just Pizarro who may have felt a final pang of guilt at what he had inflicted on the Incas. The last testament of one of the soldiers who accompanied him, Mansio Serra de Leguizamon, written in 1589 and addressed to King Philip II of Spain, shows how much the consciences of some of the conquistadors had been pricked by the horrors of what had happened. With his final thoughts turning to the fate of his soul, Leguizamon movingly wrote:

'It should be known to his most Catholic Majesty that we found these realms in such order that there was not a thief, nor a vicious man, nor an adulteress … nor were they an immoral people, being content and honest in their labour … I wish your Catholic Majesty to understand that the motive that moves me to make this statement is the peace of my conscience and because of the guilt I share. For we have destroyed by our evil behaviour such a government as was enjoyed by these natives … They were so free from the committal of crimes that the Indian who possessed 100,000 pesos' worth of silver or gold in his house left it open by merely placing a small stick across the door as a sign he was out … But now owing to the bad example we have set them in all things these natives have changed into people who now do no good, or very little, something which should touch Your Majesty's conscience as it does mine, as one of the first conquistadors and discoverers, and something that requires to be remedied.'[13]

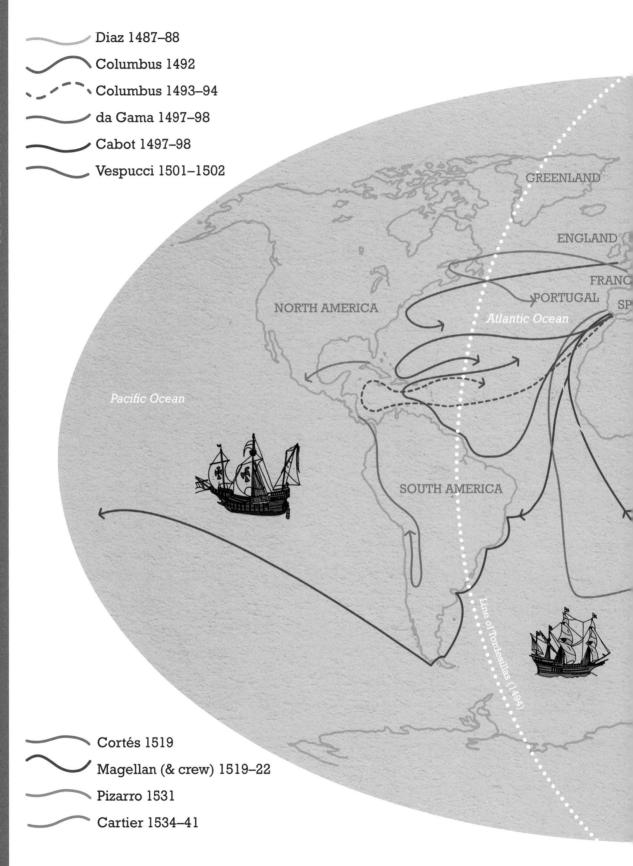

Diaz 1487–88
Columbus 1492
Columbus 1493–94
da Gama 1497–98
Cabot 1497–98
Vespucci 1501–1502

GREENLAND

ENGLAND

FRANC

PORTUGAL

SP

NORTH AMERICA

Atlantic Ocean

Pacific Ocean

SOUTH AMERICA

Line of Tordesillas (1494)

Cortés 1519
Magellan (& crew) 1519–22
Pizarro 1531
Cartier 1534–41

European Exploration

European mariner-cum-mercenaries flexed their muscles by charting the oceans and conquering previously isolated lands.

Arctic Ocean

SIBERIA

RUSSIA

CHINA

JAPAN

Pacific Ocean

ARABIA

INDIA

PHILIPPINES

FRICA

Indian Ocean

AUSTRALIA

Cape of Good Hope

23:59:59

No one can be sure how many Native American people died during those holocausts of war, disease and slavery that accompanied the Spanish invasions of Mexico and Peru, but estimates range from between two million and a hundred million. More than 90 per cent of them are thought to have been the victims of European diseases, dramatically changing the course of human history and clearing the way for new European colonization.

By 1546 the ancient Aztec and Inca sources of silver had been discovered by the conquistadors not only in Mexico, but also in Peru. The silver mines of Potosí were now being worked by slaves, and vast quantities of new wealth were being shipped back to the growing power of Spain.

Back in Europe, the effects of Spain's successes in the New World, and the Portuguese control of trade in the Far East, were beginning to make themselves felt. Charles V, grandson of Ferdinand and Isabella, came into a golden inheritance that gave him not only the kingdom of Spain, but lands in Italy, Austria, and the Netherlands. When he was elected Holy Roman Emperor in 1519, his authority stretched across Germany too. The stunning victories of Cortés and Pizarro meant that Italian bankers were more willing than ever to loan Charles money against the security of treasure from the New World. Charles could spend this money on his armies, at war across Europe where he was attempting to unite his lands into one contiguous Catholic whole.

Charles, his coffers already filled by wealth from overseas, was now boosted by another piece of good news. In a new attempt to find the spice islands of the East by sailing west across the Atlantic, Ferdinand Magellan, a Portuguese captain hired by Charles in 1519, had discovered a secret that enabled some of his crew to become the first men to circumnavigate the globe.

In 1521, after an epic three-year journey, eighteen members of Magellan's original crew of 270, including one paying tourist, Antonio Pigafetta, eventually struggled back to Spain in the ship *Victoria*, their cabins stuffed full of precious cloves and other spices – twenty-six tonnes in all.[14]

Magellan himself never made it home. After instigating an attack against a group of natives on the Pacific island of Mactan, he was hacked to death by them with spears and swords.

But the worn-out survivors of the *Victoria* also carried with them the knowledge of something far more valuable even than cloves and spices: a channel that connected the southern Atlantic to the Pacific Ocean (later called the Strait of Magellan), thus removing the necessity of sailing through the dangerous waters off the tip of South America.

Thanks to such voyages Charles now knew the full extent of the world's lands. With its secret continents firmly printed on imperial maps, he knew the earth's exact size, its fast-track channels and the direction of the trade winds that could take merchants and explorers north, south, east and west. What did this knowledge do to his ambition? Did he dream of becoming the Christian Emperor of the whole world?

Whatever his fancies, one thing was certain – for any European power to succeed in the future it was necessary to be supreme at sea. A new strategic direction for Europe was now evolving fast. Those nations with the strongest fleets had proved they could successfully import wealth from any part of the world, be it stolen from other empires, grown on foreign plantations or dug out of the ground. Natives and captives could be deployed as slaves, from the mines of Potosí to the sugar plantations of Brazil.

Here, at last, was a potential release from medieval Europe's misery – its lack of food, its need for wealth and its perilous encirclement by infidels.

Fancy a Beer?

How European merchants pioneered a new way of life overseas that fuelled a taste for growing lucrative crops, making some very rich and many more poor.

ON A COLD SPRING DAY in 1621 a band of forty-seven English pilgrims and their families were engaged in a desperate battle for survival. Six months before, their ship, the *Mayflower*, had brought them on a traumatic voyage across the Atlantic from Plymouth. But since they had landed on the coast of north-east America, disaster had struck. More than half of their company had died from disease and starvation, and the prospects for the rest looked grim. The land all around was deserted and bleak. Food and supplies were running critically low, and all they had in the world were a few wooden huts which they had built on some rocky ground to keep them dry. Not wanting to forget where they had come from, they called their new home Plymouth.

Imagine their surprise when on that bright March morning a young, tall, handsome Native American man strolled boldly into the middle of their camp and said in broken English, 'Welcome, friends! Anyone got a beer?'

What had possessed these pilgrims to flee Europe and risk their families' lives 5,000 kilometres away across the seas? How did this Native American man come to speak English? Why did he want a beer? The answers to these questions explain a great deal about what on earth happened to the world in the 200 years after Europe's first circumnavigators struggled back to the Spanish port of Seville in September 1521.

❂❂❂❂❂

England's pilgrims came across the seas because Europe had become embroiled in a series of pernicious religious wars. Savonarola's complaints about the abuses of the Papacy, that led to

revolution in Florence in 1494 (see page 276), were a limited north Italian affair. But just twenty-three years later, when Martin Luther nailed his own series of ninety-five complaints on to the door of a church at Wittenberg Castle in north-east Germany, the world was a different place. Thanks to the invention of Gutenberg's printing press, rulers across Europe, from Germany, Holland and England to France, Sweden and Denmark had by now heard of the wealth to be gleaned by trading slaves, spices and precious metals overseas.

Any excuse that could help stop Spain, with its newfound wealth, gaining supremacy was eagerly pursued. These envious rival powers shared a common objective, even if their unity was as self-interested as it was fickle: Charles V's attempt to unify his inheritance must be defeated at all costs. The key to their future prosperity lay in rivalling his power at sea and that meant preventing Spanish access to vital sea ports in northern Europe.

While northern Italy represented Europe's most ostentatious pocket of wealth, another trading network had been steadily increasing in power and influence further north. Around the Baltic Sea lay the cities of Danzig, Riga, Hamburg and Lübeck, which since the 1200s had been part of a larger trading cabal called the Hanseatic League. Gradually these ports had gained substantial wealth

from the merchandising of mundane but necessary products such as wool, timber, grain and fish. By the mid-fifteenth century merchants from Holland, based in Amsterdam, had usurped much of the former Hanseatic wealth following a war (1438–41) which broke its monopoly of trade. If seapower was vital for future financial and political success, then control of the timber supplies needed to build invincible fleets was key.

By 1520 parts of Europe were already suffering a chronic shortage of the right kind of wood. Venice was weakened by its lack of suitable timber, most of which had been stripped over centuries of shipbuilding to secure its pre-eminence on the Mediterranean. Now it was forced to trade with rulers on the Dalmatian coast, who still had supplies of the large firs it needed to make its ships.

Tapping wealth from across the Atlantic required new types of vessel. Masts had to be tall and strong enough to catch the ocean winds, ships had to be big enough to carry heavy iron cannon, and sturdy enough to survive at sea for months, if not years. Rotten boats that leaked would either sink or be sunk. Timber for making such vessels grew in the oak forests of southern England and around the Baltic Sea. But Spain had none. Even if Charles V could secure supplies from overseas, he knew his rivals were building their own fleets fast.

British immigrants prospecting the coast of North America from their ship, the Mayflower, during the winter of 1620–21.

money? He built himself a fleet. By his death in 1547 the British Royal Navy could boast fifty-seven well-armed ships.[1]

Meanwhile, popular hysteria about the abuses of the Church of Rome had quickly turned into a full-scale European religious schism known as the Reformation. It was exacerbated by a combination of Luther's protests against people having to pay Papal taxes (called indulgences) for their sins to be forgiven, and the mass printing of Lorenzo Valla's damning evidence that the Papacy's claim to its territories in Italy was based on a lie (see page 275). It was faith in Jesus alone that could save people from their sins, argued the eloquent Luther, not the payment of taxes or doing 'good works' as decreed by the Catholic Church. Other reformists like the French theologian John Calvin (1509–64) believed that a selection process had already taken place that predetermined which souls would ultimately go to heaven. No amount of taxation could possibly make a difference.

Lands controlled by German princes lay between Charles's southern kingdoms of Spain and Italy and his birthright in the Netherlands at the mouth of the Baltic Sea to the north. Technically these princes owed their allegiance to Charles, as Holy Roman Emperor, but some of them gave their support to Luther and his followers, seeing that Protestant creeds could provide useful religious legitimacy for armed resistance against increasingly threatening imperial authority.

In 1521 Pope Leo X declared Luther a heretic and excommunicated him. But he was protected from arrest by the German Prince of Saxony. Then, in 1545, the Pope, with Charles's support, opened a grand debate called the Council of Trent to try to help heal the schism by reforming some aspects of the Catholic Church, such as providing a proper education for the clergy. A new movement of pious Catholic missionaries called the Jesuits, established by a former knight, Ignatius Loyola, was created to lead this educational mission. These 'soldiers of Christ', who swore absolute obedience to the Pope, planned to spread the reformed Catholic Church's message across the globe. But by now it was too late. Despite eighteen years of talks,

Cutting them off by uniting his empire, from Italy to the Netherlands, became his biggest priority, offering the prize of additional wealth through the control of precious raw materials. Battling to unite his inherited lands both north and south became a preoccupation from which his family, the Habsburgs, and the Spanish Empire never fully recovered.

Conflict was already raging between Europe's powers during the period from 1494 to 1530, when Spain and France were locked in a series of wars over the wealth of Italy. Their conflicts culminated when, fearing that Charles's influence in the region was becoming too strong, Pope Clement VII switched sides to support France. Charles's imperial army marched on Rome in 1527, and sacked it. Clement only just escaped with his life after paying a hefty ransom. From that moment on, the Pope thought it best not to offend the Emperor ever again. So when the English King Henry VIII requested a divorce from Charles's aunt Catherine of Aragon, Clement VII resolutely refused.

It was a momentous decision. Hot-tempered Henry was outraged, and declared himself, and not the Pope, supreme head of the English Church. He then used this as an excuse to plunder ecclesiastical wealth. More than 800 abbeys and monasteries, made rich by centuries of forest-clearing, were confiscated in the famous 'Dissolution of the Monasteries'. And what did Henry do with the

23:59:59

the Council found it impossible to prevent countries and princes establishing their own breakaway Churches.

The rise of Protestant Churches reflected the growing independence and confidence of Europe's emerging powers, especially those close to precious trade links and raw materials from the north. Thanks to the early explorers, these nations could now see opportunities for enriching themselves from trade overseas. John Cabot had already impressed the merchants of Bristol with his voyage to North America, and Jacques Cartier, a French navigator, sailed to Canada in a series of expeditions for the French King Francis I in the 1530s, becoming the first European to see the St Lawrence River.

Following the abdication of Charles in 1556, Habsburg attempts to unite its Spanish and northern European possessions reached a climax. Eighty years of war raged between Spain and the seventeen provinces of Holland (1568–1648) begun by Charles's son and successor Philip II.

But thanks to the Reformation, revolt against imperial rule was easy to justify. Christianity's all-powerful God could now be deployed on all sides, each claiming for itself the one true Church. It was a fitting religious scenario for a continent already fragmented linguistically and politically. Emergent nations pitched themselves headlong into a violent contest for access to global resources. Holland, Denmark and Sweden became the most fervently Protestant nations. France dithered, swinging in the wind between Catholic and reformed, but for a while sank into in its own desperate wars of religion (1562–98), in which hundreds of thousands of people perished, and which were resolved only when the Protestant King Henry IV agreed to become a Catholic.

Although England had been early to break ties with Rome, the question of which faith to follow continued to plague its politics well into the next century. The pendulum swung first to the Catholic side with 'Bloody' Mary, the wife of King Philip II of Spain, during whose reign (1553–58) well-off Protestants were hunted down, tried and burned at the stake as heretics. But when her Protestant half-sister Elizabeth I (ruled 1558–1603) took

over, a more pragmatic approach was employed. She established a middle way, with an Anglican Church that could accommodate as many of her subjects' religious inclinations as possible. After her, the Catholic Charles I (ruled 1625–49) stirred the religious passions of the country once again, which played a key part in the build-up to the English Civil War of 1642–51. It ended with Charles's execution and a radical puritan government headed by Oliver Cromwell, a parliamentarian reformer whose zeal for religious purity saw Christmas abolished and dancing banned. His ten-year protectorate was followed by the re-establishment of the monarchy under Charles II in 1660, from which time England, now united with Calvinist Scotland, joined the Protestant gang for good.

Such wars came at an incredible cost. Silver from Mexico and Peru began to pour into Spain from the 1560s onwards. It didn't stay there long. Spread by the purchase of firearms, armadas and mercenaries, silver from the mines of Potosí underpinned a new bullion-based economy across all Europe. Pirates of all nationalities, eager to plunder their share of the New World's spoils, cruised greedily off the coasts of the Caribbean, many of them becoming national heroes at home. Sir Francis Drake – knighted in 1581 as Britain's first circumnavigator, is remembered to this day in Spain as 'el Dragón' – one of the most successful of all the pirates of the Caribbean.

So concerned was Philip II about wealth leaching away from Spain that he launched a mighty armada against Britain in 1588, reckoning that it was easier to cut off the flood of piracy through full-scale invasion of its homeland than by trying to defend his fleets in the distant Caribbean. Tradition says that in large part it was the weather that saved England from the 130 ships sent by Philip. In reality, military advantage at sea was now passing to north European nations like Britain and Holland. Smaller, more manoeuvrable ships made of superior timber could outmanoeuvre the larger, more cumbersome Spanish vessels. Military innovations such as ships' cannon mounted on four-wheeled carriages had been deployed since

Spanish ships sink off the coast of Plymouth, England, in May 1588 during the defeat of the Armada.

Henry VIII's day, making loading and reloading between shots faster and more efficient. But this improvement never caught on in Spain. Of the eight Spanish armada shipwrecks that have been studied by historical archaeologists, many have been found with substantial stocks of gunshot still on board.[2]

Eighty years of war against the Spanish gave the Dutch a powerful incentive to pioneer new tactics aimed at securing supremacy at sea. The forty-gun, 300-tonne frigates built at Hoorn in northern Holland were ideal for long-distance voyages aimed at capturing fleets of Spanish treasure. By the 1630s these ships had adopted new tactics, sailing side-on and blasting their cannon in a broadside at the opposition. Such 'ships of the line', as they came to be called, more than proved their power in 1639, at the Battle of the Down in the English Channel, where the Dutch navy sank forty of fifty-three Spanish ships. Oliver Cromwell's republican

government was quick to see the strategic importance of ships that could fight effectively in faraway seas. In 1649 Cromwell ordered seventy-seven new frigates to be built, substantially boosting the power of the British Royal Navy.

Military build-ups and religious wars shifted the balance of power in Europe following the initial successes of Spain's emerging overseas empire.

Desperate for more wealth to finance his endless European wars, and in flagrant violation of the Treaty of Tordesillas, Philip II sanctioned an expedition from Mexico westwards across the Pacific. It resulted in the Spanish settlement in 1565 of an archipelago of more than 7,000 islands called the Philippines, named after the Spanish King himself. From here came valuable spices like cloves and nutmeg, highly prized as herbal cures for European indigestion and as painkillers for rich people suffering from toothache. Spain would rule the Philippines for 333 years, from 1565 to 1898.

23:59:59

Despite such lucrative possessions, by the time Philip II died in 1598 Spain was bankrupt, its wealth scattered across the warring nations of Europe. So many loans had been taken out from banks in Genoa and Augsburg that 40 per cent of all Spanish income was being spent on interest payments alone. Prices had become hugely inflated with the influx of American silver and instead of investing those riches in local wealth-producing industries, Philip's war legacy left Spain the sinking power of Europe.

At the end of another thirty years of devastating war between 1618 and 1648, Spain's desperate attempt to secure power over northern Europe finally came to an end. The independence of Holland and the rise of Sweden as a global power were reluctantly acknowledged by Spain at the Peace of Westphalia in 1648, when an international system for securing the sovereignty of Europe's separate nation states was finally hammered out. Matters of religion, taxation, politics and foreign policy were now legally acknowledged as the province of national governments, not Holy Roman Emperors or the Catholic Church in Rome.

Europe's future strategic direction was now clear: a continent carved up into autonomous national fragments, each pursuing whatever policies suited its own individual interests in a global competition to secure wealth overseas. The Peace of Westphalia ratified this intercontinental policy for the next 300 years.

This was why the pilgrims of Plymouth Rock had sailed across the seas – their home in England was being torn apart by religious war, and they hoped to find a better life abroad. Initially they had fled to Holland, but later they sought permission to sail to America, where they hoped to preserve and extend their identity, language and culture, and to serve as evangelists, taking their reformed Protestant religion with them. Puritan followers swelled their numbers over the next twenty years as England's Civil War between Protestants and Catholics wore on, so that by 1642 about 21,000 English puritans had emigrated to North America.

The Native American who strolled into the pilgrim settlers' camp on 16 March 1621 was called Samoset. He had learned his broken English from some English fishermen who came to the shores of North America to fill their nets with cod. Samoset said he had a friend called Squanto who spoke even better English, having learned the language after twice being captured by the English. The second time he was taken by a sea captain called Thomas Hunt, who was searching for fish, corn and slaves to sell in Spain. After an epic voyage, Squanto escaped Spain and fled to England. When he eventually arrived home in 1619 he found that his tribe had been struck down by disease. Most of his friends and family were dead.

Although European settlement of North America was partly motivated by a desire to seek religious freedom, the main reason for many was, as for Hunt, the prospect of getting rich quick through trade. By 1600 British, Dutch and French adventurers were busily carving out areas of the American coast as settlements. Companies were established by Royal Charter to coordinate the allocation of territories and provide start-up capital investment in the form of ships and supplies. In return, the settlers were expected to export whatever treasure or produce that could be found, be it fish, corn or slaves.

As early as 1584, Sir Walter Raleigh, an English explorer and favourite of Queen Elizabeth I, tried to create England's first North American colony, naming the place he found Virginia in honour of his queen's celebrated chastity. But when the early settlers failed to find gold they got fed up and went back home. A second attempt at colonization three years later included more farmers and families, but within a few years it had mysteriously disappeared, falling victim either to starvation or to suspicious natives. Fortunately for the pilgrims of Plymouth, Samoset and his friends were different. They helped the English settlers plant their first crops, and showed them where it was best to catch fish and eels.

A Royal Charter was granted to the Virginia Company in 1606, with the express charge of unlocking new investment for colonizing the

Americas. Captain John Smith, a soldier from Alford, Lincolnshire, was designated leader of the first settlers, and he established a colony, Jamestown, in 1607. Settlement on Bermuda followed two years later. The question of how to make a return on investment was solved when a farmer from Norfolk called John Rolfe struck upon the idea of growing tobacco. He brought the first seeds of a sweet, palatable variety of the plant with him from Trinidad, and established a plantation on the banks of the James River. The prospect of becoming rich from the sale of tobacco attracted new investment and more settlers.

Rolfe was invited back to London, where he and his new wife Pocahontas, the daughter of a Native American chief (baptized and now called Rebecca), were paraded in public by the Virginia Company. Like a pair of modern pop stars and carrying their first child, they even secured an audience with the King. Prospective settlers could now see for themselves how attractive were the natives, and how easily they could be tamed and brought into the Christian faith. Settling the New World was advertised not just as a way of getting rich, but as a Christian duty to spread the word of God and civilize the natives. From 1619 a new scheme, called the Headright system, was introduced, in which colonists were offered fifty acres of land each and a paid passage to the colonies in return for between four and seven years of service in the tobacco fields.

Shortly after setting sail on the return journey to her homeland, Pocahontas (alias Rebecca Rolfe) fell violently ill, possibly with pneumonia, tuberculosis or smallpox. She was taken ashore and died at Gravesend in Kent, and was buried in the vault of St George's church.

Despite the inspiration and encouragement of her visit, there simply weren't enough colonists from England to exploit the tobacco plantations in Virginia and the sugar plantations in Bermuda to the full. Enslaving Native Americans proved awkward, since relations between them and the settlers were already tense, often flaring into violent skirmishes. Many of the natives fled westwards, towards the still unknown centre of the continent. Their numbers were decreasing anyway, owing to the continued outbreaks of smallpox and other European diseases.

By the end of the century a new solution to the shortage of labour, pioneered by the Portuguese in Madeira and then by Spanish settlers in Mexico and South America, was adopted by the British, French and Dutch colonies. Merchants from Europe would engage in a 'triangular trade', loading their vessels with products like sugar, tobacco, furs and timber from the Americas and shipping them to Europe for a profit. They then sailed on to the coast of West Africa, where tribal leaders sold them captured enemies as slaves, to be transported in appalling conditions to the plantations of the New World. Payment was usually made in guns or gunpowder, enabling such leaders to enslave yet more native rivals.

Enslaved Africans fetched a high price at auctions in the Americas since it was only thanks

Squanto, who learned his English after being captured by fishermen and taken back to Europe, was among several Native Americans who helped early English settlers plant their first crops.

to their unpaid labour that more plantations could be profitably established.

By 1713 there were eight slaves to every white man on the colony of Jamaica, which the British had captured from Spain in 1655. Tobacco plantations grew so quickly that annual exports to London of 22,000 pounds in 1619 had grown to twenty-two million pounds by 1700. French settlers established cotton plantations in the south of the North American continent, naming their first mainland settlement Louisiana, after their illustrious King Louis XIV. New France, as the region was called, became a vast expanse of increasingly colonized land stretching northwards up both sides of the Mississippi River. Sugar plantations were established on the French Caribbean island of Haiti, adding to fur-trading settlements further north in Quebec and along the St Lawrence River. European merchants now had exclusive access to a wide range of new commodities they could sell in their home markets and abroad without having to go through Muslim merchants.

European colonization of North America was largely a contest between France and Britain, the British position being secured by their takeover of Dutch and Swedish settlements, including New Amsterdam (which was then renamed New York) in 1674. This handover followed the third Anglo-Dutch War (1672–74) culminating in the Treaty of Westminster.

Holland's overseas empire collapsed further following the 'Glorious Revolution' that brought the Dutch Prince William of Orange and his wife Mary to the throne of England in 1688. From this time on Dutch merchants were increasingly encouraged to trade in London rather than Amsterdam. Dutch ships were now frequently captained by English seamen. Then, from about 1720, Portuguese sugar plantations in Brazil – made possible by the importation of more than two million African slaves – seriously undercut the prices that Dutch merchants could command for sugar from their trading posts in the Far East, removing a further source of wealth for Holland.

More North American settlement opportunities were scouted out by English adventurers like George Fox (1624–91), founder of a new anti-clerical religious movement called the Society of Friends, also known as the Quakers. Fox travelled extensively around the American colonies, establishing very good relationships with the Native American tribes, who he described as 'loving' and 'respectful'. It was his friend, Kentish ironworker William Penn (1644–1718), who founded Pennsylvania in 1677 by seeding 200 settlers from villages in Hertfordshire, England, centred around a new town they called Burlington.

Penn was granted this land to settle a £16,000 debt owed by the English crown to his father. He implemented a form of local government that helped shape many subsequent American institutions, with elected representatives, freedom of religious worship and trials by jury. French, German and Jewish settlers all came to Pennsylvania, attracted by the toleration written into the settlement's founding charter. It grew quickly, buying land from local Native Americans rather than forcing possession through conquest. For example, Penn is reputed to have paid more than £1,200 for land near Philadelphia in a treaty signed with the natives under the shade of an elm tree.

Shortly after Penn's death, a total of thirteen separate British colonies had been founded, the last being Georgia in 1732.

Samoset enquired if any of the pilgrim settlers had a beer because at some point during his encounter with those English fishermen off the North American coast he must have been offered a new drink which he found was rather to his liking, even if a little addictive. Contact with exotic commodities that overseas lands had to offer was also changing the palates of people in Europe, exposing them, like Samoset, to a wealth of new tastes and sensations.

Since Marco Polo wrote his famous travel diary the imaginations of explorers like Columbus had been fired by the idea of escaping Europe's misery by exploiting natural wealth found overseas. Their success created a new European taste for luxury.

Capital cities like London, Paris and Amsterdam filled up with rich merchants able to extend the cultural renaissance that had already taken place in the city states of northern Italy by buying and commissioning fabulous works of art, music, sculpture and architecture.

Fine wine and sweet foods became far more commonplace, made possible with the arrival of cheap sugar from plantations overseas. Even poor people became addicted to tobacco or snuff, turning the continent into the soft-drugs centre of the world. In 1575 a well-travelled Swiss physician calling himself Paracelsus is credited with developing laudanum, a compound based on opium imported from the poppy fields of Asia, as a painkiller. It later became nineteenth-century Europe's most popular elixir.

Hot chocolate, a drink made from cacao beans that grow naturally in Central America, was first sampled by Cortés while he was being entertained at the court of Moctezuma. By 1544 it was being drunk at the royal court in Spain, and within a hundred years people in France, England and other countries in western Europe had become addicted to its delicious flavour. French settlers established cacao plantations in the Caribbean, as did the Spanish in the Philippines. From these natural products came new tastes. Demand led to more plantations. New markets were made.

Between 1650 and 1800 the annual sugar consumption in the UK increased 2,500 per cent, to reach twenty pounds per head. By 1800, there were more than 62,000 licensed tea shops nationwide. In fact, tea was so plentiful that by 1784, on any given day, London's shops' shelves carried over 146 tonnes of it.[3]

Even though by 1800 medical science had not significantly increased human life expectancy, new wealth and better food helped European population levels to rise dramatically once again, increasing from eighty-nine million in 1600 to 146 million 200 years later.[4]

With new tastes came a confidence in the belief that perhaps humans were, after all, set apart from and superior to all other forms of life on earth. It was an old idea, reinforced by the Christian Bible, now available in mass-produced editions printed in vernacular languages for all to read. Protestant preachers zealously confirmed that from the opening Book of the Old Testament, Genesis, God made man to have dominion over all the world, its creatures and nature's bounty. After God had spent the first six days of creation making all things perfect for the arrival of humans, there was no question in the mind of John Calvin that 'He created all things for man's sake'.[5]

Religious validation that the natural resources of the earth were indeed at man's disposal was

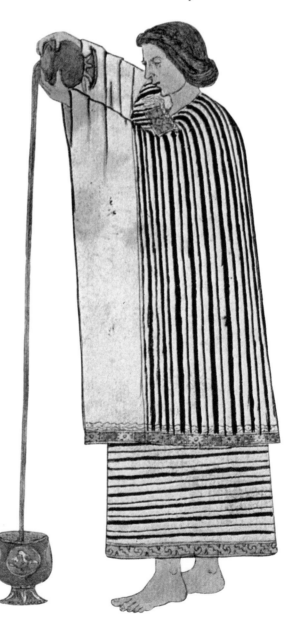

Chocolate, which originated in the Americas, being made by a sixteenth-century woman from Mexico.

ratified philosophically with the beginnings of empirical Western scientific thought. A radical book published in 1543 by Polish astronomer Nicholas Copernicus (1473–1543) demonstrated how the earth revolved around the sun – undermining traditional religious assumptions that the earth was at the heart of God's creation. No longer was mankind at the end of a long, divine chain of being. Instead, the world was part of a much larger mechanical system of planets and stars, rotating according to physical laws as first proposed by Greek philosophers like Thales and Anaximander in pre-Christian times (see pages 183, 394).

Copernicus's theories, confirmed by the observations of Italian astronomer Galileo Galilei (1564–1642), led to a revival of interest in ancient Greek scientific thought, which placed man's understanding of the universe at the centre of philosophical and scientific contemplation. Francis Bacon (1561–1626), an English philosopher regarded as one of the founding fathers of modern science, clearly expressed the new humanist view:

'For the whole world works together in the service of man, there is nothing from which he does not derive use and fruit … In so much that all things seem to be going about man's business and not their own.' [6]

Moral arguments about the rights of Europeans to exploit resources overseas were powerfully put forward by Dutch philosopher Hugo Grotius (1583–1645), whose book *Mare Liberum* (*The Free Seas*, 1609) stated that the sea was international territory, and that all nations who were capable of exploiting overseas resources had the natural right to do so.

The greatest of all philosophers who celebrated Europe's newfound confidence was René Descartes (1596–1650). This French philosopher, scientist and mathematician considered that mankind could gain insights into the workings of nature by breaking it down into small parts that could be understood through observation and reason. In a direct echo of Aristotle's treatise 'On the Universe' (see page 188),

Descartes expressed a belief that only humans have minds, reducing all other things, animate or inanimate, to mere machines that can be manipulated, dissected and exploited without feeling or scruple. 'I do not see any difference,' wrote Descartes, 'between the machines made by craftsmen and the various bodies that nature alone composes'. [7]

Descartes's confidence in man's unique capacity to understand nature slotted in perfectly alongside the views of other pioneers of modern Western science, such as the Englishman Sir Isaac Newton (1643–1727). Nearly 2,000 years after Aristotle first wrote about the constant laws of nature governing the universe, Newton finally seemed to have cracked the code of how the physical world actually worked. His three laws of motion quantified and explained the movement of all things, from the orbit of the planets around the sun to the motion of an apple falling from a tree. God, it seemed, really was the designer of a remarkable machine. Humans were uniquely blessed with the reasoning power to fathom out and understand how it worked, if only they looked carefully enough and engaged in the process of experimentation and scientific analysis.

Cogito ergo sum: René Descartes's philosophy reduced the natural world to the status of a machine and, in an echo of biblical Genesis, reinforced the idea of man's role as its master and overlord.

Europe was establishing itself as the continent of human progress. By the end of the seventeenth century France was Europe's most powerful state. Its King, Louis XIV (ruled 1643–1715), enjoyed wealth and power on a scale unknown since the most decadent days of ancient Rome. Most of Spain's riches were now in foreign hands, culminating in an enormous dowry of 500,000 gold ecus for Louis' hand in marriage to Maria Theresa that was supposed to be forked out by her father Philip IV of Spain in return for peace. In the end, Philip was too bankrupt to pay and Maria was obliged to renounce all future claims to the Spanish throne.

Under Louis' Minister of Finance, Jean-Baptiste Colbert, France's wealth grew ever greater. Colbert established a range of new industries in France in his efforts to keep wealth circulating within his country, and not to let it fritter away like Spanish silver had to rival nations. Silk manufacturing was established in Lyon, glass-blowers came from Murano in Italy, weavers settled from Flanders, ironworkers from Sweden and shipbuilders from Holland. Goods that could not be grown or made in France were imported from French colonies overseas: coffee, chocolate, cotton, dyes, fur, pepper and sugar. The enrichment of France by commerce was Colbert's single aim. Roads and rivers were improved and a 240-kilometre-long 'Canal du Midi' was constructed, connecting the city of Toulouse to the Mediterranean.

Colbert reinforced the French navy, and recruited slaves from as far afield as Turkey, Russia, Africa and North America to work as oarsmen. To help domestic manufacturers, goods imported from abroad suffered an import tax, while those made at home went tax-free. French workmen were forbidden to emigrate, and if a French seaman served a foreign nation, he paid with his life. Two of Colbert's most famous sayings sum up many of Europe's rulers' new approach to the secrets of living well:

'It is simply and solely the abundance of money within a state which makes the difference in its grandeur and power.'

'The art of taxation consists in so plucking the goose as to obtain the largest amount of feathers with the least possible hissing.'

By the time of Colbert's death in 1683, Louis XIV's annual revenues had tripled. His enormous palace at Versailles, built between 1664 and 1710 on the site of a former hunting lodge, was where the King and his entourage lived and worked. With five chapels, a fabulous hall of mirrors (mirrors were then among the most expensive items a person could possess), and even its own opera house lit by 3,000 candles, no expense was spared. The rooms were adorned with works of art from all round the world. The King's solid silver throne stood on a precious Persian carpet, and diamond-studded vases were placed alongside priceless Chinese and Japanese porcelains. So large was the palace that much of the nobility of France could live there under the watchful eye of the King. So magnificent was Louis' royal château that it became the envy of the Western world. Palaces sprang up all over Europe in an attempt to become its rival.[8] European tastes were becoming as refined as the sugar from which much of the continent's wealth was derived.

Religious wars, the enslavement of American natives by opportunistic foreign fishermen and a sea-change in attitudes and tastes all help to explain why it was that Samoset asked, in English, if his new neighbours had any beer. But while the Pilgrim Fathers of Plymouth were welcomed by friendly smiles, Europe's new global strategy was already beginning to have a dramatic impact on the rest of the non-human world.

36

New Pangaea

*How crops and creatures were farmed, harnessed,
transported and exploited for the fancies of a single,
global and mostly civilized species: man.*

QUICKLY FLIP BACK 250 million years to when the earth's land mass was just one vast continent called Pangaea, surrounded by a single enormous ocean. Life back then was very different from life on earth today. Dinosaurs ruled the globe, mammals had yet to evolve, and the origins of humanity still lay a good 247 million years into the future (see page 53).

Yet, strangely enough, the modern earth is now becoming more similar to how it was then – about one hour and fifteen minutes before midnight on the clock of earth history. The existence of a single land mass, Pangaea, allowed the dinosaurs to dominate life on earth without hindrance, from the North to the South Pole. Since the early 1500s, thanks to the success of Europe's explorers and circumnavigators, certain humans could now do the same thing. By using their ships as bridges between otherwise isolated lands, Europeans began

unwittingly to re-create conditions on earth much as they were hundreds of millions of years ago.

An artificial Pangaea was formed by the maritime adventures of European explorers, dramatically accelerating the effects of a slow natural process. In terms of the twenty-four-hour scale of all earth history, this sudden snapping together of the earth's continents has occurred in just the last *one hundredth of a second before midnight*. (If left to nature, the earth's crusts would probably take another 150 million years before they would converge to create another super-continent – that's nearly one whole hour on a twenty-four-hour scale!) Such a rapid, world-shattering development had dramatic consequences for all life on earth.

The first period of rapid change began when European people started travelling extensively across the seas and colonizing new lands from about 1550. Then, from the middle of the eighteenth century, a second, more aggressive phase began (see page 312). Today, very little of the 'pre-Pangaea' world of 1550 has been left unaffected by humans.

Man's biggest visible impact on the earth has been through the destruction of nature's woodland, a process that began when the first human civilizations appeared 10,000 years ago. Europe's strategy to develop competitive colonies and trading posts all around the world propelled the process into a violent new phase. In the period 1650 to 1749, between 18.4 and 24.6 million hectares of forest disappeared in Europe.[1] Although cooking and heating were the most immediate human demands that required a constant supply of firewood for energy and warmth, the chief reasons for such widespread deforestation were ship-building, iron smelting, salt extraction and agriculture. Compounding the problem was a lack of any coordinated replanting policy to ensure supplies for future generations.

As early as the 1640s it was becoming clear to British naval commanders that sources of the right kind of timber for building robust masts were failing fast. Sir William Monson, a revered admiral who wrote a series of historical narratives about the state of the British navy, sounded an early prophetic warning:

> 'All kinds of wood that belong to the building of ships, we do and shall find, in a little time, a great want of. For wood is now utterly decayed in England, and begins to be no less in Ireland. If seamen die, so long as there are ships and navigation, they will soon increase and make their deaths forgotten; but if our timber be consumed and spent it will require the age of three or four generations before it can grow again for use.'[2]

Part of the reason for such anxieties were the demands of shipbuilding itself. A large warship of a thousand tonnes required between 1,400 and 2,000 oak trees, each at least a hundred years old, which would take up an area of at least twenty hectares. Oaks were used because of their resistance to rot and

their naturally curved shapes, ideal for constructing a ship's hull. Fir trees were also essential, both for masts and to provide resin to make waterproof pitch and tar.[3] In Europe, they mostly grew around the Baltic Sea. Powers like Britain, Holland, Spain and France were therefore dependent on importing Baltic timber for maintaining their fleets.

By 1756 Britain was purchasing the rights to cut down and import as many as 600,000 trees a year from Russia just to supply the needs of its fast-growing Royal Navy.[4] Most of the timbers used by the Spanish navy for the Armada that sailed against England in 1588 came from Poland. By the sixteenth century the Portuguese were making new ships out of teak from Goa or brazilwood from South America.

Iron-making, the manufacture of glass, bricks and ceramics, and salt extraction all placed huge additional demands on ancient forests, as wood was needed to fuel the furnaces these processes required. Europe's annual production of iron has been estimated at 40,000 tonnes a year in 1500, but by 1700 it had more than tripled, to 145,000 tonnes a year. Following the Great Fire of London in 1666, it was decreed that to protect against future conflagrations, all new buildings must be made of brick and stone, not wood. Ironically, this substantially *increased* the amount of timber required, because of the vast quantities of firewood needed to fuel the fires for baking so many bricks.

Charcoal and potash, used to make blast furnaces sufficiently hot, consumed masses more wood. By 1500 the iron industry in Germany was burning more than 10,000 tonnes of charcoal a year, and each tonne of potash consumed a thousand tonnes of wood. By 1662 Russian production of potash used up a massive three million tonnes of wood a year. The process of evaporating water to produce salt also used huge quantities of wood. By 1720 all the forests of the Kama region of Russia had been felled, and wood had to be transported from over 200 miles away to fuel the boilers for the area's 1,200 saltworks.[5]

Fortunately, timber for making ships, charcoal and potash was in abundant supply in the North American colonies. Its value in Europe made it a

Prom.Lupi.

Bridgeheads of New Pangaea: European settlers landing on the island of Jamaica in 1591, their ships packed full of provisions brought as survival kits from overseas. The island became a bolthole for Britain's pirates of the Caribbean.

vital source of income for early colonists. European demand thus exported deforestation across New Pangaea. In 1652 the first pines were felled in New Hampshire for use as masts on British warships. Within fifty years timber had become the colony's most lucrative trading commodity. By 1696 British warships were being built in North America itself, due to the shortage of suitable timber in Europe. By 1750 most timber to the east of the Appalachian Mountains had been stripped, and by 1775 North America's supply of the tall pines needed for ships' masts had been exhausted. The Royal Navy was forced to find alternative, inferior sources from Riga.

Deforestation in North America was also driven by the first European settlers' need for pastureland for the fast-breeding domesticated animals they had brought with them as life-support systems. Christopher Columbus brought eight pigs to Hispaniola in 1492. They quickly spread to Mexico, taken there by conquistador expeditions, until their numbers in the wild became, in the words of the Spanish, '*infinitos*'.

On his second voyage, in 1493, Columbus brought with him some cows and horses. Within fifty years huge herds of cattle could be found as far apart as Florida, Mexico and Peru. The grassland pampas of South America provided perfect feeding grounds, so that by 1700 these herds totalled an estimated fifty million animals. Domesticated sheep were first taken to Mexico in the 1540s. They multiplied so fantastically that by 1614 there were 640,000 of them in the area around Santiago in Chile alone. Horses migrated in the wild from Mexico to the Great Plains, where they were re-domesticated by Native Americans, bringing a new way of life to the surviving tribes, most of which had fled to the interior of the continent to escape the encroaching European settlers.[6]

Native Americans had had no need to deforest land to make pastures for domesticated animals, because until the first Europeans arrived, there were none. Now, from the prairies in the north to the pampas in the south, the Americas became swamped by new hoofed animals: goats, sheep, mules, horses, pigs and cattle brought from Europe and Asia across the virtual sea bridges of New Pangaea. These animals also prevented tree regrowth by munching their way through young shoots and trampling the land. Within 250 years the effects of forest clearances to create pasture-land for domestic animals, combined with the devastation caused by the never-ending grazing of the huge migrating herds, meant that America's landscape was stripped bare, transformed beyond all recognition.

Diseases that wiped out so many Native American populations originated in these domestic animals that European and Asian civilizations began to live alongside about 10,000 years ago. About sixty-five different human diseases have come from dogs, more than fifty from cattle, forty-six from sheep and goats, and forty-two (including influenza) from pigs. But bridges work two ways. The venereal disease syphilis was first recorded in Europe in the French armies invading Italy in 1494. It was probably introduced to Europe by sailors who returned from Columbus's first expedition to the Americas having contracted it from over-intimate contact with Native American women. By 1498, following the voyages of Vasco da Gama, the disease had spread to India, and by 1505 it had reached as far as China.[7]

Diseases that could now spread across the entire world didn't affect just humans. In 1850 the American vine aphid (*phylloxera*), which originated in the wild vines of the Rocky Mountains, was unwittingly brought to Europe on board ships. It quickly spread throughout the vines of France and Europe, threatening the wine industry with total collapse. It was only by importing American vines, immune to the disease, that European winemakers were able to re-establish their vineyards. So one irony of New Pangaea is that for the last 150 years fine French wine has been re-rooted by an American import.

It wasn't only in the Americas that the full ecological impact of this New Pangaea was felt. European seamen first set foot in Australia as early as 1606, when Dutch navigator Willem Janszoon made landfall on the western coast of what is now the Cape York Peninsula in north Queensland –

some believe that the Portuguese may have arrived there even before that. But it was the expedition of British sea captain James Cook in 1770 that triggered the first European settlements. He named the territory he discovered New South Wales, and a penal colony was established at Port Jackson, present-day Sydney, in January 1788.

As in the Americas, there were no hoofed animals in either Australia or New Zealand until Europeans arrived. Yet within a hundred years there were more than a hundred million sheep and eight million cattle in Australia alone. The deliberate introduction of new animals for the purpose of hunting and to provide furs had even greater consequences. After a grazier called Thomas Austin released a few European domestic rabbits into the wild in Australia in 1859 the creatures, having no natural predators, bred in their millions, devastating crops and vegetation. The problem became so bad that between 1902 and 1907 a 1,600-kilometre-long fence was built from the north to the south coasts of western Australia in an attempt to control their spread. The rabbits successfully breached it by the 1920s. By 1950 there were an estimated 500 million rabbits on the loose, leading to the deliberate introduction of a disease, myxomatosis, from Brazil in an attempt to kill them off. Initially it had devastating effects on the rabbit population, but immunity soon spread, and the death rate fell to less than 25 per cent.

The introduction of new species to once-isolated continents has led to a substantial reduction in the diversity of the earth's plant and animal ecosystems. For example, when goats were brought to the South Atlantic island of St Helena in 1810, twenty-two of the thirty-three native plant species became extinct because of overgrazing. European plants often took root in New World habitats, further eradicating native species. By 1877 there were 157 different types of European plants growing in or around Buenos Aires in Argentina; fifty years later only a quarter of the plants on the pampas were of native origin.[8]

By far the biggest ecological impact of the movement of plants across the world came about owing to the deliberate introduction of new crops in new places. Growing commodities for worldwide export, such as tobacco (from the Caribbean), sugarcane (from the Americas and Asia), cotton (from North America), rubber (from South America) coffee (from Africa) and chocolate (from Mexico) provided the economic backbone of Europe's colonies. Crops like these not only meant more forest clearing, but many quickly exhausted the soil, leading to further inwards expansion to find new areas of fertility.[9] Usually only three or four tobacco crops could be sustained by a piece of land before it was abandoned, to be replanted in twenty or thirty years' time. In the meantime, more land was cleared.

Processing crops into useful products took even more forest resources. Sugar had to be boiled into syrup, which required extreme heat in order to crystallize it. Cutting down trees to fuel the fires of sugar mills destroyed more woodland than clearing forests for the planting of sugarcane. When Barbados was first settled by the English as a colony for growing sugar, an order was put out for able-bodied axemen from England with 'working tools to cut down the woods and clear the ground'. When the British started importing African slaves, each one of them was required to fell approximately 1.2 hectares of woodland a year. By 1665 the island, once covered from shore to shore with forests, had trees left only on its peaks. By 1672 the

same transformation had taken place on the neighbouring islands of St Kitts, Nevis and Montserrat.[10] The requirement for heat to cure tobacco leaves was, and still is, at least as destructive in terms of felling forests.

While crops grown for export from the Americas caused a huge change in the natural landscape, domesticated plants taken across the world to be cultivated in Europe and Asia had consequences that were just as profound. Over thousands of years native Central American women had painstakingly domesticated wild *teosinte* into maize (see page 217), and as many as 300 varieties of potatoes were grown in Peru. Now, thanks to New Pangaea, Europe and much of Asia could reap the benefits of these nutritious crops too.

Maize was brought to Europe by the Spanish, and quickly became a valuable crop grown all around the Mediterranean, primarily as fodder for livestock. The cold climate of the Little Ice Age (c.1350–1850) meant that maize didn't take off in northern Europe until after 1850, when warmer conditions were re-established. However, it was successfully grown in southern China from about 1550 onwards. Between 1400 and 1770 the population of China exploded, from seventy million to 270 million people – a rise of nearly 400 per cent. The area of cultivated land to support this enormous growth in human population rose from twenty-five million to sixty-three million hectares,

Potatoes were fodder, considered fit only for animals and Europe's poor because they originated from (and were domesticated by) New World natives.

Ireland (personified here as Erin) struggles to keep the wolf from the door as her people succumb to starvation during the potato famine of the 1840s.

much of it given over to highly nutritious maize, which unlike rice could be grown on higher, drier ground, adding substantially to the amount of land able to be cleared for growing crops.[11]

The ecological cost was immense. Without their trees, China's hillsides became vulnerable to serious soil erosion, which devastated crops when the heavy rains fell. As reported by a Ming court scholar:

'Before the reign of Cheng Te (1506–21) flourishing woods covered the south-east slopes, but then, at the beginning of the reign of Ch'i Chiaching (1522–66) the southern mountains were cut without a year's rest … and converted into farms. Now if Heaven sends down a torrent there is nothing to obstruct the flow of water – its angry waves swell in volume and break embankments. Hence the Ch'i district was deprived of seven-tenths of its wealth.'[12]

Maize also became an important crop in West Africa from about 1600 onwards, replacing the native millet and sorghum thanks to its higher yields.

But the rapid adoption of maize in new parts of the world had consequences for human health. From the early eighteenth century Western doctors and scientists were confounded by an often fatal skin disease called pellagra, that first appeared in maize-growing areas of Spain. They thought maize must be toxic, or a carrier of the disease. Only in 1926 was it discovered that eating too much maize causes a vitamin deficiency which can easily be prevented by adding an alkali such as lime, or baker's yeast. Unfortunately, by then thousands of people had died from the illness, especially in the American south-west, where its grotesque symptoms led to a revival of belief in vampires, as people with pellagra often become hyper-sensitive to light.[13]

The potato was introduced to Europe by a Spaniard returning home from South America in about 1570. It eventually transformed European diets, owing to its high nutritional value and the ease with which it could be grown and harvested. But the humble potato was exactly that – humble – and at first European snobbery seems to have prevented its rapid uptake, since it was a favourite dish of the natives of the Andes, who were widely regarded by the Spanish as savages. The idea that the potato was fit only for animals spread across Europe, with the result that it graced the dinner tables of only the poorest of farm labourers.

Attitudes began to change from the mid-seventeenth century. For example, in 1662 a farmer from Somerset wrote to the Royal Society of London suggesting that planting potatoes might protect England against a future famine. Even so, for many years potatoes were used mainly as animal fodder.

The South American potato's big European break came with its adoption amongst the peasants of Ireland, who frequently suffered food shortages. From there it made its way back over the Atlantic to the North American colonies, where from 1718 it began to be grown as fodder for animals. By 1800 potatoes had at last become a respectable food even for the rich. By then Ireland was desperately addicted to a potato-based diet. When, in the 1830s, a fungal disease struck crops on the east coast of North America, its spores quickly spread on the winds as far as Peru which is where the rot should have stopped. But such spores of disaster could now be spread by ships travelling across the Atlantic. By 1845 potato blight – possibly carried on ships carrying fertilizer, or from potatoes

Once flightless, now extinct, the dodo was an early victim of European colonization and the domestication of animals to new lands.

imported from Peru – had reached Germany, Belgium, England and then Ireland, where its progress caused complete collapse. [14]

During the great potato blight of 1846–49, more than one and a half million Irish people are thought to have died of starvation, disease and malnutrition after the total failure of three potato harvests in a row. More than a million more left Ireland, hoping for a better life in the American colonies or Australia.

Famine was not the only cause of mass emigration: by 1851 more than 25 per cent of the populations of many east coast American cities, including Boston, New York and Philadelphia, were Irish natives. Some historians view the lack of any substantial help from the British government to the suffering Irish was tantamount to genocide. Ireland had been part of Britain since the Act of Union in 1800.

Potato fields for hundreds of kilometres withered to black as if ravaged by fire, their tubers reduced to a poisonous mush. The human suffering was immense, as one witness, Captain Wynne, observed:

'Crowds of women and little children were to be seen scattered among the turnip fields, mothers half naked, shivering in the snow and sleet, uttering exclamations of despair, while their children were screaming with hunger.' [15]

Of the hundreds of different types of potatoes cultivated by the natives of South America, only as few as four varieties were ever brought across to Europe. This lack of biodiversity was one reason why a single fungus could cause such widespread devastation.

As Europeans flocked to colonies established on the other side of the world, whether as adventurers, entrepreneurs, refugees or prisoners, many animal species which had evolved over tens of millions of years also became endangered or extinct. Flightless birds were especially vulnerable. The fearless dodo, a relative of the pigeon

and native to the island of Mauritius, fell victim to Portuguese (and then Dutch) invaders who arrived in 1505. It wasn't just hungry humans who clubbed them to death for something good to eat. Dogs, pigs, cats and rats, introduced by the settlers, ate their eggs and upset their nests, too. Human deforestation finally destroyed their habitat, leading to the death of the last bird in about 1700.

Animals with furry skins have been prized by human hunters at least since Roman times as a source of warm, ready-made clothing. But after Europe's indigenous supplies of ermine, beavers and foxes were run down by Viking trappers in particular, the hunt for new game was on. It was the pursuit of these animals that led Europeans to expand both westwards across North America and eastwards over Arctic Siberia to the shores of the north Pacific. As a result of the cooling climate of the Little Ice Age, furs became an increasingly important commodity for trade in Europe from the fifteenth century onwards.

European demand for furs drove Russian Cossacks to conquer the hugely hostile and previously unexplored environment of Inner Siberia. Novgorod, in western Russia, became a centre for the export of squirrel furs, exporting 500,000 skins a year by 1400. Since several hundred skins are needed to make a single fur coat, only fifty years later supplies had fallen by half, as animals were trapped at unsustainable rates. By 1600 Spanish beavers, until then the main source of European beaver furs, were all but extinct.

Urged on by rich merchant families such as the Stroganovs, Cossack peasants from south-eastern Europe moved through Siberia, trapping furs and selling them in European markets in exchange for guns. Their descendants eventually formed the backbone of imperial Russia's armies. Russian explorers reached the Pacific coast in the 1640s. In 1648 the Russian Cossack Simeon Dezhnev chanced upon one of New Pangaea's last remaining undiscovered connections, the sea passage between Alaska and Asia. The strait was later explored by the Danish navigator Vitus Bering (1681–1741) who in the year of his death successfully landed in Alaska, adding this North American region to

Russia's imperial domain – which it sold to the United States in 1867 for $7.2 million.

Otter skins became these people's most prized trading commodities, with more than 250,000 animals killed between 1750 and 1790. Locals paid taxes to Russian Cossacks in the form of furs, which were then sent along specially constructed roads to Moscow, where they were traded for arms. By the nineteenth century severe overhunting caused the lucrative Siberian fur trade to fall into a steep decline because of huge reductions in the number of foxes, ermines and sables.

The scarcity of furry animals in Europe and western Russia at the beginning of the sixteenth century meant that trade in animal skins became the biggest business yet in the newly established colonies of North America.

From as early as 1534 French settlers in North America had traded beaver furs with Native Americans. However, the beavers' low birthrate made it hard to satisfy demand, so when hunters had exhausted one region they quickly moved on to another, travelling further inland – beavers were common around the Hudson River and New York in 1610, but had disappeared from the area altogether by 1640. Fortified trading posts were established along the St Lawrence River by rival British and French trappers, who sided with native tribes in their efforts to out hunt each other.

A series of brutal conflicts began as early as the 1630s, after Dutch and British traders traded firearms for furs, thus arming a Native American tribal confederacy called the Iroquois with European guns. As supplies of beaver ran out in the traditional Iroquois stronghold around the Hudson River, the natives attempted to push further north to the St Lawrence, where sources were still plentiful, bringing them into direct conflict with other Native American tribes such as the Algonquins and their French allies. In a grim foretaste of what was to happen in nineteenth-century Africa, disputes provoked by avaricious European traders were now settled by tribes fighting with firearms.

Thousands of Algonquins and many French settlers were killed in the Iroquois raids of the 1640s and 1650s, supported by their Dutch and

British trading allies. After a brief period of peace, war resumed in 1683, provoked by the highly aggressive beaver-hunting tactics of a new French governor, Louis de Buade.[16]

De Buade deliberately cut out the traditional role of Native American trappers by encouraging European hunters called *coureurs des bois* ('runners of the woods'), leading to massive animal exterminations. In 1743 the French port of La Rochelle alone imported the skins of 127,000 beavers, 30,000 martens, 12,000 otters, 110,000 racoons and 16,000 bears from New France (the name for the French colonies of North America from 1604 to 1763, whose capital was Quebec, on the St Lawrence River).

By the 1850s, many species had been driven to extinction by unsustainable hunting. In Canada, as in Siberia, supplies of skins fell into steep decline. Eventually, as the world's supplies of wild furs diminished, farms for the intensive breeding of furry animals took over. Today, 80 per cent of all animal furs are farmed.

The same story applied to seals, walruses and whales, which all declined rapidly from about 1500 owing to the insatiable human appetite for skins, tusks and oil. In 1456 Atlantic walruses could still be found in the River Thames; now there are only 15,000 left worldwide.

Runners of the woods were French mercenaries paid to kill wild animals, cutting out the need to rely on help from Native American tribes.

Europe's creation of a New Pangaea had the effect of reducing diversity and threatening the very survival of species. Diseases, crops, people and cultures inextricably intertwined, which triggered an enormous change in global ecology. Deforestation, which had been ongoing since the rise of the first human civilizations, accelerated into a new, highly aggressive phase. Unbeknown to people living at the time, the capacity of the world's biosystems to regulate dangerous climate-warming gases such as methane, produced by farm animals, and carbon dioxide, absorbed by trees, was now beginning to buckle. The hot, sticky, carbon dioxide-rich world that was occupied by the dinosaurs of Old Pangaea c.200 million years ago was unwittingly beginning to be rekindled by the effects of Europe's early-modern overseas explorers.

Trading the products of hunting, fishing and farming for precious metals, weapons and gunpowder was bound to have at least as big an impact on non-European human civilizations as it was having on the natural world. How these people reacted to this brave new world, in which European nations competed with each other to exploit the earth's natural resources, largely shaped the 350 years of earth history still to come – or, to put it another way, the last one hundredth of a second to midnight.

Mixed Response

How different human civilizations reacted to the arrival of European businessmen-cum-soldiers eager to trade for profit.

WHAT WOULD YOU DO if one day a Viking warrior in full battle regalia appeared out of the blue at your front door? Would you shut him out? Talk politely on the doorstep? Invite him in for a cup of tea – perhaps first asking him to remove his horned helmet, iron shield and large pointed pike? Like humans, countries respond to sudden, unexpected changes in different ways. Europe's strategy to reach out across the whole world was therefore bound to provoke various reactions with the many civilizations it touched. Some were welcoming, some dismissive, and some were downright hostile.

There two cultures closest to the bubbling cauldron of Europe lay to the east and the south-east: one was Russia, the other Turkey.

The Rus, descended from Viking traders and raiders (see page 260), had established an impressive kingdom around Kiev by about 1000 AD. Unfortunately, it was almost totally destroyed by the Mongol invasions, led by Genghis Khan (see page 252). A group of smaller states emerged in the wake of this conquest that were loosely bound into a federation called the Golden Horde. Only in Novgorod, in the north-west, did the Russians survive unscathed, by successfully establishing trade links with the Hanseatic League.

Thanks to a deep friendship struck up between the Russian Grand Prince, Alexander Nevsky (1220–63), and Sartaq Khan (died 1256), son of the Golden Horde's Grand Khan, the Russians were gifted lands just east of Moscow. With the power of the Mongols waning in the fourteenth century, Alexander's descendant, Ivan III ('the Great', ruled 1462–1505), tripled the size of these territories, merging them with the valuable trading post of Novgorod in the process. Following the fall of Constantinople to the Muslim Ottomans in 1453, a Christian monk called Philotheus wrote Ivan an apocalyptic letter telling him that it was now most certainly God's will for Moscow to become the third Roman Empire.

With Siberian expansion under way thanks to the advance of the Cossacks, precious furs were now being traded in their millions via Novgorod in exchange for European guns and powder. By this time Ivan IV ('the Terrible', ruled 1547–84) had created a powerful nation with its own armies and access to vast natural resources. In a bid to make the prophecy of New Rome come true, Ivan declared himself Tsar, a Russian modification of the old imperial title, Caesar. His power, he claimed, was absolute.

It wasn't until the reign of Peter I ('the Great', ruled 1682–1725) that a clear Russian response to the growing economic and military might of Europe's nations began fully to emerge. Peter commanded a land mass three times the size of

Europe but with less than a quarter of its population. He decided to model his empire as fully as possible on the European strategy of building mercantile bridges via the seas. To begin with he fought and won a twenty-one-year-long war against Sweden, crushing his defeated rival's overseas empire and establishing control over the northern coastal states that gave Russia precious access to the Baltic Sea. In 1703 he built an impressive new capital, St Petersburg, as close as possible to the Baltic, and thus to Europe. The city acted as a funnel for the flow of treasure from trade, boasted a modern navy, a court modelled on European lines and systems of government directly copied from the capitals of Europe.

On a grand tour around Europe, Peter studied shipbuilding in Amsterdam, toured the headquarters of the British Royal Navy at Greenwich, London, as a guest of King William III, and sent a delegation to Malta to learn the art of training knights in warfare. So keen was he for his nobles to acquire European tastes that he even imposed a tax on those who refused to cut off their beards.

Russia had chosen to join the European club of competing nations engaged in a global contest for resources. Using furs as his main source of wealth, and their proceeds to equip his military with the

Peter the Great supervises the building of his splendid new capital, St Petersburg, built in a European style.

صوف ىدكلى ىنوعظم طط : كم قا ل ه ن ه حا ول
مالك عرب ىحر وسة مطبوعه حلبى قلعه سنك ىنون سيد د كوز

Janissaries were usually Christian subjects who wore special uniforms and were responsible for protecting the Sultan and his household.

latest weapons, by 1721 Peter had transformed Russia into an Empire poised to wield substantial power within the theatre of New Pangaea.

✳✳✳✳✳

Further south, the Turkish Ottomans were less sure about how to respond to Europe's triumphant leapfrog around their traditional control of trade. By the time of Suleiman the Magnificent's death in 1566, the Ottoman Empire was conducting a series of highly successful pirate raids throughout the Mediterranean, grabbing Christians from the coasts of Italy and Spain to be sold as slaves through the Islamic ports of the Black Sea. Between 1530 and 1780 it is thought that more than a million European Christians were snatched by Muslim corsairs operating from bases such as Tunis and Algiers on the north coast of Africa.[1] Many were sent to work in quarries or as oarsmen on Ottoman galleys.

Using slaves as soldiers was an old Islamic custom. Saladin, the great conqueror of the Christian Crusaders, had armies of slave-soldiers who eventually rebelled to take control of Egypt as the Mamluk Dynasty in 1250 (see page 269). The Ottomans continued the policy after conquering Egypt from the Mamluks in 1517. Young Christian boys were brought from the Balkans into the Ottoman army to be groomed as loyal slave-soldiers at court, providing the Sultan with his own personal bodyguard. These recruits, called Janissaries, formed the backbone of the Ottoman standing army. Although forced to serve in the army, they were still paid in cash, at a time when most European forces relied on booty from successful campaigns – one of the reasons why Charles V's imperial troops sacked Rome in 1527 (see page 295). Janissaries often rose to have substantial social standing, and earned privileges for their loyalty. It became a mark of high honour for a family to have a child as a Janissary. From the 1440s these crack troops adopted firearms and became expert engineers.

Why was it, then, that the Ottomans failed in their attempts to rebuff the rise of European power?

23:59:59

Turkish tents were left lining the fields around Vienna after their siege of 1683 was repelled by Polish King Jan III Sobieski.

Poles as they approached across the mountains because he felt he had been snubbed by the Ottoman commander-in-chief, Kara Mustafa. Still trying to blast their way into the city on one side, and attacked by Polish cavalry on the other, the Ottomans were finally forced to flee, leaving their beautifully arranged camp with its 40,000 tents set out in neat, straight lines separated by narrow lanes as rich pickings for Europe's imperial soldiers.

It is a legend in Vienna that coffee first came to the city after European soldiers raided these Turkish tents, where they found sacks of coffee beans. Croissants are also thought to date from this time, as Viennese bakers working late at night sounded the alarm about the tunnelling activities of the attacking Turks. Croissants were first baked to celebrate the bakers' contribution to victory. The Islamic crescent, from which they take both their shape and their name, is a Muslim symbol comparable to the Christian Cross or the Jewish Star of David.

Within sixteen years all Hungary had fallen back into European hands. The Ottoman Turks, still unable to compete with the Europeans at sea, were never again able to muster sufficient strength to challenge Europe's forces on land.

Some historians cite the Battle of Lepanto as the moment when Ottoman seapower was finally crushed. On 7 October 1571 an alliance of Venetian, Papal and Spanish fleets sank more than 200 Turkish ships during a five-hour battle off the coast of western Greece. However, within three years the Turks had rebuilt their fleet, and they soon took back control of Tunisia from Spain. By the early seventeenth century the Ottomans made a foray into the Atlantic Ocean, capturing Madeira in 1617, even reaching as far as the isle of Lundy in the Bristol Channel in 1655. Here, for the next five years, Muslim pirates set up a powerful base for raiding the British coast and Atlantic shipping lanes.

After having come so close to taking Vienna in 1529 (thwarted only by the bitterly cold winter), in 1683 more than 180,000 Islamic troops were once again laying siege to the imperial Habsburg capital. This time they were routed at the very last minute by the arrival of European reinforcements led by the Polish King Jan III Sobieski.

The Ottomans' attempt to surge into western Europe failed partly because Turkish forces were unable to keep up with military innovations being developed in the highly charged theatre of European war. The three hundred cannon used by the Ottomans during the siege had become outdated and ineffective against Vienna's newly strengthened city walls. Delays in adopting other tactics, such as tunnelling underneath the walls, proved costly, giving time for European reinforcements to arrive.

Another handicap for the Ottomans was the increasing number of disputes between Muslims of different traditions. For example, the Khan of Crimea, who was given responsibility for defending the Turkish forces from the rear, failed to attack the

But the Ottomans could never compete with the pace and power of European military innovation. Seafaring warships of the line, pioneered by the Dutch and copied by the British and French navies, made mincemeat of the Turks' oar-powered galleys, which had evolved from a tradition of Mediterranean seafaring, and were not designed for tackling transatlantic trade winds.

Matchlock guns used a new firing mechanism introduced in Europe in about 1480 which automatically ignited the gunpowder, allowing the user to keep a firm grip on the weapon with two hands whilst firing, improving accuracy. Corning was a new method of milling gunpowder that made it more reliable and powerful.

The Ottoman's salaried slave-soldiers were often more difficult to train in new warfare techniques than the privatised mercenaries of European forces, and the spread of knowledge of the latest technology was slowed down by the reluctance of Islamic cultures to embrace Europe's mass printing revolution, pioneered by Gutenberg. Islamic calligraphers revealed the truth of God through the *written* word. Printing was often considered taboo.

But the biggest cause of the Ottomans' inability to counter Europe's relentless rise came from within Islam itself. Almost since its inception Mohammed's original vision of a religiously and politically united creed had been undermined by factionalism and rivalry. By the time Europe had found a way to circumvent the Ottomans' economic grip, the idea of a single powerful Caliphate presiding over a universe of loyal Muslims had become as outdated as it was unreal. The rise of the Safavids (1502–1722) in Persia, led by Shah Ismail, posed a new threat to Ottoman power. Rival interpretations of Islam burst out into armed conflict between Ottoman Sunnis and the Safavid Shias, each claiming political and religious legitimacy while denouncing its rivals as infidels.

◆◆◆◆◆

Unlike the Ottomans, the Safavids embraced the arrival of European culture and expertise. Shah Abbas (1587–1629) reorganized and retrained his forces along European lines, hiring the services of two British military advisers. Robert and Anthony Sherley transformed the Safavid armies, introducing salaried troops, rigorous training and firearms. By 1622 Abbas had recaptured Baghdad from the Ottomans, and secured an enduring Shia identity

for the region by restoring sacred shrines and establishing a new capital at Isfahan.[2] He built trading posts in Bahrain and Hormuz on the Persian Gulf, providing access to lucrative commercial links with the Dutch and English East India Companies.

The struggle for control of Iraq today goes back to this time, when Shia descendants of Shah Abbas settled in and around Baghdad before it was retaken by the Sunni Ottomans in 1638. Like today, Kurdish tribes were caught in the middle of the opposing Muslim forces.

The Safavids, the cultural ancestors of modern Iran, successfully latched on to Europe's lust for luxury by developing an industry making rich silk rugs. Abbas used Armenian merchants as intermediaries to sell carpets to Europe, bypassing the Ottoman Turks. He even allowed the Armenian settlers to build their own Christian cathedral, All Saviours, located in Isfahan, Iran, out of respect for their European cultural traditions. By 1602 the King of Poland had purchased a series of Persian rugs through Armenian merchants, each with his insignia specially woven into the design. Persian carpets soon became the ultimate luxury goods in the houses of European nobility – one was even to be found under Louis XIV's throne in the palace of Versailles (see page 303).

◆◆◆◆◆

While the Safavids embraced the opportunities provided by European expansion, the third large Islamic power, the Mughals of India, weren't in the least bit fazed by the arrival of Portuguese traders and Jesuit missionaries. Under Akbar the Great (ruled 1556–1605), regular gifts were exchanged between Europeans and India's Islamic rulers, including several musical instruments such as a pipe organ which, according to Akbar's biographer, caused a great deal of excitement:

'It was like a great box the size of a man. A European sits inside it and plays the strings thereof, and two others keep putting their fingers on five peacock wings and all sorts of sounds come forth. And because the Emperor

was so pleased, the Europeans kept coming at every moment in red and yellow colours, and went from one extravagance to another. The people at the meeting were astounded by this wonder and indeed it is impossible for language to do justice to the description of it.'[3]

Akbar welcomed the arrival of Jesuit missionaries to his court, inviting them to attend his weekly religious discussions at the House of Worship within the walls of his newly built capital city, Fatehpur Sikri, located just outside Agra. Representatives of all religious persuasions were invited to attend, including Muslims, Sufis, Jains, Buddhists, Hindus, Christians, Zoroastrians and Jews. Each in turn put forward their views on right religion. In the end Akbar attempted to incorporate the best elements into his own new religion, known as the Divine Faith, *din-i ilahi*, which he genuinely hoped could provide future generations with religious unity. Akbar's noble attempts were doomed to failure. Orthodox Muslims in particular found the whole idea unacceptable, and with Akbar's death his fancy for a single world religion died too.

Akbar's son, Jahangir (ruled 1605–27), gave British traders from the Honourable East India Company their first big break. A diplomatic mission led by Sir Thomas Roe in 1615 resulted in the Emperor giving him a letter for James I granting 'eternal' permission for the British to build Indian mainland trading settlements and asking him to send exotic European gifts in return:

'I have commanded all my governors and captains to give your merchants freedom answerable to their own desires; to sell, buy and transport into their country at pleasure. For the confirmation of our love and friendship, I desire your majesty to command your merchants to bring in their ships all sorts of rarities and rich goods fit for my palace and that you be pleased to send me your royal letters by every opportunity ... that our friendship may be interchanged and eternal ...'[4]

The British began by building settlements in Surat (1612) and then Madras (1639). By 1649 the Company had established twenty-three trading stations, later adding Bombay (1668) and Calcutta (1690).

Thanks to the talents of the Company's doctor, William Hamilton, Britain's fortunes further improved. In 1715 Hamilton miraculously cured a growth in the Emperor Farrukhsiyar's groin. Then, later that year, he helped the Emperor recover from a serious digestive illness. So delighted was Farrukhsiyar that he gave Hamilton an elephant, a horse, 5,000 rupees, two diamond rings, a set of golden buttons and, best of all, duplicates of all his medical instruments crafted from solid gold. More important for Hamilton's employer was the granting of privileges allowing the Company to buy thirty new villages, substantially extending its control over Calcutta, as well as permission to build its first settlements in Bengal (which later became the British centre for growing opium for export to China). To cap it all, the Emperor was so thrilled at being healthy enough to marry the daughter of the Raja of Jodhpur that he decreed that in future all British trade could be conducted rent-free.

Indian goods were now flowing back to Britain, including cotton, indigo, silk and tea. Even more significant was the discovery of natural Indian treasure in the form of saltpetre, the vital ingredient for the manufacture of gunpowder – a desperately important commodity for Britain if it was to succeed on the international stage of European war. In just two years, 1672 and 1673, Sir John Banks, a British businessman from Kent and Governor of the Honourable East India Company, supplied more than 900 tonnes of Indian saltpetre to the gunpowder mills of Surrey to supply the King's armies.

In the end it wasn't just the arrival of Europeans that caused the mighty Mughals' downfall, but rather their increasing isolation from India's Hindu majority. Riches squandered on enormous funerary monuments such as the Taj Mahal highlighted distinctions between the rival traditions of Muslim burial and Hindu cremation, differences that even the age-old caste system found hard to accommodate. During the fifty-year reign of Mughal Emperor Aurangzeb Alamgir (ruled 1658–1707), local Hindu rulers successfully reasserted their power. Aurangzeb's attempt to forcefully convert Hindus to Islam backfired, and Persian and Afghan invaders added to the misery by sacking Delhi in 1739. During this siege, led by the Iranian ruler Nadir Shah (ruled 1736–47), the Mughal Emperor's symbolic and highly precious Peacock Throne was plundered and taken to Persia.[5]

Several decades after the fall of the occupying Mughal Islamic regime, the British colonial statesman Robert Clive won a victory over the ruler of Bengal at the Battle of Plassey (1757), using trading bases in Calcutta as a bridgehead. Siraj Ud-Daulah, the Bengali ruler (commonly known as 'Sir Roger Dowlett' by the British because they couldn't pronounce his real name) submitted to Clive, despite being aided by French forces, after his army commander defected to the British side.

The Treaty of Paris in 1763, like the Treaty of Tordesillas nearly 300 years before, enshrined a new legal framework for Europe's colonization of the world, this time carved up between Spain, France and Britain. It provided international legitimacy for a new British exploitation of Hindu India through the Honourable East India Company until India's formal annexation into the British Empire took place after the failed Indian Rebellion of 1857.

Further to the east, Japan's initial reaction to the arrival in 1543 of two Portuguese adventurers blown off course on their way to China was as dramatic as it was bizarre. When the explorers pulled out their guns and shot a duck in the air, a passing Japanese nobleman bought the two firearms from them and had his swordmaker make perfect copies. By 1553 there were more guns per head in Japan than in any other country in the world.[6]

Such innovations suddenly made it possible for untrained peasants to participate as lethal combatants in Japan's feudal armies. Large-scale peasant armies with firearms transformed Japanese

Dripping with jewels: the Peacock Throne was the ultimate symbol of Mughal power until its theft by Persian and Afghan invaders in 1739.

politics. Toyotomi Hideyoshi (1536–98) a feudal ruler, marshalled sufficient forces to unite Japan under the authority of the Emperor, using the power of gunpowder. Towards the end of his life he unleashed his dream of invading Ming China, landing his forces in Korea, which was on the way. Initially Hideyoshi was successful, capturing Seoul in 1592, but his navy was sunk by the Chinese admiral Yi-Sun-sin. With his supply lines cut off and his troops on land blocked by the Chinese Emperor's forces, Hideyoshi had no choice but to retreat to Japan. But, in true Samurai style, he wasn't going to give up. His second attempt, in 1597, also encountered fierce resistance, but the next year Hideyoshi died, and his troops were ordered to withdraw.

The débâcle of Japan's first attempts at using modern European weapons to build an Asian empire dug the deepest of scars. Following the Battle of Sekigahara in 1600 a new dynasty, the Tokugawa Shogunate, rose to power. It succeeded in uniting the country from 1603 to 1867, a time known as the Edo period, named after the Emperor's magnificent castle, completed in 1636, that stood at the heart of the city that became modern-day Tokyo.

The new dynasty's first leader, Tokugawa Ieyasu, pledged to rid Japan of the cancer of violence that had caused such upheaval and devastation since the arrival of the Portuguese in 1543. Firearms were banned, their possession punishable by death. In 1614 a decree was signed expelling all Christians, and a new set of laws called the *Buke Shohatto* was introduced to purify the country's nobility and restore its morals. These laws decreed that Samurai, the military nobility of Japan, should devote themselves to the noble pursuits of swordsmanship, archery, horsemanship and literature. Expanding castles was forbidden, marriages amongst nobles had to be approved, formal uniforms should be worn, and Samurai should live frugally.

For a while limited trade with the outside world continued under a special permit system. Silver, diamonds, copper and swords were exported from Japan by European merchants, mostly in exchange for highly prized Chinese silks.

The extraordinary story of English sailor William Adams (1564–1620) provides a notable exception to the Shoguns' suspicion of European merchants. After several disasters in the South Atlantic, and more than nineteen months at sea,

Commodore Perry's ominous-looking fleet of black ships sailed into Edo Bay in 1853, forcing the Japanese Shogunate to open its markets to trade with the West.

Adams' ship landed in Japan. Here he settled, soon becoming one of the Emperor's key advisers, demonstrating to the Japanese how to build Western-style ships and helping to manage and control their maritime trading system. The Shogun later made Adams the first ever foreign Samurai.

Japanese reaction against Hideyoshi's failed flirtation with Europe's gun culture led to a determined drive towards total self-sufficiency. By 1635 the Shogun firmly shut the country's doors to the outside world. Foreigners were banned from entering the country, and Japanese nationals were forbidden to leave. Anyone who broke either law would be sentenced to death.

Self-sufficiency is the bane of those in search of global markets. The Japanese policy worked for more than 200 years, until the arrival of American commodore Matthew Perry, who on 8 July 1853 threatened the Japanese capital with his huge 10,000-pound Paixhan guns mounted on iron-clad battleships. Each one was capable of firing explosive shells a distance of more than two miles, towards Tokyo.[7]

Perry demanded that the Japanese open their country to trade with the West. A year later he returned with even stronger military support, and forced the Shogun to sign a trade treaty. In less

than a hundred years, Japan's enforced re-entry into the hectic global race for the earth's capital was to prove utterly catastrophic (see page 341).

Meanwhile, the use of Western gunpowder by Japanese armies on Korean soil led to Europeans being banned from trading in all Korean ports by the Joseon Dynasty (1392–1910) until as late as the 1880s, giving rise to Korea's Western nickname, the Hermit Kingdom.

⬖⬖⬖⬖⬖⬖⬖

Chinese reaction to the arrival of European merchants with guns was mostly one of dismissive indifference. Attempts were made by early Portuguese explorers to establish trade links (Jorge Alvarez tried in 1513, and Fernao Pires de Andrade in 1517), but imperial troops blocked their paths. In 1535 Portuguese traders were granted permission to set up a trading post at the mouth of the Pearl River, and by 1557 they had built a small trading settlement at Macao, which became the primary trading portal for the West into China until 1685.

Meanwhile Koxinga (1624–62), a naval commander loyal to the Ming Dynasty, successfully defeated the Dutch claim to Taiwan by invading the island in 1662. Despite the upheaval during the overthrow of the Ming and the establishment of the Qing Dynasty in 1644, with no significant European colony near its coast, China was now able to resist major interference by European nations.

It wasn't to last. Despite limited trading contacts, the winds of New Pangaea still blew China into uncharted waters. Its innovative paper currency, so admired by Marco Polo, had by 1600 become overprinted, overinflated and worthless. Replacement hard currency in the form of silver coins arrived from Japan on red-seal trading ships

and on Spanish galleons carrying bullion across the Pacific from the Andean mines of Potosí. A black market of pirates and smugglers in collusion with merchants from the West broke through imperial embargoes on overseas trade, ominously preparing the way for China's economy to become dependent on overseas silver.

A crisis in energy supplies brought on by massive deforestation and huge population growth, much of it driven by the extra nutritional value of South American maize, also contributed to making China dependent on the West after all. From 1685 the Chinese government opened up an official trade port for European nations (including England, Holland, France, Denmark, Sweden and Russia) through Guangzhou. By the 1830s it refused to accept payment in anything other than silver, so desperate was it to secure supplies of bullion in an attempt to shore up the country's ailing economy.

Fearing for the security of their own silver supplies, the directors of Britain's Honourable East India Company devised a new strategy to find a more acceptable form of payment. They settled on the ingenious idea of luring the people of China into addiction by dealing them a steady supply of opium, grown in India and smuggled in vast quantities across the border into China in exchange for tea. When the Qing Dynasty tried to ban the trade in 1838, Great Britain declared war (see page 361).

Despite the efforts of many cultures from the Far East to maintain their self-sufficiency by closing the door to European merchants and their Christian missionary cohorts, in the end the bridges of New Pangaea were not for disconnecting.

Kill or cure? Poppy-power has been a cause for wars, mass drug addiction (opium) and medicinal pain-relief (morphine).

23:59:59

Free Reign

How extreme inequality between people ignited rebellions in the name of freedom, and how armies were conscripted for the sake of a feeling, flag or song.

COMMAND OF THE HIGH seas was now giving European rulers and merchants the riches they had so desperately sought. Between 1650 and 1800 slave labour and the exchange of easy-to-produce arms and ammunition supplied Europe's traders with bullion, exotic foods, spices, silks, tea, coffee and chocolate. Generally, these were not available to the population at large – how could they afford them? Instead, the masses continued to suffer as much as ever from famine, poverty and disease.

The luxuries of Louis XIV's court at Versailles (see page 303) were pre-empted by the excesses of ornately dressed Catholic Churchmen such as Cardinal Richelieu (1585–1642), who did as much as anyone in history to consolidate centralized power in the name of a king. In the process he amassed for himself such a huge fortune that at his death he left one of the biggest art collections in all Europe, which he housed in an enormous palace

built in the heart of Paris. The riches of Europe's kings and cardinals were almost equalled by the new wealth of merchants and pirates.

Henry Every was one of the most successful buccaneers of all time. Born in Plymouth in 1653, he pulled off one of history's biggest ever heists. Following an epic voyage to the Indian Ocean in 1694, his ship, the *Fancy*, blasted its cannon at a treasure flagship belonging to the Indian Mughal Emperor Aurangzeb, shattering its mainmast. After two hours of hand-to-hand combat the Indian ship surrendered. Terrible horrors are reputed to have been inflicted on the hundreds of Indians taken captive. Many of the women are said to have committed suicide to avoid violation. Every's booty included a staggering 500,000 gold and silver pieces. Quite what happened to Every and the money no one knows, since he wisely took himself off somewhere beyond the reach of men. So furious was the Indian Emperor that he threatened to withdraw British trading rights.

Alarmed by the prospect of a slide in their future profits, the directors of the East India Company begged King William III to find someone who could rid the Indian Ocean of avaricious Anglican pirates. But the profits to be made from piracy proved too big a lure for the King's appointed man, the famous Captain William Kidd, who himself turned to plunder on the high seas. In May 1701 Kidd was hanged at Execution Dock in Wapping, London, on one count of murder and five counts of piracy.

Growing inequality between Europe's rulers, clerics and merchant-come-pirates on the one hand, and poverty-stricken subjects on the other, was made worse by relentless warfare both within Europe and increasingly around the world. More than half of the entire male population of what is now Germany is thought to have perished in the Thirty Years War of 1618–48, as Europe's emerging Protestant nations tussled for recognition against traditional Catholic imperial power. Only a hundred years later, a similar scale of devastation was exported worldwide as far as the Americas and India in the Seven Years War starting in 1756, dubbed by British statesman and historian Winston Churchill as the first true World War. More than one million lives are thought to have been lost in this horrific worldwide conflict.

How could a merciful, just, Christian God condone a world such as this? If Jesus Christ could perform miracles, why did He permit such inequality, wars and suffering? A fashionable new philosophy among European thinkers and politicians began to suggest that while a divine force originally created the universe, it could not be responsible for intervening in its workings. 'Deism', as this concept came to be known, rejected supernatural events, miracles and any concept of divine revelation.

Such ideas fused with the scientific observations of Copernicus, Galileo and Newton, which demonstrated nature's laws to be more rational and logical than divine and inspired. And the now widely available works of the ancient Greeks, like the *Politics* of Aristotle, fuelled a growing suspicion among many thinkers that reason and logic could also be applied to the workings of human society. Perhaps a more scientific system of social order could unlock a better, more acceptable way of life for everyone – not just kings, clerics and merchants who justified their luxurious lifestyles on the basis of either divine authority or the principle that might is right.

The concept of liberty, of giving people themselves a choice over how they are ruled, backed up by protection in the form of individual rights and laws, is almost as old as recorded human history itself. The laws of Hammurabi (see page 128), the cylinder of Cyrus (see page 176) and the edicts of Ashoka (see page 170) are just some of the foundations on which the idea of freedom and justice for all people was built. Thanks to the profound inequalities built into the system of global trade in early modern Europe, pre-Christian concepts of Greek democracy and Roman republicanism were now opened up for re-investigation.

The idea that governments could derive their authority from citizens, not gods or swords, was revived from ancient times by medieval religious scholars, such as the Frenchman Jean Gerson (1363–1429), who proposed that a council of bishops ultimately had the authority to depose a pope. Niccolò Machiavelli (1469–1527) applied the same reason to civil politics in city states such as Milan, Florence, Pisa and Venice in Italy, Europe's traditional toehold on Islamic wealth. Machiavelli is most famous for a short treatise he wrote rapidly in 1513 to gain favour with the revived Medici family in Florence. *The Prince* justified the use of whatever tactics were necessary by rulers to secure

Henry Every, master pirate, robbed the Indian emperor of 500,000 gold and silver pieces … and then wisely disappeared.

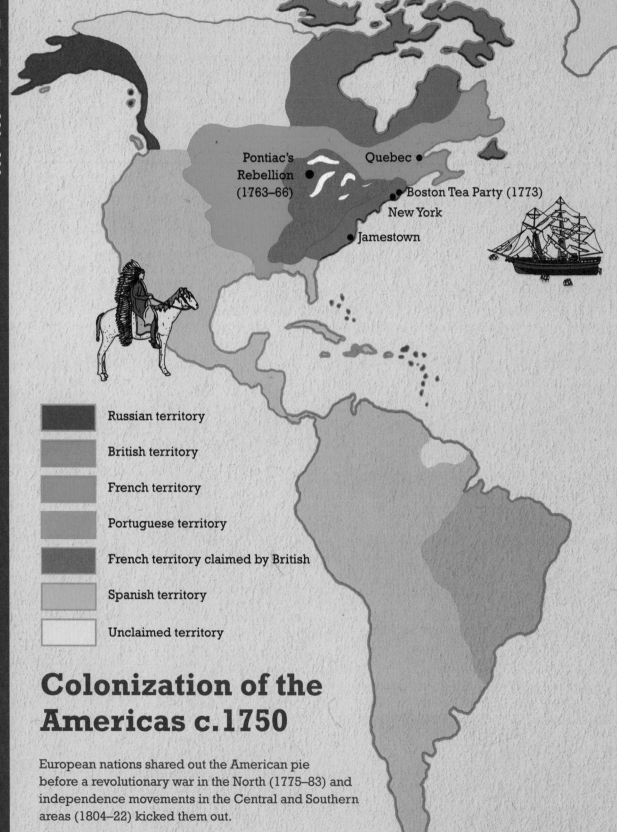

Pontiac's
Rebellion
(1763–66)

Quebec

Boston Tea Party (1773)

New York

Jamestown

Russian territory

British territory

French territory

Portuguese territory

French territory claimed by British

Spanish territory

Unclaimed territory

Colonization of the Americas c.1750

European nations shared out the American pie
before a revolutionary war in the North (1775–83) and
independence movements in the Central and Southern
areas (1804–22) kicked them out.

their hold on power. However, his biggest work was his three-part *Discourses on Livy*, a study of republican Rome inspired by Savonarola's Florentine Republic. It provided advice on how to establish a form of civic government based on popular consent and control.

The success of Martin Luther's religious grievances had also proved that change could now come from the power of the written word. With printing presses whirring in cities all over Europe and the colonies of America, words became the fuel that stoked the fires of revolutionary freedom. Could it be that from now on the pen would prove as mighty as the sword?[1]

During his country's violent Civil Wars from 1642–49, English philosopher Thomas Hobbes (1588–1679) developed the idea that political authority came not from God, but from rational laws of nature. People voluntarily delegated their authority to a ruler in the interests of their own self-preservation, said Hobbes. The lives of wild hunter-gatherers were, in his words, 'nasty, brutish and short'. However, if they were given the chance of joining a civil society, it was natural for them to acquiesce to the power of a monarch for their own protection.

Other European philosophers continued to combine theories of the universal laws of nature, such as those of Isaac Newton,[2] with new political laws for the better government of people. When the Dutch Prince William of Orange was installed by Parliament as British monarch in 1688 in place of the Catholic King James II, the time had come to decide just who had the rightful authority to appoint or dismiss a nation's rulers.

In 1689 the English Parliament passed the world's first Bill of Rights, (known as the English Bill of Rights) which set out a number of 'inalienable' natural rights, originating from laws of nature, that no one could ever take away. They included the supremacy of the law as passed by Parliament, freedom of speech and the right to bear arms for the purposes of self-defence. English Philosopher John Locke (1632–1704) proposed a form of government based on a 'social contract' between the people, whose natural rights made them sovereign, and their rulers. Most importantly, in Locke's view the rulers' authority, which came from the people, could also be taken away by the people.

Locke's *Two Treatises on Government*, which appeared in 1689, began a period now known as the Enlightenment, in which the concept of popular sovereignty became a powerful new force in the politics of Europe and its colonies. For Locke, unlike Hobbes, all people were born equal, and no one had a natural right to harm another person's 'life, health, liberty or possessions'.

What gave people a right to own property was, said Locke, the fact that they worked to acquire it. Personal liberty was an inalienable natural right of all men for the protection of their private property. If governments breached that right, it followed through reason that people were justified in taking authority back into their own hands.

Revolutionary ideas were now brewing in North America following the gruelling Seven Years War (1756–63), as a result of which France temporarily lost its colonies on the North American mainland and Britain emerged as the world's most powerful nation, sustained by its superiority at sea. British colonial taxes imposed to help recover the huge debts incurred by these conflicts were resented by American settlers, who were fighting their own battles against dispossessed natives.

Discontent grew greater when the British government refused to supply sufficient forces during major clashes such as Pontiac's Rebellion (1763–66), when natives tried to drive European settlers out of Ottawa. British officers at Fort Pitt were so desperate that they resorted to the tactics used by Jani Beg, commander of the Golden Horde at Kaffa, 400 years before (see page 265). By deliberately exposing the attacking natives to blankets smeared with highly infectious smallpox, the British hoped to wipe them out. This time, their attempts at biological warfare seem to have had little effect.

A tax on all American legal documents, newspapers, pamphlets and playing cards was

Pontiac, chief of the Ottawa tribe, consulting with his leaders in an effort to drive British forces off their traditional lands.

than by local juries, with the colonies picking up the costs. This added yet more fuel to the fire. By 1776, all thirteen American colonies felt sufficiently aggrieved to form a coalition and declare their independence from Britain.

The Founding Fathers of the American Revolutionary War succeeded in their aim of creating a separate state, independent of Britain, largely thanks to the supply of gunpowder and battleships from France. They justified their rebellion through the arguments of writers who took their lead from Locke. Fashionable Deists Benjamin Franklin, John Adams and Thomas Jefferson were the chief architects of America's declaration of inalienable human rights which, they said, had been so undermined by Great Britain that the creation of a new independent sovereign state was not only justified but an obligation under 'Nature's God'.

Jefferson, who later became the third President of the United States, even wrote his own abridged version of the Bible (published after his death, in 1895), removing all references to the miracles of Jesus and other supernatural events. For Deists like Jefferson, Jesus was a man who set a fine moral example, but (echoing Arian beliefs – see pages 258–59) was most definitely *not* the son of God.

Thomas Paine (1737–1809), an Englishman who had emigrated to America, wrote a powerful pamphlet arguing that American independence from Britain was simply 'Common Sense'. Such was the power of the printing press to spread new ideas that within three months of its publication in January 1776 more than 120,000 copies had been sold throughout the colonies of America, creating a powerful collective feeling of legitimate, indignant resistance. Among Paine's arguments in support of independence for the United States of America (it was his idea to call the new nation 'the United States') was the fact that whenever England was at war with another power, Americans were forbidden to trade with it. This limited the settlers' ability to realize the natural wealth of their land. The mother state had therefore illegally interfered with the power these individuals could exercise over their own private property, breaking the social contract. With

introduced in 1765 in an attempt by the British King George III to recoup the costs of war. So total was the resistance to this stamp duty, imposed on colonial settlers who were barely represented in the British Parliament, that within a year it had been repealed because almost no one paid up. To add insult to injury, corporations such as the East India Company successfully lobbied Parliament in London to remove the tax paid on tea sold by its agents in the colonies. That way they could undercut rival American merchants-cum-smugglers who had refused to buy the Company's tea from China.

So angry were the colonists at this latest perceived injustice that on the night of 16 December 1773 a group of protesters disguised as Native Americans boarded two East India Company trade ships and dumped forty-five tonnes of tea, worth approximately £10,000, into Boston harbour. The British government's draconian response to the so-called 'Boston Tea Party' was to close the harbour indefinitely, until all damages were repaid. Further legal restrictions meant that from now if British officials were ever charged with misdemeanours they had to be put on trial back in London rather their natural liberties usurped, revolt was legitimate, and political authority reverted to the people.

Thomas Jefferson combined the ideas of Paine and Locke in his draft of the United States Declaration of Independence, which was published on 4 July 1776. In its preamble, before condemning the abuses of British rule, it boldly declared:

> *'We hold these truths to be self-evident, that all men are created equal, that they are endowed by their Creator with certain inalienable Rights, that among these are Life, Liberty and the pursuit of Happiness.'*

The United States Bill of Rights (1791), which was issued following the Revolutionary War, enshrined the principle of personal liberty in a series of ten amendments to the new nation's Constitution. For example, the First Amendment prohibits Congress from making any law whatsoever 'respecting an establishment of religion'.[3] The Second Amendment gives individuals the right to 'keep and bear arms'.[4] The Fifth Amendment prevents an individual from having to incriminate himself in court, or being 'deprived of life, liberty and property without due process of law'.

Individual rights for the protection and exploitation of private property now formed a cornerstone of American liberty and personal freedom. But some issues that should have been dealt with then were instead brushed under the carpet. Despite the powerful opening statement of the Declaration of Independence that 'all men are created equal', American society was not yet ready to practise what it preached.

Several of the Founding Fathers came from states for which slave labour was the axis of their wealth. Mount Vernon, George Washington's huge Virginia estate, deployed more than a hundred slaves. At the end of his two terms as the first US President (1789–97), their number had tripled to more than 300. Thomas Jefferson suffered many moral qualms about slavery, although even he had an estate on which dozens of slaves toiled in the fields for his profit. When, in his first draft of the Declaration of Independence, Jefferson condemned the British sanctioning of slavery as against 'human nature herself', the words were struck out by delegates from South Carolina and Georgia.

A huge gap remained between the rhetoric and the reality of what 'all men are created equal' really meant. Women, slaves and Native Americans, who traditionally had never even considered the idea of individual land ownership in the first place, were not, in reality, equal in the eyes of America's Founding Fathers. The stark omission of a ban against slavery later contributed to the outbreak of a bitter civil war that cost nearly a million lives (see page 348).

Negro slaves turn to look up to their lord and master, George Washington (first President of the United States from 1789–97) on his estate at Mount Vernon, Virginia.

However, flush with the exhilaration of its newly won independence, the United States of America turned its attention westward, and its citizens, whose right to bear firearms was now an inalienable part of the country's constitutional freedom, began to force themselves on to new lands. In 1804 Meriwether Lewis and William Clark, with financial support from the American Congress, set off on an overland expedition to reach the Pacific coast. Their discoveries would bring about the westward expansion of the United States, eventually leading to the concentration of natives in special reservations (see page 350) and devastating battles with rival Spanish-cum-Mexican settlers.

Back in France, passions about the excesses of the King, his court and the privileges afforded by the Catholic Church provoked a revolutionary zeal just as impassioned as that in the United States. Within just three years of the Americans declaring their independence from Britain, protesters in France had decided it was time to put into practice their own version of what liberty had to offer.

French ideas of freedom were rather different from those of the Americans. Tying natural rights to parcels of land was far less important to the masses who liberated the prisoners from the Bastille in Paris on the night of 14 July 1789. After all, theirs wasn't a fight against either a distant colonial power or fierce bands of dispossessed natives. Their struggle was more to do with making sure everyone got fed, and ridding their country of excessive royal, noble and clerical privileges. Liberty, Equality and Fraternity became their rallying cry, and everything that stood in their way was swept aside.

The champion of these high ideals was a man whose powerful philosophy was truly radical for its day. Jean-Jacques Rousseau (1712–78) was not a Frenchman, but 'a citizen of Geneva' who converted to Catholicism so he could qualify for a good education. He converted back to his original Calvinism in 1754, once his education was complete.

Unlike Hobbes, Rousseau believed that man in his natural state was caring and compassionate, his goodness deriving from an instinctive sense of self-sufficiency. It was society (which Rousseau suggested emerged as a result of population growth) that had corrupted man. His desire to own property, and his comparisons between himself and others around him, had turned his natural, positive self-love (*amour de soi*) into artificial pride (*amour-propre*). Primitive humans who lived in the anti-social wild were free of greed and envy, but with the development of agriculture, private property and social status this had given way to inequality and conflict. Society was an unjust contract, made, said Rousseau, by the rich to secure their unequal advantages and disenfranchise the poor. Inequality, therefore, was an inherent ingredient of human society, not part of mankind's original, natural condition.

Rousseau went on to describe a revised contract as the basis of a new society that would correct this artificial inequality. His work, called *The Social Contract*, published in 1762, explained how people could collectively liberate themselves from the tyranny of the few by becoming sovereign, ridding themselves of the supposedly divine authority of kings and clerics. Laws made by the people would, he argued, set them free. By suggesting that private property was the root of all evil, Rousseau's concept of liberty could not have been more different from the ideas adopted by the United States of America. What's more, he specifically argued against representative assemblies, saying that laws had to be made by the people directly – although it's hard to see how this concept could have been put into practice in a society as large as France. Perhaps with the rise of mobile phones and instant text messaging Rousseau's ideas have at last come of age …

Nevertheless, there was more than enough moral ammunition here for the disaffected peasants of France to justify rebellion under the mantra of '*liberté, égalité, fraternité*'. By 1789 many had been driven to the verge of starvation by a series of bad harvests, followed by a severe drought.[5] A punishing tax on grain called the *dîme*, imposed by the all-powerful Catholic Church, only added to their fury. The contrast between the lifestyles of the peasants and those of the nobles at Louis XVI's lavish court at Versailles eventually became too much to bear. By the time even more taxes were heaped on the peasants by Louis XVI to pay off the cost of his wars, popular anger meant that Colbert's goose was completely plucked and its hissing had reached fever pitch (see page 303).

When the King agreed to summon the ancient Estates General, a supposedly representative assembly that hadn't met for over 150 years, popular grievances spilled over into revolution. Jacques Necker, Finance Minister and leader of the self-styled Third Estate (the First being the clergy, the Second the nobility, and the Third everyone else) demanded the assembly draw up a new constitution for France. Despite Louis XVI's efforts to close the meeting down (he had the doors of the hall locked), delegates swore an oath on a nearby indoor tennis court that their National Assembly, as they called themselves, would not break up until a new constitution was agreed.

Events unfolded fast. After Necker was sacked by the King, on 14 July 1789 a mob stormed the Bastille, a symbol of royal oppression which, conveniently for the rebels, held a considerable cache of gunpowder and weapons. In August the Assembly abolished all special rights and privileges of the nobles, clergy, towns, provinces, companies and cities. Despite the 'Declaration of the Rights of Man and of the Citizen', published on 26 August, being modelled on the American Bill of Rights, the idea of liberty in France couldn't have been more different – abolishing inequality between classes of people was not the same as providing protection for landowners.

As in America, the issue of women and slaves was fudged. Not until 1946 was the French Declaration's opening statement, 'Men are born and remain free and equal in rights', legally extended to include women. French slaves working in the sugar plantations of the Caribbean found some solace in its words, though, and although slavery was never explicitly outlawed, the Declaration was quickly adopted by African workers in Haiti (then Saint Dominigue), who at that time produced as much as 40 per cent of the world's sugar thanks to the labour of some 500,000 African slaves. After a thirteen-year struggle, they finally liberated the island from French oppression. Their victory in 1804 created the first black republic in modern history and the second liberated colony in the world, after the United States of America.

Freedom in the name of liberty, equality and fraternity spread further across New Pangaea. In 1794 a Colombian, Antonio Narino, translated the French Declaration into Spanish and distributed copies in Central and South America, a crime for which he was sentenced to ten years in prison by the Spanish Inquisition. Two years into his sentence he escaped from jail in Cádiz, fled to France and London and finally back to South America, all the while spreading ideas about Latin American independence. On 20 July 1810 the citizens of Colombia (then called New Granada) declared their independence from Spain. After their leader Simón Bolívar secured victory over Spanish forces at the Battle of Boyaca in 1819 (with support from the British), Colombia's declaration became a reality, and the capital Bogotá fell to the

French anti-government rebels storm the Bastille prison in Paris, during a revolution against aristocratic and royal privilege.

23:59:59

Slaves picking cotton on the island of Haiti before they won their independence from colonial rule following the French Revolution.

independence fighters soon after. Waves of American countries were now declaring, and soon securing, their own independence. Venezuela in 1811, Argentina in 1816, Mexico, Guatemala, El Salvador, Honduras, Nicaragua and Costa Rica in 1821, and Brazil in 1822.

Independence in the name of freedom spread across the Americas almost as fast as smallpox brought by the first Europeans more than 300 years before. Nothing natural or human could now happen on one side of the global continent of New Pangaea without having serious consequences on the other.

Meanwhile, back in France, idealistic aspirations of human equality soon turned sour. For ten years a bitter civil war raged, including the infamous 'Reign of Terror' (June 1793–July 1794), notorious for the extensive use of the relatively humane form of mechanical execution proposed by, among others, Dr Joseph-Ignace Guillotin,[6] which chopped off its first victim's head on 25 April 1792.

It was a lot more humane than the Viking Blood Eagle ritual (see page 260) or being burnt, or disembowelled, or having your head inexpertly hacked off by several strokes of an axe.

Anyone suspected of harbouring anti-republican views was sentenced to death by the guillotine. Louis XVI and his Austrian Habsburg wife Marie Antoinette were among its victims, in January and October 1793 respectively. Under the auspices of the Committee for Public Safety, headed by Maximilien Robespierre, more than 18,000 people met their deaths, many under the guillotine, mostly for 'crimes against liberty'. Those sentenced were taken to their executions in an open wooden cart through jeering crowds. Two groups of liberal radicals, the Girondists and the Jacobins, fought for control, each executing as many of its rival's leaders as possible whenever the pendulum of power swung their way.

One man who found himself on the wrong end of the guillotine's blade was Antoine-Laurent

Lavoisier, a chemist and aristocrat. For the previous twenty years he had revitalized France's ability to produce gunpowder by dramatically increasing domestic supplies of saltpetre, making up for ground lost to the British with their plentiful sources from India. Lavoisier pioneered the addition of potash, which intensified the power of the powder, and provided financial incentives for local farmers to increase their output of saltpetre (made from mixing manure, earth, straw and compost with woodashes) from the barnyards of France.[7]

It was largely due to Lavoisier – later to be dubbed the 'father of chemistry' for his discovery of the role of oxygen in combustion – that liberty in the United States was won in just a few years of conflict. French saltpetre supplies, now plentiful thanks to his work, may have made the crucial difference between success and suppression, since American colonists had almost no gunpowder-manufacturing capacity of their own. But if Lavoisier's services helped further the cause of liberty on one side of the world, they were condemned on the other. As a privileged aristocrat who had been prominent in the King's *ancien régime*, he was a marked man. Lavoisier's wooden cart set out across the cobbled streets of Paris on 8 May 1794.

A new French parliament, consisting of 500 elected representatives and 250 senators, was inaugurated in 1795. A committee of five directors was given executive power. But, like so many rulers, they soon met with popular hatred, and were able to achieve their ends only via unconstitutional means. In the end it was the army who, on 9 November 1799, finally intervened with a dramatic coup. Their general had met with some success repelling invasion attempts by other European powers. Now, having seen enough with his own eyes of what ideals of liberty and equality meant in practice, he intervened to restore order

and rebuild French national pride. His name was Napoleon Bonaparte.

Kings, nobles, clerics and merchants in Europe's other big powers were appalled at what was happening in France. For the sake of the continued attachment of their own heads to their bodies, they wanted to ensure that whatever form liberty took in their countries, it didn't end in a French-style revolution. But still, it didn't take long for Europe's ruling classes to figure out how to make use of the trinity of liberty, equality and individual rights to boost their own grip on power.

For a start, it now became easier for rulers to recruit much larger armies. Since the 1530s, firearms had made it possible to turn peasants into foot-soldiers with only a few days' training. But armies stayed small, because the bigger they were, the more they cost. Standing armies were a wasted expense when a country was not at war, and mercenaries fought for one thing only: money. Once people became infected with the idea that ultimately *they* were sovereign, it became much easier for rulers to galvanize their masses into taking up arms. Individual

A satirical political cartoon in which French revolutionary leader Robespierre gets to play the part of executioner.

liberty proved a sufficiently powerful concept to persuade ordinary people that it was their duty to die for the sake of an idea, or for the colours on a national flag or the melody of an anthem. If rulers, be they aristocratic, mercantile or elected, could convince those they ruled that *their* country stood for freedom, then it became the patriotic responsibility of everyone to defend their nation from attack. In this way, ideas of national service and conscript armies were born.

In 1798 the French Republic passed a law declaring in its first article that 'Any Frenchman is a soldier and owes himself to the defence of the nation.' Napoleon was a genius at rallying France to the cause of defending its new principles, and was the first European statesman to introduce the idea of a conscript army. It quickly caught on. At the height of the Napoleonic Wars (1804–15), the French Emperor – as Napoleon became – managed to raise as many as 1.5 million troops, mostly from France, to his ranks. At the end of the wars, some 900,000 Napoleonic troops faced forces of about a million from a coalition of other European countries. Until now, no wars had ever been fought on such a scale.

In 1812, when Napoleon invaded Russia, he took with him 600,000 men, nearly half of them French. During his winter retreat, 370,000 died, mostly from exposure, disease and starvation, and 200,000 were captured. Even so, within a

year, Napoleon had managed to rebuild his army from 27,000 to 400,000 men.

The birth of fighting for the causes of good versus evil, freedom versus tyranny, ushered in a new age of propaganda and popular heroes. After Admiral Horatio Nelson defeated the French fleet at the Battle of Trafalgar in 1805, the British government built a statue and a square in the heart of London as a permanent reminder to its people of this hero who had led them in the defence of their country against tyranny. National champions like Nelson and Wellington, the British General who finally crushed Napoleon at the Battle of Waterloo in 1815, were given state funerals and buried in the nation's most sacred places.[8]

Besides the fear of a French-style revolution, another problem for Europe's rulers was how to recover from the economic dislocation caused by the loss of their colonies, on which much of their wealth was based.

France had lost Haiti and ceded most of its western colonies in a desperate bid for cash. The French colony of Louisiana was sold by Napoleon to the United States in 1803 for $15 million to help fund French wars against the British. Unlike the state of Louisiana today, the area conceded stretched from the south coast all the way up to the border with Canada.

Thanks to Latin America's independence movements (see page 329), Spain's overseas empire began to crumble during the Peninsular War when Napoleon's armies invaded Spain (1808–14). By this time, Britain could no longer rely on exclusive supplies from the plantations of the newly independent United States of America.

One solution was to find new colonies elsewhere. Napoleon got off to a quick start. In 1798 he occupied Egypt in an attempt to protect French trade interests and, if possible, block British routes to India. But after also invading Syria and Lebanon, he was halted by a British and Ottoman alliance (it was during this campaign that his soldiers found the Rosetta Stone, see page 124).

Hungry for power and determined to restore order and control, Napoleon pioneered conscript armies in Europe and used the ideologies of patriotism and freedom to take warfare to a new scale.

Meanwhile, the British consolidated their position in India, extending the rule of the Honourable East India Company across most of the subcontinent by 1813. The conquest of Burma enabled Britain to compensate for its loss of North American timber supplies by gaining access to thousands of square kilometres of precious hardwoods. Singapore became a British colony by agreement with the Dutch in 1824, and from there began a series of expansions up the coast of Malaya in 1826. Finally, no longer able to send convicts to North America as indentured servants, Britain's gaze turned southwards to find an alternative place for the disposal of its large numbers of poor people locked up in the country's overcrowded jails. From 1788 to 1868 regular shipments of British convicts turned Australia, which had been home to an estimated one million previously undisturbed Aboriginals, into another satellite of Europe's New Pangaea.[9]

Such strategies didn't help resolve the problem of what to do about trade with former colonies, such as America, which could no longer be exploited on an exclusive basis. What's more, with its millions of African slaves, the United States was at a huge economic advantage now that it was free to export its goods to the highest bidder anywhere in the world.

Thanks to the tireless efforts of Quaker campaigners and evangelistic Christians such as William Wilberforce (1759–1833), the British Parliament voted in 1807 to outlaw slave trading throughout the British Empire. By 1833 Britain had declared not just slave trading but slavery itself illegal throughout its Empire. It provided a total of £20 million in compensation to slave-owners such as the Bishop of Exeter, who received £12,700 for his 665 slaves – putting the price of freedom at just under £20 a head. The British Parliament now positioned itself as liberty's true champion, and in a determined effort to level the economic playing field, British foreign policy was now directed at persuading its trading partners to follow suit.

In 1815 the British government paid the Portuguese authorities £750,000 to abolish slave trading, and the Spanish £400,000 to cease shipping slaves to Cuba, Puerto Rico and Santo Domingo two years later. Between 1807 and 1860 the Royal Navy's West Africa Squadron helped to enforce the policy by seizing more than 1,600 ships and freeing as many as 150,000 African slaves.[10]

The restoration of profitable trading relationships with former colonies was also helped by the pen of a Scottish professor called Adam Smith. In *An Inquiry into the Nature and Causes of the Wealth of Nations* (1776), Smith proposed a new economic philosophy, arguing that some countries were naturally more efficient at making certain goods than others. Take a country with a warm, dry climate suitable for growing raw materials like cotton. If it freely traded with a hilly country that had plenty of rainfall to power watermills to manufacture finished clothes, then, said Smith, both countries would be better off.

Smith took the idea of liberty and extended it to mean free trade between nations by the removal of monopolies, taxes and tariffs. In such an economy, if demand for a particular product outstripped supply, then its price would rise, increasing profit margins for producers. That, said Smith, would act as an incentive to people to make products that customers wanted. The same theory worked in reverse: if a product is in too much supply, its price falls, profits are squeezed, and producers will switch to manufacturing other goods. Smith's 'invisible hand' of free market forces acted as a self-correcting system designed to regulate the demands of the market.

Smith's economic liberty became the cornerstone of a new science called economics. Capitalism based on free trade between separate nations was a direct challenge to the protectionist concept underpinning the empires of Europe in which wealth came exclusively from a nation's private colonies. In a world where some empires were beginning to break apart, and for new nations like the United States that had few colonies of their own, this idea of free market Capitalism had considerable attractions.

Capitalism's eventual global predominance can be accounted for not just through revolutionary wars of freedom, but also thanks to a different type of conquest that was just starting to unfold. The scientific and industrial revolution was no less violent than the American or the French, but its sights were set firmly on the conquest of nature, not men.

Monkey Business

How the human species freed itself from nature's limitations by mastering its own source of transportable power and how its populations increased beyond all reasonable measure.

ARE HUMANS fundamentally different from other animals? Today people seem rather divided on the issue. Some feel secure in their belief in man's separateness, even superiority, perhaps persuaded by the fact that no other animal ever tried, let alone succeeded in, sending its own species to the moon and back. Others think that Charles Darwin's discoveries, first published in 1859, proved beyond all reasonable doubt that humans evolved from animals, and that the difference between man and monkey is only a matter of degree – or of brain size, to be precise. Humans, they say, may be good at inventing things that other animals cannot; but then, was making an atomic bomb really such a smart idea? And

what's all this recent anxiety about the climate and global warming? If humans are to blame for melting ice caps, rising sea levels, changing rainfall patterns and dreadful new droughts which may mean that soon hundreds of millions of human beings, not to mention other species, will perish, perhaps they're not so clever after all.

This dispute is at the core of the current debate about man's relationship with nature. Will the future of life on earth be best secured by man using his ingenuity to tackle and take responsibility for the great problems of climate change, energy supplies and shrinking biodiversity? Or should mankind revert to a simpler, less frenetic way of life, restoring to nature its traditional role of regulating

rise of at least 500 per cent in less than 2,000 years. The biggest success was Asia, where the introduction of a new variety of rice (c.1100 AD) that could be harvested twice a year, and the sowing of South American maize (see page 308), had led to a huge population boom: approximately 64 per cent of the world's population lived there. Europe had about 21 per cent, Africa 12 per cent, South America 2.3 per cent and North America 0.7 per cent.

But in 1798 an English economist called Thomas Malthus (1766–1834) published an essay that predicted imminent disaster. Within fifty years, he claimed, the human race would have increased to such an extent that there would not be sufficient food to sustain it. The problem, wrote Malthus in *An Essay on the Principle of Population*, was that the human population was growing far more quickly than the supply of food that the earth could produce. If disasters resulting from human vices, such as war, didn't cut humanity's numbers sufficiently, then nature would take over, in the form of disease, famine and starvation, until the proportions of population to food supply regained a proper balance:

'The power of population is so superior to the power of the earth to produce subsistence for man, that premature death must in some shape or other visit the human race. The vices of mankind are active and able ministers of depopulation. They are the precursors in the great army of destruction, and often finish the dreadful work themselves. But should they fail in this war of extermination, sickly seasons, epidemics, pestilence, and plague advance in terrific array, and sweep off their thousands and tens of thousands. Should success be still incomplete, gigantic inevitable famine stalks in the rear, and with one mighty blow levels the population with the food of the world.'

and balancing the planet's life-support systems, regardless of the success or failure of any individual species? Just about every key decision facing modern scientists, politicians and civilized populations at large depends on answers to these questions.

The foundations of this mother of all debates dates back about 200 years. Until then most civilized people were clear in their minds that in terms of living species mankind was a cut above the rest.

Most followers of Judaism, Christianity and Islam believed that God had given man dominion over all other life,[1] and with the advent of guns and gunpowder he had the means to back it up. Once Europeans had connected the earth's previously separate continents, using their ships to transport artificially bred crops and animals from one part of the world to another, man's command over nature must have seemed almost complete.

Large human populations covered most of the globe, and even where they had been all but wiped out by disease, as in the Americas, stocks were rapidly replenished by fresh supplies from Africa and Europe. Numbers tell the story. Despite wars, plague and other natural disasters by 1802 the world's population had risen to a staggering one billion people – that's a

Malthus's essay came as a great shock to many people in Europe and America who believed, mostly for religious reasons, that man's place among the species of the earth was special thanks to the gift of a superior intellect and a divine soul that made him different from all other living things. But what if Malthus was right? It was almost as if his dire warnings were a direct challenge to the most creative minds of Europe and its American offshoots to make sure that they couldn't come true.

For centuries China had led the world in technological and scientific innovation. Printing, gunpowder and the compass, three inventions which according to the philosopher Francis Bacon helped shape the modern world, all came from the East. But it wasn't from the East that man's newest bouts of invention designed to overcome the limits of nature were to come. After all, who in China had ever heard of an Englishman named Malthus?

Since the mid-fifteenth century China had shunned approaches from the West. Its government's instincts were not directed towards the expansion of markets overseas. Following the rise of the Qing Dynasty in 1644,[2] its strategy was simple: to keep its borders secure from invasions and other foreign influences, to keep tribute payments from neighbours flowing in, and above all to prevent its prodigious numbers of peasants from clamouring for regime change. Without the pressure of external enemies, which had plagued the Song, the Qing's message to China's rural poor was as consistent as it was conservative: learn the works of Confucius, who teaches obedience to the family and the state, and if you're lucky you may win a well-paid job as a minor official in the enormous civil service. Otherwise, stick to growing rice.

The contrast with Britain in 1800 could not have been greater. This small island-nation, naturally protected by the sea, was now at the centre of a worldwide system of fleets, markets and colonies. A knock here or there, such as the loss of its American colonies, was simply redressed by the acquisition of new imperial territories elsewhere (e.g. Burma, Malaysia, Singapore, Australia). When Britain was threatened by a chronic energy crisis after its supplies of wood had been depleted by the demands of shipbuilding, iron-making, beer brewing, glass production, brickmaking and salt extraction, an alternative energy source was found in the form of coal. By 1800 most industries in Britain had converted from burning wood and charcoal to coke or coal.[3] With apparently infinite supplies of the magical black fuel under the ground in Tyneside, Yorkshire, Lancashire and South Wales, Britain's temporary energy crisis was well and truly over.

Another incentive to be creative came from the ban on slave trading, introduced in 1807, which stretched the nation's most ingenious minds to find ways of replacing slave labour with machines. This was considerably boosted by a system that gave individuals who came up with new ideas the chance to grow rich. Patents, which gave inventors a monopoly over sales of their ideas for a number of years, were first introduced in England by James I in 1623.[4] By 1714 it was made mandatory for inventors to publish their designs in exchange for their monopoly entitlement, encouraging creative minds to share their ideas as well as devise new and useful inventions.

It was within this innovative, entrepreneurial environment that Bolton barber Richard Arkwright (1732–92) came up with a revolutionary way of manufacturing cloth, inventing the world's first water-powered cotton mill. Arkwright's mill, opened in 1781 in the village of Cromford in Derbyshire, transformed the process of textile-making by spinning thin, strong threads which were then fed into an automatic, water-powered loom. In spite of a series of disputes over the rightful claim to the invention, by the time Arkwright died in 1792 he was a very wealthy man, with a fortune estimated at £500,000. From nothing, he had become one of the richest men in the world.

Despite the ingenuity of Arkwright's loom, his invention still relied on a natural source of power. It could work only next to suitably fast-flowing rivers or streams. Indeed, from the start of human history until about 1800, everything to do with man's relationship with the rest of the world had been subject to nature's own operating systems. Energy came directly from the natural, solar-powered processes of the earth – either from life itself, be it animal or human, or from nature's life-sustaining cycles of wind and water.

Power to move across land came from the feet of either humans or animals. An alternative was to navigate rivers on boats, or to pull barges using animal power along man-made canals. A huge canal-building project was undertaken in Britain from the middle of the eighteenth century, initially to transport coal from mines to markets around the country. A horse could tow ten times more weight by pulling a barge on a canal than it could by pulling a cart on a road. Between 1760 and 1820 more than a hundred canals were built across Britain.

Travel on the seas depended on natural wind power. In 1800, if sailors were becalmed in the Doldrums, an area of frequent calm near the equator, they simply had to stay there until either help came or the weather changed, even though being stuck in such hot, muggy conditions could mean death.

Wind power on land had been successfully harnessed using windmills, from ancient Rome to imperial China. As much as 60 per cent of the land in Holland today is below sea level largely thanks to its reclamation from the sea using wind power. By the sixteenth century Holland had more than 8,000 windmills, used for everything from draining the land to lifting equipment from mines, purifying (fulling) cloth, dressing leather, grinding grain, corning gunpowder and rolling copper plates.[5]

Power from a source as natural as free-flowing water downhill was successfully exploited in areas blessed with high rainfall and swift rivers. The Romans used watermills for pounding grain, the Chinese for operating the bellows of blast furnaces, and Europeans for sawing wood. By 1800 mechanical inventions like Arkwright's loom led to the industrialization of areas close to fast-flowing rivers. Among them was Manchester, situated at the foot of the Pennine hills, which became the world's first industrial city, dubbed the cottonopolis of England.

Indeed, it was entirely within this traditional worldwide context, one in which natural power was more abundant in some places than others, that Adam Smith's radical ideas about free market Capitalism were originally conceived (see page 333).[6]

Automation made possible by water power led to mass-production assembly lines, with teams of labourers trained to do the same repetitive tasks day in and day out. Products could be made quicker, more cheaply and in greater volumes than ever before. Whereas a craftsman had to be skilled in many disciplines to make a finished item, bustling around his workshop, wasting time finding the right tools for different stages of the job in hand, the assembly-line worker simply waited for the arrival of the next half-made object as it passed down the line, ready to do his one simple task again and again and again. Little skill required. No time wasted.

So huge was the impact of the transformation from skilled labour to mass production that people whose families and neighbours had enjoyed a traditional way of life for generations felt their world had come to an end. In Britain those who resisted such changes came to be known as 'Luddites', after their mythical leader Ned Ludd. Beginning in 1811 in Nottingham, skilled workers resenting their replacement by machines went on the rampage, doing battle with the English army. A mass trial of Luddites took place at York in 1813, leading to seventeen executions for crimes against machines (industrial sabotage was a capital offence). Dozens more were shipped off to the penal colonies of Australia.

It wasn't just in Britain that the idea of mass production had caught on. Eli Whitney (1765–1825) was a mechanical genius from Massachusetts, USA, who became familiar with the ideas of mass production after having worked in his father's nail-making workshop during the American Revolutionary War.

By 1798 the newly independent United States, anxiously watching the rapidly escalating warfare between Britain and Napoleonic France, had recognized its need for armaments to secure its fledgling liberty. To the amazement of the American Congress, Whitney demonstrated how he could assemble ten separate guns from a single heap of interchangeable parts. They were so impressed that they awarded him a contract to produce 10,000 muskets.

By reducing the number of components, Whitney's system not only substantially lowered the

costs of manufacture, but it meant that if one mechanism in a piece of artillery failed, it could simply be replaced by another from military stores.

Templates applied by semi-skilled labourers using machine tools quickly replaced craftsmen who used their skills and their hands to make bespoke products. An era of mass standardization was born. Whitney's water-powered armoury would later produce guns under the direction of another American inventor, Samuel Colt (1814–62), who had patented a design for a new type of repeating firearm called a revolver. Thanks to this invention American soldiers fighting in the Mexican-American War (1846–48) and prospectors in the Californian Goldrush (1848–55) could now shoot storms of bullets at the same time as riding full-speed on horseback. By 1850 Colts were being mass produced using the same interchangeable-

High-pressure steam released people from nature's power limits, provoking a race to unlock underground energy supplies.

parts technique pioneered by Whitney. By his death in 1862 Samuel Colt had also become a very rich man, with an estate valued at around $15 million.

The advent of mass production marked the beginning of modern consumerism and waste. Strong emotional attachment between humans and many of the tools and objects they used in their everyday lives had existed since *Homo habilis* crafted his first Stone Age implements (see page 90). But with the emergence of mass production all this began to fade. Instead of treasuring possessions because they were hand-crafted and perhaps passed down through generations, or made locally by some well-known artisan, people had little lasting attachment to individual objects that had been mass produced by machines. As a result, and contrary to mankind's hunter-gathering instincts (see page 213), people developed a new habit of throwing away an object before the end of its useful shelf-life and purchasing replacements for broken items rather than bothering with their repair.

Ships, windmills, waterwheels, canals and mass production in newfangled factories were all still, however, products of man's ability to harness the forces of nature. Even the first steam engines, pioneered by Englishman James Watt in 1769, used the weight of atmospheric air to provide a limited source of power for pumping water out of mines.[7] If mankind was to prevent Malthus's catastrophic predictions coming true, it was essential to find ways of breaking free of nature's constraints and tapping into the resources that she herself used to create and sustain all life.

When Cornish inventor Richard Trevithick (1771–1833) turned up the pressure on his 'Puffing Devil' steam engine in 1801, the truly transformative effects of man's potential power over nature could at last be seen. This machine did not rely on the earth's natural forces at all. High-pressure steam meant that Trevithick's engine could be mounted sideways on a track and be made to pull a wagon without the help of gravity, atmosphere, wind or water. All it needed were the energy-rich raw materials of the earth itself, the leftovers of life once lived (coal, oil and natural gas

TREVITHICKS,
PORTABLE STEAM ENGINE.

Catch me who can.

Mechanical Power Subduing Animal Speed.

are all products of leftover life). Simply by burning wood or coal in the oxygenated atmosphere, water could be heated in a high-pressure kettle to produce a fully independent source of portable power.

From now on, nature and humans began to compete for the earth's finite resources. While nature used them for the purposes of creating, sustaining, and evolving and recycling habitats, species and life forms, humans now exploited them to maintain a comfortable lifestyle and to increase their numbers far beyond natural limits.

◄▮◄▮◄▮◄▮►

On 14 October 1829 a competition was held at Rainhill, near Liverpool, to find the steam engine best able to transport people and goods on a newly constructed railway that connected the cotton-producing city of Manchester to the sea.

The winner, Robert Stephenson's *Rocket*, used a multi-tubular boiler to maximize its high-pressure steam power, allowing it to run continuously for fifty miles and transport loads of up to thirteen tonnes at twelve miles per hour, and to reach speeds of up to twenty-nine miles per hour when running light.

Railways transformed the landscapes of Britain, Europe and America. By 1890 they criss-crossed the whole of Britain, providing a convenient, fast, reliable transportation service that ran according to a timetable of human needs regardless of natural conditions. Clocks were synchronized across Britain from 1847 as railways standardized their schedules according to Greenwich Mean Time.

Until the mid-nineteenth century it was customary in the UK to add about fifteen minutes every twenty-four hours when travelling west to east to compensate for differences of local time. *Bradshaw's Railway Guide* for January 1848 was the first to show synchronized schedules in Greenwich Mean Time.

In the United States, rail transport allowed the proper exploration and exploitation of the vast inner continental area, opening up new lands for plantations and mining. In 1869 the first transcontinental railroad was opened, linking the Atlantic seaboard in the east to the Pacific in the west. It was considered by many as the crowning achievement of the presidency of Abraham Lincoln, who was assassinated four years before its completion (see page 349). By 1890 the US railroad system stretched some 230,000 miles. Networks in Russia, Europe and in other Western territories such as India and South Africa developed in parallel, providing mass overland transportation across the trading world.

Steam power allowed electricity, pioneered by a clutch of American, French, Italian and German scientists, to be carried across national grids following the development of the world's first electrical transmission system by American inventor Thomas Edison. Having perfected the design for a cheap, mass-produced lightbulb, Edison went on to establish America's first distribution system in 1882, linking fifty-nine customers in New York with mains electric power. In the same year, a steam power station came into operation in London, supplying energy to streetlights and private houses nearby.

Still today, as much as 86 per cent of all electricity is generated by steam-powered turbines heated by coal, oil, natural gas or nuclear fuels. Electricity gave rise to the first system for worldwide telegraphic communications, initially based on chemical battery power, which was trialled in France as early as 1810. In 1866 a transatlantic cable was laid on the seabed, linking Britain with the United States, thanks to the development of steam-powered ships, which had by then entered service in the British Royal Navy.[8] By 1900, instant electronic communications networks linked every inhabited continent on the globe.

The power of steam was matched by the development of another artificial transportable power-production system, in the form of the internal combustion engine. In 1876 German scientists Nikolaus Otto and Gottlieb Daimler developed a four-stroke engine that used oil, in its refined state of petroleum, as an energy-rich explosive mixture from which to produce a lighter, faster system for motive power. Within three years Karl Benz produced the first engines that powered a four-wheeled automobile.

23:59:59

Ransom Olds built the first American automobile factory in 1902, and Henry Ford applied mass-production techniques, beginning in 1910. Road networks were upgraded by governments keen to promote trade and transport, and urged on by people's desire to express their personal liberty by being able to travel, independently, where and when they chose. Less than a hundred years later there were approximately 590 million cars spread over the entire world, 140 million of them in the United States and fifty-five million in Japan.[9]

Small, highly efficient engines that could power cars were soon mounted on wings, further defying nature's limits on man's mobile capabilities. In 1903 Orville and Wilbur Wright successfully piloted the first controlled, powered flight near Kitty Hawk, North Carolina.[10] Aircraft became a vital component of military forces, providing reconnaissance and the possibility of delivering explosive bombs from on high. Man-to-man aerial combat was pioneered in the First World War (1914–18), making modern knights out of fighter pilots. Germany's Manfred von Richthofen, 'the Red Baron', was the ultimate flying ace, credited with more than eighty victims, each one shot down during dogfights using a front-mounted machine gun. War needs meant that the dynamics of mass production were quickly applied to aircraft manufacture. During the Second World War (1939–45) the United States built almost 300,000 military planes in the space of just five years.[11]

English inventor Frank Whittle's patent for a new type of propulsion system for aircraft in 1930 led to the development of the jet engine, which gradually replaced internal combustion engines after the Second World War. Now the continents were so close that it became possible to travel to the other side of the world in less than a day.

Like coal, oil deposits usually lie underground, formed by the remains of once-living things crushed over tens of millions of years by the weight of sediments overlaid by the movements of the earth's restless crusts. Much of the earth's oil naturally seeps to the surface, where it is broken down by bacteria that feed off it as part of nature's systemic recycling. The first modern oil wells were dug in the 1850s to provide a source of fuel (kerosene) for lamps. Slow production continued throughout the nineteenth century, until the invention of the internal combustion engine transformed the industry. In 1879 the US was producing nineteen million barrels a year. By 2005 worldwide production was almost eighty-three million barrels *a day*.

The ultimate form of power, which does not occur naturally on earth, but is nevertheless created by man out of the building blocks of the universe, was pioneered by one of the most famous scientists of all time, Albert Einstein (1879–1955). In 1905 this German-born genius spelled out a number of new theories that revised man's basic understanding of the laws governing the physical universe. Isaac Newton's theories of motion and gravity (see page 302), although apparently correct for large objects, broke down at the atomic level, said Einstein. Most critically of all, he calculated that the enormous quantity of energy stored in an atom to hold its

Orville Wright mounts his heavier-than-air aircraft that successfully accomplished the first powered flight in 1903. His brother, Wilbur, looks on.

constituent parts together was a massive source of untapped power. The amount of energy inside a single atom was, Einstein worked out, equivalent to its mass multiplied by the speed of light (186,000 miles per second), *squared*, i.e. E = mc².

In August 1945 the residents of Hiroshima and Nagasaki in Japan were the first people to feel the power of this awesome, supra-natural force when atomic bombs were dropped on their cities by the American air force, each one killing as many as 70,000 people in an instant. Thousands more died in a deluge of invisible but lethal radioactive energy that lingered for years to come. In 1951 a means of harnessing this energy for peaceful purposes was demonstrated by a test reactor in the US state of Idaho. It showed that nuclear energy could provide another source of heat to evaporate water into steam for generating electricity.

Modern transport systems, global communications and electricity on demand transformed man's ability to produce and transport goods and services all over the world. In the process new material wealth and unbridled personal freedom were created for millions of humans on an unprecedented scale. Most of this progress, however, was made entirely without regard for nature or other forms of life.

Just as significant was the discovery by German scientist Friedrich Wöhler (1800–82) that chemicals produced by life itself could be created artificially in a laboratory. While trying to concoct the compound ammonia cyanate in 1828, Wöhler, quite by accident, synthesized something else. When he discovered what it was, he could scarcely believe his eyes, reporting to his friend and mentor the Swedish chemist Jöns Jakob Berzelius: 'I cannot, so to say, hold my chemical water and must tell you that I can make urea without thereby needing to have kidneys, or anyhow, an animal, be it human or dog.'

The scientific world was astonished. Until then people had believed that a fundamental 'vital' force separated animate from inanimate matter. The artificial creation of a chemical of nature, such as urea, out of inanimate substances in a laboratory had been considered quite impossible.

The birth of organic chemistry, triggered by Wöhler's discovery, signalled the opening of a second front in man's knowledge of how to use the same materials as nature for his own ends. Life's 'modelling clay' is constructed from two main elements – carbon and hydrogen – that can combine with traces of other elements and oxygen in an almost infinite variety of chains, curls and rings to produce the diverse stuff of living things.[12] One of the richest sources of such ingredients is crude oil. Wöhler's discovery meant that it was now possible for mankind to learn how to model with this clay too – not yet for making life itself, but for synthesizing new useful but unnatural materials.

In the 180 years since Wöhler's accidental synthesis of urea, mankind has modelled countless new substances and materials using oil or its aerated equivalent, natural gas, as ingredients. Among the most significant are artificial fertilizers. After British forces cut off German supplies of Chilean saltpetre in the First World War, it seemed that Germany might be forced to surrender due to a lack of gunpowder. However, two German scientists, Fritz Haber and Carl Bosch, developed a new artificial process for producing ammonia from nitrogen and hydrogen. Not only could ammonia be used as a substitute for saltpetre in making gunpowder, but they found it could also be used as an artificial fertilizer. Today, more than a hundred million tonnes of artificial fertilizers are produced each year for spreading on fields all over the world. The Haber–Bosch process is now reckoned to be responsible for sustaining up to 40 per cent of the world's population – that's as many as three billion people.[13]

Almost as important was the development of artificial pesticides and herbicides. Although Dichloro-Diphenyl-Trichloroethane (DDT) was originally synthesized by the German chemist Othmar Zeidler in 1874, its power to inflict a holocaust on insect life was only discovered in 1939 by Swiss scientist Paul Hermann Müller.

DDT quickly became the most widely used insecticide in the world. Then, in 1962, American biologist Rachel Carson wrote a book called *Silent Spring* which exposed the devastating environmental effects of this chemical on wildlife. The long-term damage DDT might have done to humans now it had been incorporated into the food chain as a crop spray was a prospect too horrific for many governments to ignore. Despite the lack of hard evidence, Carson's image of a springtime when there would be no birds left to sing was alarming enough to launch the modern environmental movement and resulted in a ban on the use of DDT in eighty-two countries, although other artificial chemicals have since taken its place.

A huge fightback was orchestrated by American chemical companies which accused Carson of hysterical scaremongering. One spokesman for the industry, Dr Robert White-Stevens, said in a CBS news documentary: 'If man were to follow the teachings of Miss Carson, we would return to the Dark Ages and the insects and disease and vermin would once again inherit the earth.'[14] The debate between those who supported man's attempts to conquer nature and those who believed he should live within its operating margins was by now well and truly raging.

In 1921, Thomas Midgley, an engineer from Beaver Falls, Pennsylvania, developed a new organic compound called tetra-ethyl lead (TEL). The company for which he worked, General Motors, found that it could be added to petrol to make car engines run more smoothly. But the effects of lead poisoning quickly became apparent to Midgley, who in 1923 felt that he had to have a break from his job. 'After about a year's work in organic lead,' he wrote, 'I find that my lungs have been affected and that it is necessary to drop all work and get a large supply of fresh air.'

Lead poisoning has been shown to cause insomnia, weight loss and learning difficulties – or, as the Romans found out, premature senility (see page 201).

More than fifty years later, in 1976, US petrol companies finally began phasing out the use of leaded petrol, a project that was largely complete in America and Europe by 1986. A few years later a research report found that levels of toxic lead in human blood had dropped 78 per cent as a result.[15] In many parts of South America, Asia and the Middle East, however, leaded petrol is still made and legally sold today.

Midgley may have been history's least lucky inventor. After the débâcle of TEL he was shuffled, apparently out of harm's way, into looking for a new refrigerant for household appliances. He discovered how to synthesize chlorinated fluorocarbons (CFCs), which quickly became widely used in household fridges and aerosol sprays all over the world. But then, unknown to Midgley, who died in 1944 after being strangled to death by a harness he had invented to help him get out of bed, it was found that these man-made chemicals wreaked havoc in the earth's upper atmosphere.

CFC molecules are responsible for destroying the earth's ozone layer, allowing potentially lethal cancer-producing ultraviolet radiation from the sun to flood down on to the earth's surface. So serious was the issue that as a result of the Montreal Protocol of 1987, most countries have now agreed to stop CFC production. Nature is left to heal the hole above the earth's polar caps which Midgley's chemicals made. (Ironically, the ban on CFCs is now thought to have exacerbated the issue of global warming since these chemicals at least provide some protection against infra-red radiation in the upper atmosphere.)

Plastics are also synthesized by man from the oily clay of life. Radios, televisions, kitchenware and jewellery were some of the products once made from Bakelite, named after the Belgian scientist Leo Baekeland, who invented it in 1907. Strong, easily moulded into any shape, cheap to produce and perfect for use with electrical appliances because of its insulating properties, Bakelite ushered in a new era of plastic materials that quickly came to dominate the modern consumer world. Between the First and Second World Wars huge numbers of innovations were made in the range, variety and uses of plastics and other artificially synthesized products: polystyrene, PVC, nylon, synthetic rubber and plastic explosives to name just a few.

One of the biggest problems with man-made materials, however, is that because they are designed by humans, for humans, they have no intrinsic place or function in the world's natural ecosystems. Nature's products degrade and decay over time, so that she can re-use the planet's raw materials *ad infinitum* without clogging up the earth's environment with waste. Most synthesized materials and artificial plastics do not biodegrade, and often they can't be recycled.

The problem of unnatural waste is well demonstrated by a process pioneered in 1839 by American inventor Charles Goodyear. By applying intense heat to natural rubber and mixing it with sulphur he concocted a new substance, vulcanized rubber, not found in nature, which was perfect for providing air-tight elastic seals in machines like steam engines to make them run more efficiently. Later, this springy but durable material proved ideal for use as tyres on the wheels of bicycles and motorized vehicles. But because vulcanized rubber does not biodegrade, there is now such a mountain of worn-out tyres that no one knows how they can be got rid of. By 2007, stockpiles of waste tyres had reached three billion in Europe and more than six billion in America, with 300 million more added each year.

Direct artificial intervention into the processes of life is the third frontier in man's push over the last 200 years to break free from the natural limits endured by his ancestors. Building on the pioneering work of Englishman Edward Jenner (1749–1823), Frenchman Louis Pasteur (1822–95) devised vaccinations that worked by injecting people with artificially weakened forms of diseases to provoke their bodies into building defences to ward off potential infection. By 1979 such vaccinations had eradicated smallpox from the world, possibly man's biggest triumph over one of nature's most persistent and devastating historical plagues.

Direct biological interventions were spurred on by Rosalind Franklin (1920–58), a British biophysicist whose X-Ray images of DNA helped James Watson and Francis Crick to model nature's ultimate polymer, DNA, in 1953. Their work unlocked the control system inside every living

A 1932 magazine advertisement portrays the power of universal transportation, crowned by the use of Charles Goodyear's vulcanized rubber to make cheap pneumatic tyres.

cell that directs the process of evolution itself, thereby leading to the rise of the modern science of genetic engineering.

Since then scientists have replicated life in a test-tube by cloning cells; genetically modified crops to make them grow fitter, faster and larger; and fought diseases with drugs and therapies that lengthen human lifespans. To cap it all, the Human Genome Project, launched in 1990, has codified the complex genes that make up humans themselves, giving modern scientists the keys with which to unlock the inner secrets of evolutionary mechanics.

Modern man's achievements would seem to suggest that he is indeed rather different from any other species that has ever lived. After all, what other living things have managed to interrupt, modify, compete with and even usurp nature's age-

Man's manipulation of nature has led to a runaway rise in world population.

old operating systems? As for the dire warnings of Thomas Malthus: when he died in 1834, the world's population had just passed the one billion mark. By 1928 it had reached two billion; by 1961, three billion; by 1974, four billion; by 1987, five billion; and by 1999, six billion. The growth rate is such that for each extra billion humans added, it takes roughly half as long to add a billion more.[16] Dramatically falling child death rates have been accompanied by substantially longer life expectancy. Artificial fertilizers, medicines, improved hygiene, fossil fuels, industrialized cities, mass production, organic chemistry and vaccinations are the cause. As of 2006, the world's net population grows by approximately 211,090 *every day.*[17]

Malthus was right. None of this would have happened if nature had had her way.

Boom! The Rise in World Population

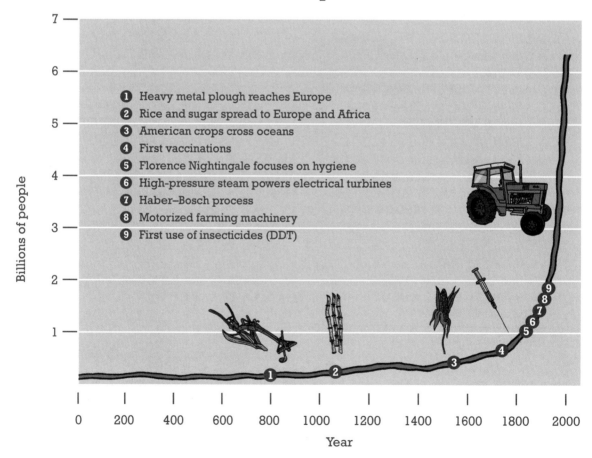

1 Heavy metal plough reaches Europe
2 Rice and sugar spread to Europe and Africa
3 American crops cross oceans
4 First vaccinations
5 Florence Nightingale focuses on hygiene
6 High-pressure steam powers electrical turbines
7 Haber–Bosch process
8 Motorized farming machinery
9 First use of insecticides (DDT)

Billions of people

Year

White Man's Race

How people from the West became convinced they were superior to all living things, believing it was their duty to subjugate the globe to their way of life.

CECIL RHODES was a British empire-builder, mining magnate and politician in southern Africa whose ambition was totally unfettered. He recognized that the advance of railways, steamships and mass-produced manufacturing had ushered in a new age of human supremacy over nature. He dreamed of building a railway line all the way from Cape Town, on the southern tip of Africa, to Alexandria on the Mediterranean coast.

But, unlike Alexander the Great, Rhodes didn't just want to conquer the world, its people and its riches. In a prophetic foretaste of the space race that was to begin half a century after his death, his goals reached out much further – even beyond the earth itself. As he wrote in his will:

'The world is nearly all parcelled out, and what there is left of it is being divided up, conquered and colonized. To think of these stars that you see overhead at night, these vast worlds which we can never reach. I would annex the planets if I could; I often think of that. It makes me sad to see them so clear and yet so far.'

Between 1850 and 1930, imperial ambitions like these transformed the world. Now unshackled from the limitations of the earth's natural forces, steamships could travel anywhere regardless of prevailing winds, railways could transport goods at high speeds without the need for human or animal

Cecil Rhodes, the British imperialist who wanted to colonize the world, announces plans to straddle a telegraph line from Cape Town to Cairo in 1892.

still hollow words. Real power lay in cabals of aristocrats and hungry merchants, who dominated parliamentary politics, or with imperialist tsars and autocratic kings. The story of how these countries 'parcelled out' the rest of the world between them is as extraordinary as it was traumatic.

Until about 1850, Africa was 90 per cent untouched by European nations. But then a new, industrial-scale method for manufacturing the drug quinine sourced from the cinchona tree of South America (the name was derived from '*quina*', the Inca name for the bark), dramatically reduced the hazards for Europeans of the deadly disease malaria, which was widespread throughout Africa. Large-scale use of quinine by Europeans began in the 1850s.

In addition, railways now made inland mining for raw materials such as copper, diamonds and gold logistically and financially viable. European powers found that Africa's natural wealth could provide a hat-trick of opportunities: capital riches to pay for their new machines (diamonds and gold); raw materials to feed into them (cotton and rubber), and markets for their finished goods (clothes, tea, coffee, chocolate and armaments).

Large areas of Africa's mysterious interior were unmasked by the Scottish missionary and explorer David Livingstone between 1852 and 1856 – the first European ever to see the Victoria Falls, which he loyally named after his reigning monarch. Following in Livingstone's footsteps, European explorers, settlers, prospectors and merchants, supported by their governments, flocked to the continent. By 1939 there was hardly a single African country that was not ruled by, or run for the benefit of, a European (or, in the case of Liberia, American) power.[1]

During the same period the United States successfully reached across North America to the Pacific Ocean, battling with Mexico in 1846–48 to bring into its Union new states such as Texas and California. In Australia, early British settlers had by 1859 overrun the continent, forcing the remaining Aboriginals into its hot, dusty interior. A new European colony was also established in

power, and industrialists could manufacture limitless numbers of cheap finished goods using artificial as well as natural materials, unskilled labour and automatically powered machines.

Britain's industrial lead didn't last long. By 1870 France and the United States of America had caught up, and their factories were manufacturing everything from pots and pans to clothes, ships and weapons. New countries such as Italy (united in 1860) and Germany (formed in 1871) also joined the club of quickly developing nations. Each followed a similar pattern: merchants grew rich off the profits of new enterprise, while their workers stewed with discontent about low pay, poor working conditions and the lack of a real political voice. Liberty, equality and fraternity were

New Zealand in 1840, following the Treaty of Waitangi negotiated with the Maori inhabitants, originally Polynesian explorers who colonized the islands between 800 and 1300 AD.

The treaty, signed on 6 February 1840 by representatives of the British crown and Maori chiefs, granted land-owning rights to both British settlers and Maoris. Its various multi-lingual versions mean land disputes still rage to this day, with Maoris arguing that the British crown has repeatedly breached the Treaty's second clause granting Maoris 'exclusive and undisturbed possession of their Lands and Estates, Forests, Fisheries and other Properties'. A New Zealand Crown Commission is still trying to settle the issue.

Back in Europe and America it hadn't escaped the notice of some late nineteenth and early twentieth century commentators that the inventions that were changing the world all seemed to originate from the genius of white Caucasian people – British, French, German, Italian or American. Over the course of a lifetime Samuel George Morton (1799–1851), Professor of Anatomy at the University of Pennsylvania, collected hundreds of human skulls from all over the world in an attempt to understand the source of Europe's apparent intellectual superiority. He claimed to be able to determine the respective mental abilities of different human races based on their average cranial capacity: the larger the brain, he said, the greater the intelligence.

In his book *Crania Americana* (1839) Morton claimed that the average skull capacity of whites was eighty-seven cubic inches, whereas in African Blacks it was just seventy-eight. At eighty-two cubic inches, Native Americans came somewhere in between. Later in the century Paul Broca, founder of the Anthropological Society of Paris, claimed to have confirmed Morton's findings after conducting his own study by weighing brains of members of different racial groups after autopsy.[2]

Despite Charles Darwin's theories, which suggested that all humans derived from a common African ancestor (see pages 98–99), supposedly scientific theories about the natural superiority of the white man continued to appear. In 1899 William Ripley, a social anthropologist based at the highly influential Massachusetts Institute of Technology, suggested there were three separate European racial stocks: Teutonic, Mediterranean and Alpine. It was this line of research that later inspired American lawyer Madison Grant to write his best-selling book *The Passing of the Great Race* (1916). Grant proclaimed that European racial superiority originated from a process of natural selection amongst Nordic people that enabled them to survive the harsh north European winters. He went on to recommend a systematic programme of state sterilizations to combat the spread of black African stock into North American cities:

'The individual himself can be nourished, educated and protected by the community during his lifetime, but the state through sterilization must see to it that his line stops with him, or else future generations will be cursed with an ever increasing load of misguided sentimentalism … A rigid system of selection through the elimination of those who are weak or unfit – in other words social failures – would solve the whole question in one hundred years, as well as enable us to get rid of the undesirables who crowd our jails, hospitals, and insane asylums.'[3]

By 1937 Grant's book had sold more than 1.6 million copies in America alone.

Voyeuristic Europeans and Americans had ample opportunities to see the apparent inferiority of other races for themselves. During the 1870s human zoos displayed African Nubians and American Inuits alongside wild animals. They could be found at imperial exhibitions held in Hamburg, Antwerp, Barcelona, London, Milan, New York and Warsaw, which drew as many as 300,000 members of the public. The 1889 Exposition Universelle in Paris was visited by a remarkable twenty-eight million people. One of its main exhibits was a 'Negro Village' featuring 400 native Africans. Other exhibitions in Marseilles (1906 and 1922) and Paris (1907 and 1931) displayed natives in cages, often naked.

In his capacity as head of the New York Zoological Society, Madison Grant advised the Bronx Zoo on its display in 1906 of a Congolese pygmy man called Ota Benga. He was placed in a cage with an orang-utan to demonstrate the missing evolutionary link between apes and real humans. The sign outside his cage read:

'The African Pygmy, "Ota Benga". Age, 23 years. Height, 4 feet 11 inches. Weight, 103 pounds. Brought from the Kasai River, Congo Free State, South Central Africa, by Dr. Samuel P. Verner. Exhibited each afternoon during September.'

Deeply felt racism was one of the factors that lay at the heart of America's devastating Civil War (1861–65), which almost caused the Union to break up irrevocably. The status of African slaves had been under scrutiny ever since the Declaration of Independence's statement in 1776 that 'all men are created equal' (see page 327). When Britain abolished slave trading in 1807 and then slavery throughout its Empire in 1833, pressure for change elsewhere grew greater. America's Northern states had little to lose from the abolition of slavery, having quickly industrialized with help from Britain. Not so the cotton-, tobacco- and sugar-plantation-owners of the South, where free slave labour underpinned the white way of life. Although the US Constitution was unequivocal about the equal rights of men, the legality of slavery was deemed a state matter, not a federal one. Eleven Southern states refused to outlaw slavery, using some ingenious mental gymnastics to justify their stance.

'Freedom is not possible without slavery!' was the slogan of the plain white folk of the old South, for whom the prospect of a life without slaves was tantamount to becoming slaves themselves.[4] By the 1850s American politics was almost entirely dominated by this single issue. The Northern states, which mostly abhorred the concept of slavery, were spurred on in their self-righteousness by books such as *Uncle Tom's Cabin*, written in 1852 by Harriet Beecher Stowe, a schoolteacher from Connecticut, which described the humiliations and injustices suffered by slaves in the South. The novel became one of the nineteenth century's best-selling books.

With populations growing in the Northern states, pressure on the South to outlaw slavery grew more intense. The questions as to whether new states should be admitted as 'Slave' or 'Free' further inflamed debate. When Abraham Lincoln, a Northerner, became President in 1861, eleven Southern states, led by South Carolina, formed their own Confederation and declared themselves separate from the Federal Union.

The Texas Declaration of Causes for Secession, published in February 1861, accused Northern states of 'proclaiming the debasing doctrine of equality of all men, irrespective of race or colour – a doctrine at war with nature, in opposition to the experience of mankind and in violation of the plainest revelations of Divine Law'. The African race was 'rightfully held and regarded as an inferior and dependent race'.[5]

Stowe's book struck a chord with Americans from the Northern states, most of whom thought the Southern states' use of slavery was unlawful, sinful and evil.

135,000 SETS, 270,000 VOLUMES SOLD.

UNCLE TOM'S CABIN

FOR SALE HERE.

AN EDITION FOR THE MILLION, COMPLETE IN 1 Vol., PRICE 37 1-2 CENTS.
" " IN GERMAN, IN 1 Vol., PRICE 50 CENTS.
" " IN 2 Vols., CLOTH, 6 PLATES, PRICE $1.50.
SUPERB ILLUSTRATED EDITION, IN 1 Vol., WITH 153 ENGRAVINGS,
PRICES FROM $2.50 TO $5.00.

The Greatest Book of the Age.

When a new government was established for the Confederacy, its Vice President Alexander Stephens boldly declared:

'They [the Northern states] rested upon the assumption of the equality of races. This was an error ... Our new government is founded upon exactly the opposite idea; its foundations are laid, its corner-stone rests, upon the great truth that the negro is not equal to the white man; that slavery – subordination to the superior race – is his natural and normal condition.' [6]

Between April 1861 and the spring of 1865 almost one million people died in a most bitterly fought civil war, including President Lincoln himself, assassinated at point-blank range whilst attending a theatre on 15 April 1865 by incensed Confederate John Wilkes Booth. In the end the Northern states secured victory thanks to their superior industrial technology. Four million black African slaves were freed, and three new amendments adopted into the US Constitution: outlawing slavery (the Thirteenth Amendment), granting black citizenship (the Fourteenth), and granting all men – but not women – the right to vote regardless of race, colour or creed (the Fifteenth). By 1877, after a painful period of 'Reconstruction' in the South, all the former Confederate states had been sworn back into the Union.

But ideas of white supremacy endured. In 1883 the US Supreme Court ruled that Congress did not have the power to outlaw racial discrimination by private individuals or businesses, causing many former black slaves to flee from the South.[7] Rulings in 1896 and 1908 allowed legal discrimination under the doctrine of providing public facilities (called accommodation) that were 'separate but equal'. In reality this meant that until the Civil Rights movement of the 1960s many states in America ran a system of apartheid, in which blacks were forbidden by law to attend the same schools as whites. They were not allowed to share a taxi with whites, or enter public buildings by the same doors. They had to drink from separate water fountains, use separate toilets, be buried in separate cemeteries, and even swear in court on separate Bibles. They were excluded from public restaurants and libraries, and barred from public parks with signs that read 'Negroes and dogs not allowed'.

A soldier fighting to preserve the Union plants a flag during the siege of Vicksburg, Mississippi, during the American Civil War.

Ku Klux Klan hoodies march with the American flag in 1923 in a display of solidarity aimed at defending racial segregation.

Vigilante whites, congregating in secret, such as the Ku Klux Klan, quickly grouped following the abolition of slavery in 1865 and carried out campaigns of violence and lynching to intimidate blacks, particularly to prevent them from voting.

The Ku Klux Klan was re-formed in 1915 inspired by D.W. Griffith's *The Birth of a Nation*, the most profitable film of its day, in which the original Klan was romanticized as the saviours of white victims of black violence following the Civil War.

Native Americans who had survived the diseases brought over by early European settlers fared little better under the ever-expanding commercial and agricultural interests of the United States. Americans of European descent came to believe that by expanding their territories westwards they were fulfilling the will of God, an idea known as the 'Manifest Destiny'. The concept was revived during the Spanish-American War (1898), which resulted in Puerto Rico, the Philippines and Guam coming under US control.

In 1862 Congress passed the Homestead Act, giving any white American the opportunity to own the freehold of up to 160 acres of undeveloped land in the West just by cultivating it, living there for five years and building a house at least twelve feet by fourteen feet. By 1900 white settlers had filed some 600,000 claims, amounting to more than thirty million hectares of land. Large rectangular blocks, surrounded by dirt tracks and with a single house in the middle, are still a common feature of the landscape in many central and western areas of the United States.

These determined settlement activities inevitably pushed indigenous people further into the pockets and corners of society. Special reservations were set aside for Native Americans where they could practise their traditional lifestyles, which were often diametrically different from those of white Europeans.

Samoset, the friendly native who stumbled across the Pilgrim Fathers in 1621 (see page 298), was a member of the Wampanoag tribe, among whom work was organized on a familial basis. Family groups gathered together in the spring to fish, in the early winter to hunt, and in the summer to cultivate crops. Boys were schooled in hunting and survival skills in the woods, while girls were trained from their earliest years to work in the fields and to erect or dismantle the family *wetu*, a portable round or oval house. Women were often made tribal leaders because they were responsible for the group's most vital activity–food production. Monogamy, polygamy and divorce were all acceptable, because tribal and clan allegiances were considered more holy than marriage.

With no need for slaves, private property or legal institutions, these native people soon became like aliens in their homeland. After the discovery of gold at Sutter's Mill in California in January 1848, a flood of new settlers from all over the world descended on the western US coast. Within ten years 300,000 prospectors had arrived hoping to make their fortunes from American gold. San Francisco's population exploded, increasing from 1,000 to 25,000 in

1849 alone. Steamships, railroads, agriculture, schools, towns, churches and roads transformed the landscape. Native Americans on the west coast suffered most of all, through disease, starvation and vigilante attacks. Their population, estimated at 150,000 in 1845, fell dramatically to less than 30,000 just thirty years later. Meanwhile, gravel, silt and toxic chemicals, which were used to extract gold, killed the local fish and destroyed their habitats.

Native Americans further to the east fared little better. In 1830 President Andrew Jackson, in collusion with Congress, passed the Indian Removal Act, encouraging natives to sign treaties removing them to reservations west of the Mississippi River. In theory, these were voluntary arrangements, but if the tribes declined to move, military conflicts such as the three Seminole Wars in Florida (1817–58) invariably followed.

By the end of these conflicts there were estimated to be only around one hundred Native Americans left in Florida. US government forces used a mixture of bribes and forcible removals to achieve their eventual eviction.

In 1835 the Treaty of New Echota enforced the removal of more than 17,000 Cherokees from Georgia in return for a cash payment. Despite a petition in which 15,000 of them pleaded with Congress to overturn the treaty, in May 1838 federal troops forcibly rounded them up into camps for eviction to the west. Four thousand died on their long trek to Oklahoma. Their passage is now called the 'Trail of Tears'.

Many tears were also shed by the Aboriginals of Australia when British settlers further consolidated their positions on the other side of the world. Between 1788 and 1900 the Aboriginal population is estimated to have fallen by as much as 90 per cent owing to a combination of disease, land appropriation and violence. In his frank *History of New South Wales*, written in 1834, Dr John Lang, a Presbyterian minister, was clear about what was happening:

'There is black blood at this moment on the hands of individuals of good repute in the colony of New South Wales of which all the waters of New Holland [Australia] would be insufficient to wash out the indelible stains.'

Lang's radical ideas for a free, independent, democratically governed Australia were far ahead of his time. All such concepts were put firmly on the back burner once Edward Hargraves discovered gold in Australia in 1851. The news provoked a new wave of European colonization which increased Australia's population from 431,000 in 1851 to 1.7 million two decades later. Roads, railways and telegraph lines quickly followed. Native people, as in America, found themselves in the middle of a confusing, unsympathetic world. European illnesses such as smallpox, influenza, measles and venereal diseases did most of the killing, although it is estimated that some 20,000 Aboriginals who got in the way were variously massacred by European whites.

Traditional hunter-gathering ways of life became less and less viable thanks to the development of massive pastoral enclosures and the monopolization of precious inland waterholes by imported cattle, rabbits and sheep that ate the lush vegetation which had previously sustained the continent's indigenous fauna. Aboriginals became dependent on jobs as miners, pearl divers or on cattle and sheep farms, and were often paid only in food and clothes.

Many white European settlers hoped that over time Aboriginal culture would disappear altogether. Between 1869 and 1969 Christian missionaries, supported by the Australian government, systematically made Aboriginal children wards of the state, forcibly taking them to internment camps and orphanages where they were raised independently of their parents as agricultural labourers or domestic servants. In an effort to stamp out their native culture, they were forbidden to speak except in English. According to a government inquiry published in 1997, more than 100,000 Aboriginal children were forcibly removed from their parents between 1910 and 1970. In a

The Scramble for Africa (1880–1914)

How the riches of a continent were sliced up
like a melon to feed foreign fortunes.

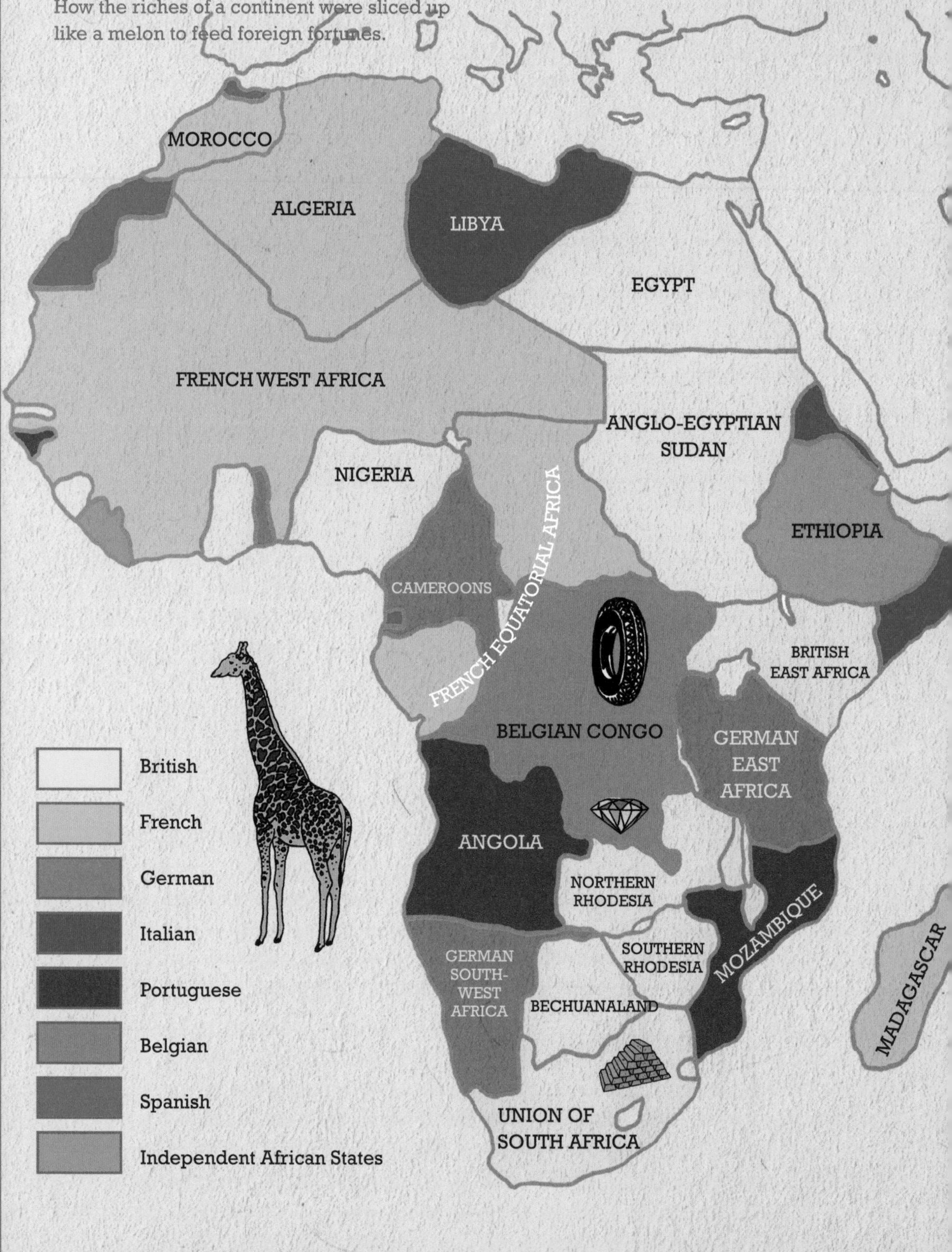

MOROCCO

ALGERIA

LIBYA

EGYPT

FRENCH WEST AFRICA

ANGLO-EGYPTIAN
SUDAN

NIGERIA

ETHIOPIA

CAMEROONS

FRENCH EQUATORIAL AFRICA

BRITISH
EAST AFRICA

BELGIAN CONGO

GERMAN
EAST
AFRICA

ANGOLA

NORTHERN
RHODESIA

MOZAMBIQUE

GERMAN
SOUTH-
WEST
AFRICA

SOUTHERN
RHODESIA

MADAGASCAR

BECHUANALAND

UNION OF
SOUTH AFRICA

British

French

German

Italian

Portuguese

Belgian

Spanish

Independent African States

candid admission of what happened to these 'Stolen Generations', the report concluded:

> *'These violations continue to affect Indigenous people's daily lives. They were an act of genocide, aimed at wiping out Indigenous families, communities and cultures, vital to the precious and inalienable heritage of Australia.'* [8]

Now, largely thanks to better immunity from European diseases, Aboriginal populations have risen from a low point of about 90,000 in 1930 to an estimated 458,920 in 2005.[9] But their torrid history has taken its toll. Unemployment, alcoholism, discrimination, drug abuse, crime and depression are huge problems facing these communities today.

※※※※※

Europe's conquest of inland Africa was no less traumatic. At the beginning of the nineteenth century sub-Saharan Africa was still a patchwork of as many as 3,000 distinctive human groups speaking over 1,500 different languages. Most of the continent was dominated by Bantu farmers and herders who, over several centuries, had spread out from their ancient homeland in West Africa, displacing hunter-gathering bands such as the bushmen (Khosians) and Pygmies. These people were now living a Neolithic-style village existence, sustained by crops such as sorghum and pearl millet alongside domesticated cows and sheep. In some sub-Saharan areas, such as Mali, Islam had taken hold thanks to the overland transport of salt, slaves and gold.

France began the first major European incursions in 1830, when its troops invaded Algeria, then a vassal state of the weakening Ottoman Empire. The Algerian government was furious with France for refusing to pay back enormous debts owing for grain shipped to France during Napoleon's Italian campaigns back in 1796. In his frustration, the Algerian ruler Hussein Dey assaulted the French ambassador with a fan. The insult was regarded as sufficient pretext for a French invasion that lasted seventeen years.

By 1834 French troops controlled a population of three million Muslims, and Thomas Bugeaud, the colony's first Governor-General, began constructing new roads for transporting goods and materials for export to Europe. In parallel with European emigration to North America, by 1848 more than 100,000 French people had settled in the territory, cultivating the land to make a living through the export of valuable but cheap cotton harvested by forced African labour. A French Lieutenant-Colonel serving in Algiers revealed his true feelings about the purposes of the occupation in a letter he wrote to a friend in 1843:

> *'All populations which do not accept our conditions must be despoiled. Everything must be seized, devastated, without age or sex distinction: grass must not grow any more where the French army has put the foot. I personally warn all good militaries … that if they happen to bring me a living Arab, they will receive a beating with the flat of the sabre. This is how, my dear friend, we must do war against Arabs: kill all men over the age of fifteen, take all their women and children or send them to the Marquises Islands or elsewhere. In one word, annihilate all that will not crawl beneath our feet like dogs.'* [10]

By 1843 Bugeaud's 100,000 French troops had finally crushed Arab resistance, turning Algeria into one of Europe's largest colonial suppliers of grain and raw materials until its eventual independence in 1962.

Following the success of this first major invasion by European forces into the heart of Africa, Europe's other nations found themselves in a competitive race to grab as much land and labour, and as many commodities and markets as possible. By the 1870s a renewed fervour for colonization – known as 'the Scramble for Africa' – was fuelled by the demands of European industrialization.

The scramble was whisked up further by the arrival of a new power on the European stage – Germany. On 1 September 1870 France was humiliated in the Battle of Sedan by German-

Prussian forces that surrounded and captured the French Emperor Napoleon III and his army. Following this Franco-Prussian war (1870–71), Germany arose as unrivalled master of the River Rhine, around which lay prime land for industrialization, with excellent transport links and power from natural fast-flowing rivers.

Flushed with victory, Germany's Chancellor Otto von Bismarck united twenty-five separate German states into a single mighty nation, headed by a new all-powerful ruler in the person of Wilhelm I of Prussia. He was crowned Kaiser of the new German Reich on 18 January 1871. As his title suggests ('Kaiser' is the German translation of 'Caesar'; 'Tsar' is the Russian equivalent), Wilhelm intended to follow directly in the footsteps of his antecedents in ancient Rome.

Africa's fate as the next theatre of European colonization was sealed by a conference in Berlin in 1884–85, orchestrated by Bismarck, at which the major powers of Europe agreed the terms under which the continent was to be sliced up between them like a melon. To prevent colonization in name only, claims of occupation had to be accompanied by proper colonial administrations that pressed the land into economic use. Without these, it was agreed that another European power could legitimately usurp control for itself.

With rules like these, the stage was set for the rapid and comprehensive creation by European powers of what has since become called the Third World. This term was first coined by French economist Alfred Sauvy. It followed the term for the Third Estate (Tiers État) in which the commoners during the French Revolution sought to redress the balance of power between themselves and the nobles (Second Estate) and the Church (First Estate). Like the Third Estate, wrote Sauvy, the Third World has nothing but 'wants to be something, too'.

Putting the land into 'economic use' meant either cultivating cash crops such as coffee, cacao, rubber, cotton and sugar to generate mercantile profits, or extracting minerals such as copper, diamonds and gold to pay for and feed Europe's

appetite for industry. In both cases, as European countries occupied Africa, its people, soil and society were consistently pressed into producing goods mainly suitable for export to other countries. Africans were also denied the chance of investing in strategies to feed and sustain their own growing civilizations. The legacy of this mostly European policy (although the US controlled Liberia, its commercial exploitation of rubber plantations was no less ruthless), which lasted in some cases for more than a century, is the poverty and hunger suffered by most of Africa today.

There was so much land and wealth to be plundered from Africa that even small countries like Belgium were able to get rich quick. King Leopold II (ruled 1865–1909) established the Congo Free State as his own private estate, a claim that was ratified by the powers at the Berlin conference. In 1881 Leopold added substantially to his domains by seizing the mineral-rich region of Katanga after his forces assassinated its King, Msiri, whose head was cut off and hoisted on a pole. Needless to say, his successor willingly signed a peace treaty. This new territory alone added an area sixteen times larger than Belgium to Leopold's private estate.

Between 1885 and 1908, Leopold's regime inflicted utter terror on native African populations, which were forced to produce rubber and ivory for export. His army, the Force Publique (FP), terrorized the people, ensuring that each individual delivered a set quota of rubber or ivory at a fixed price. Those who chose to hunt wild elephants to secure their quota of ivory were ruthlessly exploited. Ivory bought from Africans for eighty-two centimes a pound could be sold in Liverpool for 12.5 francs, a profit for Leopold of over 1,500 per cent.[11]

Armed with modern weapons shipped from England, Leopold's FP tortured the natives, took hostages and raped the women. Black soldiers recruited by the FP were permitted to cut off the right hands of natives whose rubber quotas weren't filled. They could even harvest baskets of smoked human hands, which were used as bargaining chips by local officials in lieu of meeting official rubber quotas. Bonuses were paid on the basis of how many hands they collected.[12]

A huge increase in demand for rubber from the 1890s boosted Leopold's profits, thanks to the advent of inflatable pneumatic tyres, invented by John Dunlop, who opened his first factory in Dublin in 1889. However, thanks to the 'invisible hand' of Adam Smith's market forces (see page 333), fresh demand encouraged the cultivation of new plantations across the colonized world, including South America and the British-controlled territories of Malaya. As a result, rubber prices fell, increasing pressure on the population of the Congo to produce more latex at a lower cost. Brutality increased. Finally, in the early 1900s, despite a series of elaborate cover-ups, Liverpool shipping clerk Edmund Morel raised the alarm when he noticed that ships arriving from the Congo laden with rubber were returning to Africa loaded with vast quantities of guns and ammunition. It has since been estimated that in the twenty years that the Congo was ruled as a personal estate for King Leopold, as many as five million natives were killed. At his death in 1909 Leopold left a personal fortune of US$80 million.

After the Belgium government took over control in November 1908, the Congo continued to be one of the world's richest sources of precious materials in the form of diamonds, copper, cobalt, tin, zinc and uranium. Its mines paid for extensive railways to be built throughout central Africa, with profits channelled back to Europe through the Belgium company the Union Minière du Haut Katanga, established in 1906. By 1919 more than 22,000 metric tonnes of copper were being exported a year. The local people of Katanga, the Luba, had once ruled a mighty kingdom, and were famous for their skill as woodcarvers. Now they worked as forced labourers in the Belgian-owned copper mines.

Little Belgium's African enterprise was as nothing compared to the scale of Bismarck's ambition now that he was at the helm as Chancellor of a united Germany. Colonies were needed to bolster the new nation's prestige. By 1900 Germany had become the third-largest European power in Africa, with an empire comprising fourteen million African subjects. Its biggest territories were in Rwanda, Burundi, Botswana, Cameroon and Togoland. German South-West Africa (now called Namibia) was added in 1884, and Germans settled there in large numbers, attracted by the prospects for mining diamonds and copper, and growing crops. By 1917 about 10,000 Germans controlled approximately 160,000 African bushmen, divided between the Herero and Namaqua tribes. Between 1904 and 1907 revolts against German rule led to the deaths of 50 per cent of the Namaqua and 80 per cent of the Herero populations, most of whom starved to death after fleeing into the desert. Alternatively, their wells were poisoned by occupying troops.[13]

※✕✕✕✕✕※

By the mid-nineteenth century, the precarious balance of power in Europe was beginning to shift, partly as a result of Germany's entry into the scene. The Russian Tsar Nicholas I was determined to extend his own sphere of influence south-wards, towards the strategic trade routes of the

Mediterranean, by defeating the Ottoman Turks and capturing Istanbul. Britain, petrified of losing control of trade routes to its territories in India, launched a pre-emptive strike by besieging the Russian Black Sea fleet in Sevastopol, on the Crimea, in 1854. After a year-long siege supported by France, the Ottomans and Italy, the port was captured, and in 1856, after the end of the war, the Russians were forced to accept a humiliating peace treaty that banned all navies from the Black Sea.

Among the many battles fought during the Crimean War was a much celebrated human war against nature. Teams of nurses headed by Florence Nightingale, with help from the Jamaican-born Mary Seacole, began to understand that through meticulous cleanliness and attention to hygiene, the rate of bacterial infection among wounded soldiers could be dramatically reduced.

Italy's reward for its loyal support of the British, French and Austrian allies was recognition of its own unification under the Kings of Sardinia. Europe now welcomed yet another new nation that also showed itself eager for a fiery colonial baptism in Africa. Italy annexed Eritrea (in 1882), Somaliland (1899), Libya (1911) and finally Abyssinia (present-day Ethiopia, 1936–41).

Anxiety over trade routes that provoked the Crimean War lay at the heart of the British government's policies towards African colonization. In 1875 Prime Minister Benjamin Disraeli purchased shares in the Suez Canal, which had been rebuilt in 1869 by French engineers on behalf of Egypt's Islamic rulers, so that steamships could travel to and from the Far East in a fraction of the time it took to sail the long, arduous route around the Cape of Good Hope.[14]

Control of southern Africa was Britain's other big interest, originally in order to protect the shipping route to India and China. Dutch settlers, called the Boers, had first arrived in 1652, establishing the colony as a resupply stop for their ships travelling to the Far East. In 1806, following the Battle of Blaauwberg, the British took over the Cape Colony (modern South Africa) from the Dutch, to ensure that it couldn't fall into French hands during the Napoleonic Wars. The Boers

were none too happy with their new overlords, particularly when slavery was abolished throughout the British Empire in 1833, as the enslavement of the natives had become both a tradition and a way of life for the Boers. Within two years 12,000 Boers began the so-called 'Great Trek' inland to create their own independent slave states (Natal, Orange Free State and the Transvaal) separate from British control, although only after much blood was shed in wars with native Africans. At the Battle of Blood River in 1838, Boer firearms slaughtered 3,000 Zulus. The blood of their dead, accounts said, turned the river red.

British settlers started arriving in their thousands during the 1820s, many hoping to establish lucrative sugar plantations. They also ran into conflict with the native populations. So effective were the Zulus as a fighting force that even without firearms they inflicted a humiliating defeat on the British in 1879 at the Battle of Isandlwana, when over 1,400 British-led soldiers were surrounded and slaughtered. But within six months, British supremacy was restored (at the Battle of Ulundi) thanks to the deployment of Gatling guns, an early type of revolving-barrel machine gun invented in America which could fire an almost never-ending stream of bullets.

The inventor, Richard Gatling, wrote to a friend that he hoped his invention would promote peace:

'It occurred to me if I could invent a machine – a gun – which could by its rapidity of fire, enable one man to do as much battle duty as a hundred, that it would, to a great extent, supersede the necessity of large armies, and consequently, exposure to battle and disease would be greatly diminished.'

But even the collective might of 50,000 Zulu warriors with their short-stabbing spears and cowhide shields were no match for such firepower.

In 1886, when gold was discovered in the Transvaal, tensions between the Boers and the British turned explosive. Almost overnight a new city, Johannesburg, sprang up out of the arid South African bush. It was soon flooded with prospectors like Cecil Rhodes, whose British South Africa Company became one of the richest mining empires of all time.[15] Backed by Royal Charter from London, by 1895 Rhodes had carved out his own country, called Rhodesia in his honour, comprising what is now Zimbabwe and Zambia. His ambition was driven by a single-minded ruthlessness to profit from Africa:

> *'We must find new lands from which we can easily obtain raw materials and at the same time exploit the cheap slave labour that is available from the natives of the colonies. The colonies would also provide a dumping ground for the surplus goods produced in our factories.'*[16]

But British ambitions in southern Africa were frustrated by Paul Kruger, the Boer President of the Transvaal, who taxed dynamite – essential for prospectors – in the gold and diamond fields he controlled, and denied foreigners the right to vote on local affairs. When British officials protested in 1899, the Boers declared war.

The Boer War raged on until April 1902, forcing the British to deploy 250,000 soldiers. More than 22,000 were killed in action, along with 7,000 Boers and about 20,000 Africans. In addition, 28,000 Boer civilians are thought to have died in the appalling conditions of the British concentration camps in which they were confined, most of them from starvation, malnutrition and disease.

Cecil Rhodes's dream of building a railroad all the way from Cape Town to Alexandria typified the attitudes of Europeans, whose confidence in their own supremacy over nature, including other races, was now so ingrained that to question it was almost unthinkable. After a long struggle with illness Rhodes died in Bulawayo on 26 March 1902, leaving this thought in his will:

> *'I contend that we are the finest race in the world and that the more of the world we inhabit the better it is for the human race.'*

Nineteenth-century Europe's financial and material gains were made at the expense of the native populations of Asia, America, Africa and Australia. Grand buildings like the Royal Palace of Brussels still stand proud, refurbished in splendour by King Leopold II with his profits from Congolese

Zulu warriors successfully overwhelm British forces at the Battle of Isandlwana in 1879, but victory was short-lived following the arrival of Richard Gatling's machine gun a few months later.

Hundreds of British bodies fill a trench following the battle of Spion Kop during the Second Boer War (1899–1902).

ivory and rubber. Yet by the time most African countries received their political independence in the second half of the twentieth century, their land had been exhausted, their raw materials removed, their economies sucked dry by loans, and trade agreements locked their populations in poverty. Worse still, their people had been ripped out of their traditional tribal and ethnic groupings, rearranged into new colonial territories and armed with Western guns. In reality, European colonization just extended the misery of Africa that began with the demand for slaves in the Muslim world to the east and in America to the west (see pages 269, 299).

To cap it all was Europe's imperial morality, epitomized by Rhodes himself, which cultivated a sense of supreme self-righteousness. Only in such a spirit could Rudyard Kipling, one of Britain's most celebrated poets of the age, write:

Take up the White Man's burden –
Send forth the best ye breed –
Go bind your sons to exile
To serve your captives' need;
To wait in heavy harness,
On fluttered folk and wild –
Your new-caught, sullen peoples,
Half-devil and half-child. [17]

Back to the Future

How some people tried to resist the advance of Western civilization, wishing instead to return to a more natural order, but whose attempts often met with catastrophic consequences.

KARL MARX, the Jewish German social philosopher and economic theorist, wrote his famous *Communist Manifesto* in the heady climate of a Europe swept by revolutions. His testament directly challenged Europe's ruling Capitalist elites. Marx proposed a new type of society that would finally put an end to centuries of pernicious inequality between rich people and poor.

In 1848 a series of disastrous crop failures, most importantly the potato blight that crippled Ireland, Belgium and Germany, led to popular rebellions in Italy, France, the German states, Poland and throughout the Habsburg Empire. The cause, according to Marx, was plain to see. Human history, he said, was a long series of

struggles between rich and poor. As a result of industrialization that struggle was now being waged between Capitalist businessmen (the bourgeoisie) and impoverished factory workers (the proletariat). But the ideology of constant economic growth around which Europe and America had staked their strategies was now tottering on the brink. Capitalism's imminent collapse, said Marx, would lead to a new social order across the whole world in which equality and true freedom for the masses could be attained, if only the workers of the world would unite.

Marx explained that man's industry based on naked profit was as physically brutal as it was morally bankrupt:

'In place of the numberless indefeasible chartered freedoms, [the Bourgeoisie] has set up that single, unconscionable freedom – Free Trade. In one word, for exploitation, veiled by religious and political illusions, it has substituted naked, shameless, direct, brutal exploitation.'

Capitalism had caused the breakdown of traditional family values:

'The bourgeoisie has torn away from the family its sentimental veil, and has reduced the family relation into a mere money relation.'

Its requirement for constant economic growth had created an insatiable appetite for global conquest:

'The need of a constantly expanding market for its products chases the bourgeoisie over the entire surface of the globe. It must nestle everywhere, settle everywhere, establish connections everywhere.'

Such globalization, he said, undermined man's natural self-sufficiency and created products no one needed in order to satisfy desires that had to be artificially created:

'In place of the old wants … we find new wants, requiring for their satisfaction the products of distant lands and climes.'

Cheap, mass-produced goods manufactured on the other side of the world forced other cultures irretrievably into Capitalism's net, creating a spiral of human dependency and natural exploitation:

'The cheap prices of commodities … compel all nations … to adopt the bourgeois mode of production; it compels them to introduce what it calls civilization into their midst … In one word, it creates a world after its own image.'

Industrialized towns had swamped the countryside, massively increasing human populations and obliterating traditional, rural ways of life:

Karl Marx was convinced that free market Capitalism was just a passing phase and that a fairer system would eventually take its place.

'The bourgeoisie has subjected the country to the rule of the towns. It has created enormous cities, has greatly increased the urban population as compared with the rural … as it has made the country dependent on the towns …'

Capitalism's accomplishments were, said Marx, the work of mankind's conquest of nature:

'Subjection of nature's forces to man, machinery, application of chemistry to industry and agriculture, steam navigation, railways, electric telegraphs, clearing of whole continents for cultivation, canalization of rivers, whole populations conjured out of the ground.'

Today's debates on globalization, Third World poverty, the unequal distribution of wealth, social breakdown, the obsession with consumerism and the environmental damage caused by the relentless exploitation of the earth's natural resources all lead directly from a battle of ideas that dates from Karl Marx.[1] His adversaries were the imperialists and Capitalists of Europe and North America, whose deep-seated belief in scientific and industrial

progress, rooted in the philosophies of ancient Greece and Rome, made them determined to spread their economic formulae across the entire world.

✕❮❯✕❮❯✕

Shortly after Marx wrote his prophetic manifesto, humanity's largest ever civil war broke out in China. Between 1850 and 1871 an estimated twenty million people perished in the Taiping Rebellion – that's twenty times more than in the American Civil War of the same period.

Despite the efforts of the Qing Empire to isolate itself from foreign interference, by 1850 European influences in China had become very profound indeed. On the one hand Western missionaries were busily converting sectors of Chinese society to the gospels of Jesus Christ, while on the other merchants forced their way through imperial trade embargoes with a mixture of illegal smuggling and gunboat diplomacy.

Tea, silk and porcelain were highly sought-after commodities in Europe. But there was a problem. Chinese society was built on a philosophy of self-sufficiency. Since the mid-fifteenth century, China had been a civilization independent of overseas fleets and trade with far-flung vassal colonies. Food and luxury goods were all manufactured in the home market. The Chinese Emperor himself explained as much in a letter he wrote to King George III of England in 1793, in response to a British request for trade:

> 'You, O King, live far away across the mighty seas ... The difference between our customs and moral laws and your own is so profound that our customs and traditions could never grow in your soil ... I have no use for your country's goods. Hence there is no need to bring in the wares of foreign barbarians to exchange for our own products ...'[2]

Such self-satisfied sufficiency provoked the most extreme imperialist reaction. If the Chinese didn't want Western goods, then something had to be done to make them want them.

Officials in Britain's Honourable East India Company came up with the rather less honourable solution of drug trafficking. An elaborate system was established whereby British traders would buy Chinese tea in Canton and issue credit notes to Chinese traders, who could then redeem them against opium smuggled across the border by Bengalese agents from Calcutta. Between 1750 and 1860 thousands of tonnes of opium grown in the poppy fields of Bengal were smuggled into China in exchange for silk, tea and porcelain. The trade was a masterstroke of ingenuity. Rather than the British paying for goods in valuable silver, locally grown opium could be used as currency instead. And the problem of China's self-sufficiency was solved by a freshly cultivated dependency on highly addictive drugs.

By 1810 Chinese imperial court officials were themselves increasingly under the influence of opium despite repeated bans on the trade. In desperation the central Qing government issued a decree ordering customs searches at all ports in an attempt to stamp out the drug's devastating effects:

> 'Opium has a harm. Opium is a poison, undermining our good customs and morality. Its use is prohibited by law. Now the commoner, Yang, dares to bring it into the Forbidden City. Indeed, he flouts the law!'[3]

It made no difference. By the 1820s more than 900 tonnes of opium a year were flooding into China from Bengal. In 1838 the imperial government introduced the death penalty for anyone caught trading the drug. When the British refused to stop shipments, the Chinese government imposed a trading embargo on them. Two years later a British fleet arrived with the object of forcing the Chinese to revoke the ban, and its cannon wreaked havoc on towns and villages along the fertile banks of the Yangtze. After the British seized the Emperor's tax barges, by 1842 the Chinese government was forced to sue for peace. A treaty re-established trade links, ceded Hong Kong to Britain and allowed Christian missionaries to preach unfettered on Chinese soil.

In hindsight these unwitting agents of God did more harm than all the poppy fields of India combined. In 1850, Hong Xiuquan, an unorthodox Christian convert (another revolutionary who had repeatedly failed his imperial examinations), was so entranced by missionary teaching that he claimed to be the long-lost brother of Jesus Christ. Blessed with a charismatic personality, Xiuquan raised a giant peasant army to challenge the Qing government, which was, largely thanks to Britain, now impoverished. Between 1853 and 1864 Xiuquan and his rapidly swelling number of followers established a rival civilization across southern China, with its capital at Nanjing. The 'Heavenly Kingdom of Great Peace' replaced the teachings of Confucius with the Christian Bible. Women were treated for the first time as equal to men. Opium, gambling, tobacco and alcohol were all banned.

But once installed in his new capital, the movement's divine leader lost his zest for politics, instead choosing to spend more time with his extended family in his private harem. In 1856 Britain used China's civil war as an opportunity to make more military mischief in a conflict known as the Second Opium War, and attacked the port of Guangzhou. This conflict ended in 1860 with the occupation of Beijing by Western forces, who compelled the Chinese government, still battling Xiuquan's rebels in the south, to sign a new treaty which legalized the import of opium, authorized foreign warships to sail along the Yangtze River, established eleven new ports for trade with Britain, France, Russia and the United States, and paid a large indemnity in silver to compensate Britain for its recent loss of profits. How times had changed since the Chinese Emperor confidently penned his letter to George III in 1793! Meanwhile, French and British forces supported the struggling Qing government and together they finally put an end to the rebellion in 1864 but not before it had become one of the deadliest wars in human history in which between twenty and thirty million people died.

Japan's reaction to the rude awakening of Commodore Perry's gunships (see page 320) was no less dramatic. During a brief civil war (The Boshin War, 1868–69), a clique of disaffected nobles, unhappy at the ruling Shogun's acceptance of American trade terms, restored imperial power in a military coup. The new Meiji regime was determined not to suffer the same humiliation as China, its old rival and master. Its five-charter oath, established in 1868, spoke of an 'international search for knowledge to strengthen the foundations of imperial rule'. In an astonishing turnaround the Japanese government recruited 3,000 foreign experts to teach its people English, science and engineering. It dispatched students around the world to learn Western ways, and formed its first great industrial companies, Mitsui and Mitsubishi. Production started with the finishing of textiles, and a kind of Japanese Lancashire was set up, which began a brave new world modelled on Western Europe but located in the Far East.

By 1895 Japan's rulers adopted an aggressive, expansionist foreign policy supported by their new military strength gained through industrialization. In 1894–95 Japanese forces successfully fought a war against China in Korea, despite China's own attempts at 'self-strengthening' following its humiliation by the Western powers. Japan now played the Western game. It established Korea's independence from China, forced China to relinquish control of Taiwan, secured itself a large payment in silver, and won precious trading and manufacturing rights along the Yangtze.

Just ten years later, Japan stunned the world by defeating Russian forces as they surged southwards in search of an all-weather sea port on the Pacific coast. At the Battle of Tsushima on 27 May 1905 the Japanese navy annihilated the Russian fleet in a surprise night-time attack. Japan had now truly proved herself a great power. When it formally annexed Korea in 1910 hardly a murmur of disapproval was heard from the mighty imperialist club of nations.

Meanwhile, China, exhausted by war and still reeling from the effects of addictive opium, was easy prey for colonial vultures. By 1887 the French had won their own Far Eastern empire, luring Vietnam and Cambodia (known as Indochina) away from China's sphere of influence. Ten years later German forces occupied the strategically important mainland coastal region of Jia Zhou. By this time the Japanese had established control over Korea.

All of which accounts for why another devastating Chinese rebellion broke out between 1899 and 1901. Traditionalist rural peasants, calling themselves the Boxers, wanted to rid their homeland of the pernicious influence of the West and its alien Capitalist culture. Their forces invaded the imperial capital Beijing in June 1900, killing tens of thousands of Chinese Christians and taking thousands of Western foreigners living in the city hostage. An international force of 20,000 troops from eight nations (Austria, France, Germany, Italy, Japan, Britain, Russia and the United States) scrambled to the rescue. By August the force had defeated the Boxer rebels, but then, in an eerie echo of the disaster of the Fourth Crusade (see page 263), this cabal of mostly European nations went on to plunder Beijing itself, setting fire to its palaces and forcing the Emperor and his Dowager Empress to flee.

Brutality was sanctioned by the famous words of German Kaiser Wilhelm II, who on 27 July ordered his troops to 'Make the German name remembered in China for a thousand years so that no Chinaman will ever again dare to even squint at a German.'[4]

The price of international 'rescue' was set at reparations of some £67.5 million, to be paid in precious silver by the imperial government of China and split amongst the members of the eight-nation alliance. Such a sum could be met only by punishing new taxes on the rural population. As a result, within ten years the imperial Chinese government had become so weak, and so loathed by its people, that a popular revolution finally succeeded in throwing out the 2,000-year-old imperial institution, replacing it with a republic in January 1912. After decades of civil war and invasions, this fledgling republic itself succumbed to Communist takeover on 1 October 1949 under its chairman Mao Zedong.

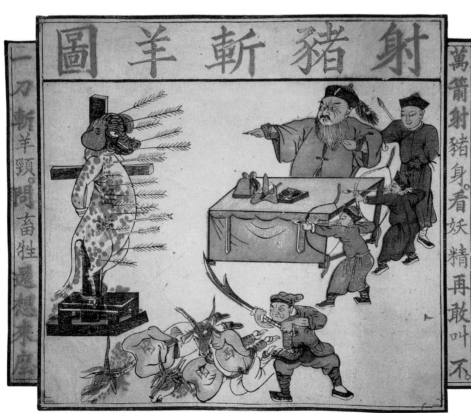

Shooting the pig (Jesus Christ) and beheading the sheep (Christians) epitomized years of anti-Western feeling that led up to the Boxer rebellion of 1900.

In the Far East Marx's predictions as to the course of Capitalism and the rise of the suppressed proletariat had proved uncannily accurate. In Europe his ideas were stoking similar revolutions, whose effects turned out to be no less profound.

✕◗◖◗◖◗✕

Popular zeal for a return to a simpler, fairer way of life boiled over during the First World War. The Great War, as it became known, was fought mostly in Europe between 1914 and 1918. It was only slightly less devastating than China's Taiping Rebellion, with eighteen million people left dead, twenty-two million wounded and the slaughter of some eight million horses.

It began after the heir to the imperial Habsburg throne, Archduke Franz Ferdinand, was murdered by a Serbian student in Sarajevo on 28 June 1914. What had been intended as a brief incursion by the Kingdom of Austria-Hungary into Serbia to seek revenge, turned into a power-struggle between the nations of an imbalanced Europe newly enriched with German industrial, colonial and military might.

Against the combined forces of imperial Germany, Habsburg Austria and the Ottoman Empire was pitched an alliance of Britain, France and Russia, and from 1917 the United States, in a titanic struggle for the control of Europe and its overseas colonies. In such a climate, popular revolutions quickly gathered pace. First to fall was Russia, in two separate revolts during 1917. Military humiliations inflicted by Japan in the east and Germany in the west had severely weakened the authority of Tsar Nicholas II. A lack of reforms and an imperial obstinacy against sharing power resulted in a total breakdown between his government and the people of Russia. Inflation, food shortages and a battered economy ripened the conditions for revolt, while rapid industrialization increased urban populations but failed to provide them with a better quality of life.[5]

When the women of Petrograd revolted over the shortage of bread in February 1917 they unleashed a massive uprising of popular discontent. In March Tsar Nicholas was forced to abdicate his throne. The political radical Vladimir Lenin now chose his moment to smuggle himself back to Russia from Switzerland, where he was living in exile. The Germans let him pass through their country on board a sealed train, anxious to make sure his revolutionary ideas couldn't leak into their own country, itself teetering on the brink of breakdown after three years of devastating war.

By early November Lenin and his army of Bolshevik revolutionaries led by Leon Trotsky had seized formal power in Russia. Inspired by the anti-Capitalist ideology of Karl Marx, Lenin established peace with Germany on terms that many regarded as highly unfavourable to Russia.

The Treaty of Brest-Litovsk was signed between Russia and the Central Powers (Germany, Austria-Hungary, Bulgaria and the Ottoman Empire) on 3 March 1918. It confirmed the independence of Finland, Estonia, Latvia, Ukraine, Lithuania and Poland, removing roughly a third of the total number of people under Russian control.

In May 1918 the Tsar and his family, under arrest since the abdication, were executed in cold blood – whether on local or central orders, no one knows. Now civil war broke out between the Marxist Bolsheviks' 'Red Army' and the monarchist 'Whites' backed by Britain, France, the USA and Japan. In June 1923 the Bolsheviks finally gained control of the country, leading to the world's first experiment with Communism in which new technology promised to provide the magic ingredient to make Marxist dreams of class equality at last come true. Lenin wrote:

'We must show the peasants that the organization of industry on the basis of modern, advanced technology, on electrification which will provide a link between town and country, will put an end to the division between town and country, will make it possible to raise the level of culture in the countryside and to overcome, even in the most remote corners of the land, backwardness, ignorance, poverty, disease, and barbarism.'[6]

Universal healthcare, equal rights for women and education for all formed the backbone of Lenin's socialist ideals. But Russia, being a predominantly peasant, rural society, did not yet have the industrial capacity to achieve full-blown socialism. So a New Economic Policy had to be introduced in 1921. Lenin allowed surplus agricultural yields to be sold by farmers as an incentive for them to produce more grain to help build up Russia into a rich enough society to afford the technology it needed to make the dream of classlessness come true. By 1928 agricultural and industrial production in Russia had fully recovered from the devastations wrought by world and civil wars.

But Lenin had died in 1924, and his successor Joseph Stalin (1878–1953) had different ideas. His purge of the Communist Party in the 1930s revealed that his was a dictatorship as total as any in imperial times. Although his successive Five Year Plans, begun in 1928, transformed the Soviet Union from a backward, peasant society into a major world industrial power, countless millions died of starvation owing to his confiscation of grain and food from farmers between 1932 and 1934, after declaring that all agricultural produce was the property of the state.

Stalin's goal of industrialization was far removed from Lenin's ideal of creating a classless and equal society. Under Stalin the Communist Party became a brutal ruling class, suppressing anything and everything that threatened its grip on power. Those who resisted were either executed or dispatched to labour camps in Siberia. By 1939 an estimated 1.3 million people had been interned in such camps, called gulags.

Further to the south, another man's ideals centred around trying to turn back the clock of history, as a wave of African and Asian movements towards political independence from European colonial rule gathered pace. Mohandas Karamchand Gandhi (known as 'Mahatma' – great soul – 1869–1948) pioneered the art of civil resistance using the philosophy of *ahimsa* (non-violence), one of the five Jain vows (see page 169). After campaigning against racial discrimination in South Africa, where he worked as a lawyer, Gandhi returned to India to take over leadership of the Indian National Congress. His philosophy was based on equal rights for women, grants of land to the poor and the supremacy of self-sufficient local vegetarian communities (*ashrams*) that resisted imported goods from overseas and the export of raw materials to colonial powers. Gandhi's ideas led to one of the boldest ever attempts to turn back the tide of New Pangaea.

Gandhi's leadership was successful in that it led to Indian independence, eventually granted by Britain following the Second World War in 1947. But his search for racial harmony and a society built on non-violence was in vain. Muslims and Hindus, it seemed, could not live side by side without an oppressive master to keep them in check. The British partition of India, in which

Lenin, as depicted in this revolutionary poster of 1924, was passionately devoted to the idea of redressing the gap between rich and poor.

more than half a million people died as a result of intercommunal violence, gave birth to a new Muslim country, Pakistan. Racial and political hatred has plagued these nations ever since, a throwback to the Mughal / Hindu divisions of pre-British India.

On 30 January 1948, while Gandhi was walking in the grounds of his house in New Delhi, a Hindu radical named Nathuram Godse, who blamed him for making too many concessions to the newly formed Pakistan, shot him dead. Hindu ideals of non-violence and a return to self-sustaining local communities seemed to die with him, although recently there has been a revival in the concept of *ashrams*, local food production and ecological living in response to concern about global warming and carbon emissions (see page 377).

While Gandhi was orchestrating his campaign for a return to traditional Indian ways of life and Lenin was fighting to build a new classless Russia, the Islamic Ottoman Empire finally collapsed after Western European powers invaded Istanbul in 1918. The Middle East now became a source of booty for victorious France and Britain, which grabbed the lion's share between them following a secret agreement they had made during the First World War.[7] France took control of Syria and Lebanon, while Britain took over Iraq and Palestine, stating its support in the Balfour Declaration of November 1917 for the re-creation of a national home for the Jewish people based in Palestine, to which it encouraged Jewish settlers.

As far as the core Anatolian home of Ottoman Turkey was concerned, Greek, Italian and Armenian nationalists all claimed it, leading to a bitter international and civil war which was finally settled in 1923, after a victory by Turkish nationalists led by Mustafa Kemal (1881–1938, now known as Atatürk, meaning Father of the Turks). After a bitter struggle his forces successfully established Turkish national independence, formally recognized by world powers in the Treaty of Lausanne in July 1923.

Mohandas Gandhi championed non-violent political resistance and attempted to rekindle the human appetite for hand-crafts, but he fell victim to the stronger forces of religious intolerance.

In today's Turkey, Atatürk is regarded as a national hero. Mirroring the Meiji restoration of Japan, he embraced Western styles of dress, industrialized the country, promoted women's rights and secularized politics to protect against the rise of Islamic fundamentalism. But in neighbouring Greece, school textbooks describe him as a murderer and traitor who denied Greece its ancient heritage, founded on the historic Christian city of Constantinople (Istanbul). In Armenia, Atatürk's reputation is even worse, because they believe he was complicit in an act of genocide in which 1.5 million Christian Armenians were forcibly deported from the eastern Ottoman Empire between 1915 and 1917, leading to an estimated 500,000 deaths.

The Ottomans' forcible removal of an entire ethnic group to enhance its religious and racial purity wasn't lost on a man whose rise to power not only came close to ridding the world of Communism, but nearly toppled the entire Capitalist system too. Adolf Hitler's aim wasn't to usher in a Utopia of Marxist equal rights, nor was it to build up a mighty global empire based on overseas trade.

Instead he wanted to restore society to its 'natural principles' which, he believed, had been severely corrupted by the rise of Western culture.

Adolf Hitler (1889–1945) was a veteran of the First World War who felt deeply let down by his country's leaders who, in the opinion of many in the German army, had accepted a humiliating armistice agreement in 1918 which had burdened the country with impossible war reparations. According to the terms of the Treaty of Versailles, signed on 28 June 1919, not only must Germany accept full responsibility for the Great War (article 231) but it had to pay reparations of a staggering *269 billion* Reichsmarks in gold (£11.3 billion).

Whilst recovering from an attack by British forces at Ypres, Hitler described his reaction to the news of Germany's surrender:

> *'I broke down completely ... Because the war was lost and we were at the mercy of the victor ... The Fatherland would have to bear heavy burdens in the future ... Darkness surrounded me as I staggered and stumbled back to my ward and buried my aching head between the blankets and pillow.'* [8]

Waves of workers' strikes that had crippled the munitions factories of the German Fatherland, turned what Hitler had believed was almost certain victory into a humiliating surrender. Those responsible were, in Hitler's eyes, Socialist Jewish Marxists. People like Rosa Luxemburg, the Polish-born Jewish revolutionary who led the Communist Party of Germany, and her fellow socialist Karl Liebknecht, who tried to ferment a Marxist revolution in Germany in January 1919. Both Luxemburg and Liebknecht were murdered by right-wing soldiers that same month. Hitler set down his ideas in a book he called *Mein Kampf* ('My Struggle'), written while he was serving a prison sentence for an unsuccessful coup attempt in 1923:

> *'The Jewish doctrine of Marxism repudiates the aristocratic principle of Nature ... it impugns the teaching that nationhood and*

race have primary significance and takes away the very foundations of human existence and civilization.' [9]

It wasn't just Jewish meddling in Germany's domestic affairs that enraged Hitler. Jewish bankers, he believed, were responsible for the rise of the Capitalist powers themselves, with all their money lending and pursuit of profit. Thanks to them Germany had been dragged into a war with an ignominious end, after which it was forced to admit guilt and pay enormous reparations that could be honoured only by borrowing capital from Jewish bankers in the United States. Under such a crippling burden, Hitler wasn't alone in thinking that the German economy stood no chance of recovery:

> *'The struggle against international finance capital and loan-capital has become one of the most important points in the programme on which the German nation has based its fight for economic freedom and independence.'*

But what, exactly, was Hitler proposing instead?

In *Mein Kampf* it is clear that the same strand of racial supremacy that fed Europe's colonial enterprise, with its 'White Man's Burden', also led to Hitler's system of eugenics, a philosophy that advocated selective human breeding, and the Nazi adoption of a policy of ethnic genocide:

> *'Just as Nature concentrates its greatest attention not to the maintenance of what already exists, but on the selective breeding of offspring in order to carry on the species. So in human life also it is less a matter of artificially improving the existing generation – which, owing to human characteristics is impossible in ninety-nine cases out of a hundred – and more a matter of securing from the very start a better road for future development.'*

These were the very principles argued by American zoologist Madison Grant at the beginning of the twentieth century (see page 347). Hitler's grand plan was to force back the frontiers of Europe's New

Ja!

Führer wir folgen Dir!

Adolf Hitler: charismatic war-leader who wanted to restore what he believed was the world's natural order.

Pangaea, with all its consequences, and go back to the philosophy of ancient Sparta, with its pure racial stock providing national security and social welfare (see page 184). But to wind back the clock would require intervention on a massive scale. Hitler took it upon himself to bring this about.

The German government that ruled between the end of the Great War and 1933 (since called the Weimar Republic) was an attempt to establish a social democracy which failed in large part due to economic hardship and attempts by conservative economist Heinrich Brüning (German Chancellor from 1930–32) to strengthen the economy by cutting workers' benefits. Elections in July 1932 gave Hitler's Nazi Party an overall government majority and by 30 January 1933 Hitler was sworn in as Chancellor. Almost immediately the reality of his ideas when put into practice began to come clear.

Hitler knew that his country's population, like many others', was rising fast. He believed that Germany needed more space in the long term, since industrialization had led to overcrowding of the homeland.[10] Eastward expansion became his top priority. Domestic economic revival and the provision of new land could both be accomplished through a massive programme of rearmament, designed to intimidate, and if necessary to force, Germany's neighbours into territorial concessions.

To put his plans into effect, Hitler introduced a totalitarian regime. He used the country's constitutional emergency powers to suspend all democratic elections, and then he banned opposition parties. Next, he introduced a secret police force to enforce conformity to his creed. More surprising, and less well known, was his introduction of laws against cruelty to animals. The strong affinity between Nazi ideology and the natural order of the world, in which only the fittest survive, led to new laws that gave animals their own official legal status.

Hitler was an advocate of vegetarianism, as was his henchman Heinrich Himmler, who professed to hate hunting. Laws introduced in 1933 abolished the distinction between wild and domestic animals, and vivisection was prohibited. In what was a direct affront to western Europe cartesian tradition (see page 302), anyone found treating animals as if they were lifeless objects, Hermann Göring warned, would be dispatched to a prison camp.

Nazism saw itself as an alliance between people of pure Aryan stock and the forces of nature, in which the superior race worked on behalf of nature to restore the balance of a world corrupted and diverted by subhuman races' mismanagement of global affairs.[11] Such policies were designed for the long term. Planning for future human generations lay at the heart of Hitler's political creed. Re-establishing natural racial supremacy would require persistence and the acceptance of suffering today, but it had to be done for the sake of generations to come.

Beginning in 1933, Hitler inaugurated a mass sterilization initiative with the cooperation of the leaders of Germany's medical establishment. By 1945 more than 400,000 people had been neutered against their will so as to eliminate them from the chain of heredity. Physically weak, homosexual, religious, ethnically mixed and criminal stock were being bred out of the system.

In an extension of the programme, between 1939 and 1941 an estimated 75,000 to 250,000 people with 'intellectual or physical disabilities' were slaughtered through a system of forced euthanasia called 'Action T4'. Later this was extended to what is now called the holocaust during the latter part of the Second World War, when between nine and eleven million people were executed in gas chambers, most of them Jews, but also Christians, homosexuals and prisoners of war, as well as Polish and Romany people. Systematic genocide was at the apex of the Nazi regime's efforts to cleanse the racial stock of its Fatherland, to turn back the clock to a time when its people were genetically pure.

The playing out in reality of the theories of eugenics and racial hygiene, many of which were popular in the pre-Second World War United States of America, left deep, lasting scars across humanity's increasingly globalized civilizations owing to the unprecedented scale of suffering they caused. More than sixty-two million people are thought to have died in the Second World War, making it the most devastating human conflict in all history.

But in reality these deaths were as a result of two wars, although arguably they both stemmed from the same inexorable rise of global Capitalism. Hitler's European war was a revolt against what he saw as a Jewish plot to rape the world for profit which must, he believed, be stopped from perverting the natural order of humankind for ever. The second war took place in the Far East, where Japan's enthusiasm for and brilliance at copying Western-style industrialization was equalled only by China's inability to implement effective reform.

Japan's aim was to colonize the Far East in much the same way that Europe had conquered Africa. It wanted to secure permanent access to the raw materials required to further its economic growth, independent of the meddlesome powers of the West, whose colonies were still scattered all around. That meant controlling the vast agricultural and mineral wealth of China, which, largely thanks to Soviet and Japanese intervention, had become embroiled in another long, drawn-out and bitter civil war starting in 1927.

In 1931 Japan invaded the north-eastern Chinese province of Manchuria and installed a puppet regime. In 1937 it launched a large-scale invasion of China with 350,000 soldiers, and began a series of aerial bombing raids on cities all over the country. But Japanese advances had stalled by mid-1938, as Chinese resistance grew stronger under its nationalist leader Chiang Kai-shek. By 1940 the war in mainland China had become a stalemate, and Japan was increasingly being strangled by economic sanctions from Western powers which controlled its energy and oil supplies via their colonies in India, Burma, the Philippines, Malaysia, Indonesia and Singapore.

With Europe's major powers distracted by the struggle against Hitler, Japan waded into the World War with a surprise attack on the American Pacific fleet at Pearl Harbor, Hawaii, on 7 December 1941. It hoped that, in their desperation to avoid the opening of a second front, Western nations would lift their sanctions, giving Japan a free rein to complete its conquest of China. Instead, it found itself surrounded on all sides as the United States declared war on Japan, ultimately securing victory through its use of atomic weapons in the summer of 1945.

By then Hitler's hegemony was already over. His long-term plan failed in part because time does not run backwards. His experiment to reverse generations of interracial mixing solved nothing. It just provoked yet more war, violence and slaughter. The scale of devastation was immense. Its effects dramatically altered the final fraction of our history still to come.

Witch Way?

How the whole world was bound into a single system of global finance, trade and commerce, sustained by relentless scientific endeavour. Can the earth and its living systems sustain humanity's ever-increasing demands?

'**WHEN SHALL** we three meet again? In thunder, lightning or in rain?' Three weird witches dance around a bubbling cauldron casting spells that let them gaze into the future.[1] A windswept, treeless, barren moor is the setting for this opening scene of Shakespeare's *Macbeth*, about a man who would be king. It also works well as a backdrop for the final *one thousandth of a second to midnight* that represents the extraordinary sixty years that have followed the Second World War when seen on the scale of earth history as a twenty-four-hour clock. In that tiny fragment of time so much changed in both the human and the natural worlds that no number of books could hope to tell the whole story.

To help us see the job through, these witches will now conjure up the ghosts of three thinkers from the past whose prophetic insights bring into sharp focus dilemmas of today that may yet determine the future for all living things.

The ghost of Karl Marx haunts anyone who believes in the supremacy and wisdom of the human system of economic organization called Capitalism. Marx believed that Capitalism was merely a necessary phase which would one day be replaced by an altogether fairer society.

In the first three weeks of July 1944 more than 700 bankers, representing the forty-four nations allied against Hitler's Germany, met in secret at Bretton Woods, deep in the forests of New Hampshire. Their goal was to devise a robust financial system that would not only repair the damage done by two devastating World Wars, but would minimize the risks of such conflicts ever happening again.

The bankers agreed about the two wars' main causes. Industrialized nations had ignored the sacred mantra of free trade, as originally spelled out in the eighteenth century by Adam Smith (see page 333), and had instead engaged in a pernicious cycle of mercantile protectionism, exploiting their colonies as security for raw materials (for making goods) and consumer markets (for selling them). A series of economic blocs competed for global supremacy leading inevitably to conflict, especially when times got tough.

The stock-market crash of 1929, which led to a period now known as the Great Depression of the 1930s, was one such time. Instead of a global system of central banks co-ordinating a rescue, each colonial empire was confined to the limits of its own trading system – such as the Sterling Area in the British Empire. If Britain was struck by a slump, its colonies suffered appalling hardships as the one-way flow of produce to their mother nation stalled. Workers in the factories of the industrialized motherlands suffered too, since few consumers in their overseas markets could now afford to buy finished goods. Spiralling unemployment led to social unrest which could be settled only by massive state intervention programmes, often established through military rearmaments, potentially leading, as in interwar Germany, to global war.

The Bretton Woods system was designed to prove Marx wrong. Capitalism was not doomed, but required a single global system that allowed the free flow of capital and goods without exchange controls and government taxes. In the changed conditions following two World Wars, imperialist nations that had previously relied on their colonies for economic strength were now encouraged to relinquish them and establish an international collaborative financial framework instead. Free trade agreements would allow Smith's 'invisible hand' of market forces to regulate the supply and demand of goods, taking away from individual nations the power of economic blackmail as well as the threat of being sunk without trace.

It was also thought that governments should create a safety net for the welfare of their citizens to enable them to provide payments in the event of unemployment, healthcare for the poor and state pensions for the elderly and infirm. Welfare states were designed to alleviate economic hardship sufficiently to eliminate the likelihood of French-, Chinese-, Russian- or German-style popular revolts. Further reforms came in the shape of universal adult suffrage, adopted between the wars by countries like Britain which, after protracted lobbying by female 'suffragettes', extended the right to vote to all women over the age of eighteen in 1928. The huge nation of India followed in 1950 after gaining its political independence. The voting rights of blacks were at last enforced in the United States following the Civil Rights Act of 1964, and women finally got the vote in Switzerland in 1971. South African blacks were eventually enfranchised after the racist apartheid system collapsed in 1994.

Sixty years after the Second World War, free trade, welfare states and democratic governments underpinned by universal suffrage had come to dominate the politics of the human world. The principles of Capitalism and free trade were enshrined in a system of global exchange championed by the World Trade Organization, established in 1995. By 2007 it had 123 nations as members, with most of the world's remaining economic powers – for example Russia, Libya, Iran, Iraq, Ethiopia, Algeria and Afghanistan – waiting in a queue to join.

It took this long because the Soviet Union and China were not party to the Bretton Woods Agreement. After the surrender of Japan in 1945, China's political independence was quickly usurped by a Communist/Marxist regime supported by Stalin's Soviet Union. In October 1949 Mao Zedong's 'People's Republic of China' declared itself a one-party state. The Communists' defeated political rivals, Chiang Kai-shek's Kuomintang, fled to the small island of Taiwan, where they remain to this day, as Capitalist converts firmly tied to the Western camp.

Between 1945 and 1991 China and the Soviet Empire formed economic blocs that rivalled the

Capitalist West. These were closed, centrally managed systems backed up by massive armament programmes. Thanks to the doctrine of Mutually Assured Destruction ('MAD'), atomic weapons ultimately helped keep the peace between the ideologically divided Communist East and the Capitalist West during what is called the Cold War.

A number of flashpoints arose during this period that could have led to global war. They include the Cuban Missile Crisis of 1962, the Korean War (1950–53), the Vietnam War (1964–75) and the Soviet invasion of Afghanistan (1979).

But then, starting in 1985, the Soviet Union (USSR) began to disintegrate through economic stagnation, and its satellite countries in Eastern Europe and the Baltic States (Lithuania, Latvia and Estonia) snatched at the chance of greater political and social freedoms offered to them by reforming Soviet leader, Mikhail Gorbachev (last head of state of the USSR from 1985 to 1991). Following a failed coup in 1991 the Soviet empire finally collapsed as former Soviet republics declared their independence – by December 1991, 14 out of the 15 Soviet states signed the Alma Ata Protocol effecting the dissolution of the Soviet Union. On Christmas Day President Gorbachev resigned as President of the USSR – declaring the office extinct and even Russia became a Capitalist democracy of sorts. On 1 May 2004 seven former Soviet bloc countries formally joined the European Community (Estonia, Latvia, Lithuania, Poland, the Czech Republic, Slovakia, Hungary). Romania and Bulgaria joined later, on 1 January 2007.

China is still a Communist one-party state. Economic reforms introduced by Party leader Deng Xiaoping in 1978 have brought it increasingly within the Capitalist system of commerce and trade, culminating in its accession to the World Trade Organization in 2001. Communist China is modern Capitalism's biggest growth area. With the benefit of cheap labour provided by its almost limitless supply of manpower, the Chinese economy has risen to become the second-largest, and the fastest-growing, in the world. With an average annual growth rate of 9 per cent, China is now on a path

to take over from the United States as the kingpin of global Capitalism.

Meanwhile, the imperial powers of Europe let their former colonies go. Sometimes the process happened peacefully, sometimes not. It took years for some countries to find their feet – especially those, like French-controlled Algeria, whose populations included thousands of settlers from their colonial motherland.[2]

Nations in the Middle East were transformed not only by their newfound political independence, but by the discovery beneath their deserts of huge reserves of crude oil.

In the 1930s disaffected British diplomat Jack Philby handed the United States a most precious gift by milking his close friendship with Arab ruler Ibn Saud. Following the collapse of Ottoman power at the end of the First World War, Ibn Saud conquered the Muslim holy cities of Mecca and Medina, and formed a new kingdom which he named after himself – Saudi Arabia, recognized internationally by the Treaty of Jeddah on 20 May 1927. Largely thanks to Philby's influence, US oil companies were granted exclusive rights to prospect the Saudi desert for oil.

By 1938 their searches had borne fruit. Aramco, the company formed by the American oil industry to exploit the reserves of Saudi Arabia, soon became the largest oil company in the world. By 1950 its profits were so huge that Ibn Saud demanded a 50 per cent share, or else he threatened to nationalize the firm. In the end the US government compensated Aramco's shareholders for their lost profits by providing them with a tax break called the 'Golden Gimmick', equivalent to the amount siphoned off by the Saudi regime.[3]

Oil money pouring into the Middle East transformed the influence of the region's autonomous, monarchical rulers, whose grip on power was solidified by their sponsorship of a strict form of puritan Sunni Islam. Wahhabism was an eighteenth-century Sunni reform movement, founded by Muhammad ibn Abd-al-Wahhab, that sought to purify Islam. It was subsequently adopted by the House of Saud, which had historically ruled the region of Najd in central Arabia, where Wahhabism first took hold.

Peak Oil Pipeline

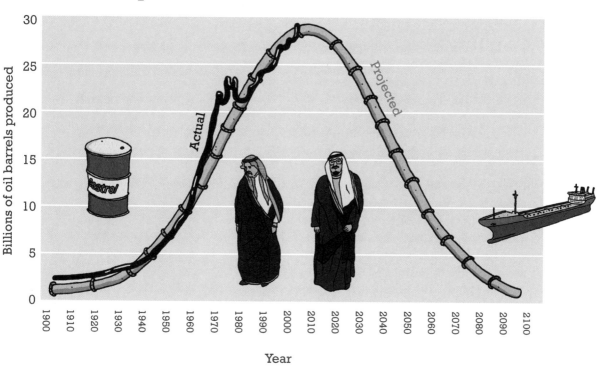

The rise and fall of annual oil production, past, present and predicted.

Backed with money from the oil-hungry West and with tight control of Islam's holiest cities, this creed spread quickly throughout the region via religious schools, newspapers and outreach organizations. When Western powers officially created the State of Israel in 1947, carved out of Palestinian lands where the majority of the existing population were Muslims, a potent mix of fundamental Islamicism and anti-Western sentiment exacerbated the long-standing hostility between the Jewish and Middle Eastern peoples that had its origins back in the pre-Christian world (see page 173). Racial war, bitter hatred and international terrorism still pour out from this Middle Eastern conflict, which remains unresolved.

Between them, cheap Chinese labour and the rich oilfields of Saudi Arabia underpin today's global economy. They demonstrate that Capitalism has no intrinsic requirement for democracy, despite some of its leading advocates declaring otherwise.[4]

The desire for personal enrichment through the constant, never-ending acquisition of material possessions – which Marx called 'commodity fetishism' – has, since the end of the Second World War, provided a common incentive for people to collaborate in business regardless of colour, politics, race or creed.

So far the system has proved remarkably robust. Despite a succession of crises such as the Middle Eastern oil shock of 1973, the Latin American debt crisis (1981–94), the great global stock-market crash of 1987, the East Asian financial meltdown of 1997 and the long-term capital-management hedge-fund crisis of 1998, the scheme of central bank interventions, first proposed by the Bretton Woods bankers in 1944, has managed to recalibrate and rebalance the global economic system to keep underlying growth on track. Since 1983 the United States has avoided serious recession, with only two brief interruptions in 1987 and 2000, leading some experts to believe that the traditional Capitalist boom-and-bust cycle, one of Marx's main criticisms of the system, may finally have been put to rest thanks to modern economic management.[5]

Symbols of American financial power and capital wealth are destroyed on 9/11 by Islamic terrorists determined to make sure their voices are heard.

Unfortunately, Capitalism has a dark side. Hugely improved living standards for millions in some parts of the world hasn't rescued millions more from impoverishment elsewhere. By the end of 2001, the richest 2 per cent of adults in the world owned more than half of global household wealth, with the super-rich 1 per cent owning more than 40 per cent. Conversely, the bottom 40 per cent owned less than 1 per cent between them. It has been estimated that the richest 10 per cent of the world's adults now own more than 85 per cent of its total wealth. North America alone houses just 6 per cent of the world's population, but accounts for 34 per cent of all household wealth.[6]

Today's inequality of wealth between individuals is mirrored by the differences between the world's richest and poorest nations. The scale of poverty in the Third World, particularly in post-colonial Africa, shows no sign of lessening. The system of free trade, proudly trumpeted by advocates such as Ronald Reagan (US President 1981–89) and Margaret Thatcher (British Prime Minister 1979–90) has proved in reality to be far from free. Massive subsidies for farmers in Europe under the Common Agricultural Policy (CAP), introduced in 1958, meant that the playing field was uneven. Poor countries in Africa, whose economies had been shaped by their colonial past, depended on income from food exports to feed their people. But the rich countries artificially depressed prices through subsidies and import taxes, claiming that they had to do so to protect the livelihoods of their own farmers, who produced as much food as they wanted for an inflated fixed price.

Mountains of European butter, lakes of wine, and stocks of cheese and grain that far exceeded what Europe itself could consume were dumped at rock-bottom prices abroad, putting Third World farmers out of business. Although many of these market abuses have since been addressed, they have left a deep legacy of mistrust. Lacking an industrialized base, many poor countries have been forced to borrow funds from First World banks. Unable to pay back their loans, they have been caught in a vicious spiral of dependency. To add to their economic woes, many colonies won their political independence only for their governments to fall into the hands of corrupt, despotic rulers, who, with weapons sold to them by developed nations, greedily clung to power. In this way, ethnic and tribal disputes still predominate today even after the tyranny of arbitrary colonial rule.[7]

These are some of the reasons why Capitalism has failed to make amends for the colonialism of the past. Desperate people are resorting to desperate measures. While refugees flood from the barren wastelands of places like the Democratic Republic of Congo and Somalia in central and eastern Africa, terrorist groups recruit suicide bombers to their cause in the repressed oil-rich states of the Middle East. Fourteen of the nineteen terrorists who smashed three American civilian airliners into the World Trade Center in New York and the Pentagon in Washington on 11 September 2001 came from the Kingdom of Saudi Arabia. Their actions were an extreme example of a devastatingly effective new way of making desperate voices heard. With such a dramatic increase in the stakes of international terrorism, fuelled by the politics of envy, race and inequality, the spectre of Marx's warnings lives on in the twenty-first-century 'War on Terror'.

◆◆◆◆◆

Witch number two conjures up a ghost no less haunting. Thomas Malthus was so concerned about the rising levels of human population that he prophesied a time when nature would take revenge.

The massive increase in the population during the twentieth century was directly linked to rising levels of economic wealth. At 78.8 years, life expectancy in Britain is now thirty years higher than it was in 1900 – and more than thirty-two years higher than in sub-Saharan Africa today.[8] By far the biggest recent increases in population have been in Asia. China and India between them host nearly half the world's overall population of 6.7 billion – a number projected to rise to more than nine billion by 2050.

The fact that so many more people are now crowded on to the same-sized planet has dramatically changed life on earth in the last sixty years. Natural habitats, long since vulnerable to human settlement, have been devastated by rapid industrialization and the growth of new towns and cities. Deforestation, mining, deep-sea trawling and intensive agriculture are some of the main causes of the massive decline in the number of species on earth.

A sixth extinction event may turn out to be no less profound than the five previous mass extinctions that are known to have occurred in prehistory (see page 52). Human activities over the last few hundred years are now thought to be responsible for increasing natural rates of extinctions as much as 1,000 per cent, with some experts estimating that two million different species of plants and animals may already have fallen victim to habitat loss, increased farming, pollution and infrastructure projects such as the building of dams.[9] The rate of extinctions today is reckoned to be between one hundred and one thousand times greater than the historic norm known as the 'background' rate. According to the World Conservation Union Red List of Threatened Species, as many as 52 per cent of all major living species are in jeopardy; plant extinctions head the list, with 70 per cent of species reported to be at risk.

Deforestation accelerated dramatically during the twentieth century, especially in the tropics, as demand for natural hardwood products grew to new heights. Between 1920 and 1995 nearly 800 million hectares of tropical forests were cleared, an area approaching that of the United States of America.[10] Between 1980 and 1990 roughly 15.4 million hectares – an area almost double the size of the United Kingdom – were felled each year.[11]

Such destruction has been driven by economics, usually regardless of the human or natural cost. Poor people in post-colonial countries desperately need a crop they can easily and cheaply trade for cash. Gang violence has become synonymous with trade in illegal logging. There have been more than 800 land-related murders in the Amazon region over the past thirty years. Sometimes things got personal. Sister Dorothy Stang was an American nun who devoted her life to helping educate people living in the Amazon rainforest on how they could extract natural forest products without resorting to cutting down trees. After reporting illegal loggers to the Brazilian authorities she began to receive death threats from gangs who faced prosecution. On 12 February 2005, as she walked to a meeting in her village, she was shot by two gunmen at point-blank range. They then emptied another five bullets into her dead body.[12]

Chopping down trees destroys more than just animals, insects and humans. It also sterilizes the earth itself. Soil quality is severely compromised in areas with no trees, because the ground is exposed to erosion by the weather. Deforestation is also thought to have a significant effect on rainfall patterns, since it is through the natural process of transpiration that much of the world's water, locked up in the ground, ends up seeded in clouds (see page 45).

The annihilation of other species by humans through overhunting is another major cause of extinctions, as the Cossacks of Siberia and the Iroquois of North America found to their cost (see page 311). One extraordinary example, which occurred in the late nineteenth century, is the case of the North American passenger pigeon.

Fatal flight: the last male passenger pigeon died in captivity in 1912.

These birds were once so numerous that their flocks regularly stretched more than a mile across the skies during springtime migrations from the south to their breeding areas in New England. Human hunting began in earnest in the 1860s and 1870s to provide a source of cheap meat for the growing cities on the east coast of the United States. In 1869, Van Buren county in Michigan sent more than seven million of the birds to markets in the east. Such extreme levels of hunting meant that by 1914 the passenger pigeon, a species which once numbered more than five billion individuals, was added to the list of the extinct.

The same story may soon be repeated for thousands of other species, some as common as cod. A study released in 2006 concluded that one third of all fishing stocks worldwide have now collapsed to less than 10 per cent of their previous levels, and that if current fishing trends continue, the seas will be virtually empty of edible fish by the year 2050. Bottom-trawling, the practice of dragging long trawl nets along the sea floor, churns up seabeds so severely that the damage caused to deep-sea ecosystems is far greater than any amount of man-made pollution that leaches into the oceans. A detailed global map, published in February 2008, illustrated the impact on the seas of 17 different human threats. It showed that only

4 per cent of the world's oceans remain undamaged by human activities. Even the small amount of ocean left untouched was preserved mostly thanks to the protection of polar ice, itself threatened by global warming as the ice sheets melt.[13]

Pollution caused by the huge rise in human population is another reason for the rapid decline in the diversity of living things. Air pollution comes from the burning of fossil fuels, causing rainwater to become acidic. Metal foundries and petrochemical plants are sources of poisonous contaminants that destroy delicate ecosystems. Landfill sites release methane and harmful chemicals like cadmium, found in discarded electronic products, which poison the surrounding soil. In 2007 Britain had the worst record for landfill use in Europe, discarding some twenty-seven million tonnes of waste into dumps that now extend across 227 square kilometres.

A report published in 2007 by the Blacksmith Institute (a New-York based environmental group) shows the enormous environmental cost behind the rising aspirations of countries that have historically benefited least from economic development, as they desperately try to catch up with the lifestyles enjoyed by those who have prospered most. Not one site in Blacksmith's 30 most polluted world sites (the 'dirty thirty') is in a 'First World' country. Rather they are

Chinese coal-fired power stations, like this one in the Hebei Province, provide the energy needed to power the world's biggest industrial nation, but at what environmental cost?

located in Russia, China, Zambia, Peru, Azerbaijan, India and the Ukraine.

If the citizens of every country were to have a lifestyle such as that enjoyed by the average westerner, as many as five planet earths would be required to deliver sufficient natural resources in terms of energy, food and water.[14] The effects of the dash by poorer countries to industrialize fast are exemplified by China, which has the world's largest rural population and the fastest-growing economy. Since 1978 its leaders have jumped on to the Capitalist bandwagon, perhaps believing that by increasing their population's prosperity they may prevent a repetition of China's history of violent peasant revolts.

Today the number of households in China is increasing at twice the rate of its population due to increasing divorce rates and more families living apart as young people seek employment in cities. If everyone in China led a lifestyle similar to that of people in Europe and America, it would require roughly double the amount of raw materials currently used by the world's entire population.[15] Just to keep up with China's huge demand for power its government is currently commissioning two new coal-fired power stations *every week*.

The need for resources to power the economic growth driven by the Capitalist system has led to an increase in global oil production to almost eighty-three million barrels a day in 2005.[16]

The consequences of burning fossil fuels are now well understood. Atmospheric carbon dioxide levels have increased dramatically since the 1800s, when fossil fuel deposits ignited man's first wholly independent source of power, in the form of high-pressure steam (see page 338). Between 1832 and 2007 levels have risen from 284 parts per million to 383. Carbon dioxide, like methane, is a gas which has a major impact on the earth's temperatures, by absorbing infra-red radiation. Its increasing levels in the earth's atmosphere are reckoned to be the most likely cause of a recent rise in global temperatures that has already led to the erosion of many of the world's major glaciers, the melting of the ice caps, and changes in sea levels and patterns of rainfall.[17]

During 2007 the fabled North-West Passage sought by generations of explorers (see page 283) was fully clear of ice for the first time in known history, allowing ships to navigate between the Atlantic and Pacific Oceans by the Arctic route. Since the absorption of carbon dioxide by trees is one of the chief mechanisms through which the earth naturally regulates levels of the gas, massive human deforestation may now have neutered the planet's capability to keep global temperatures at an optimum for life.

The geo-political effects of global warming are already starting to unfold. A war in Darfur, a region of western Sudan the size of France, began in February 2003. It was triggered by decades of drought and soil erosion, probably caused by changing rainfall patterns as a result of global warming.[18] In a desperate bid for survival, camel-herding Arab Baggara tribes moved from their traditional grazing grounds to farming districts further south in search of pastureland and water. As a result of their attacks on the non-Arab population more than 2.5 million people are thought to have been displaced by October 2006, of whom approximately 400,000 have died of disease, malnutrition or starvation.

Further south in Africa, the HIV virus is destroying the human immune system, causing the deaths of millions of people who have been left defenceless against common infections. First diagnosed in 1981, the virus somehow jumped the species barrier from monkeys to humans. Since then it has killed more than twenty-five million people, mostly Africans, and infected as many as forty-six million more. There are currently more than a million orphaned South African children, most of them infected themselves, since their parents died from the disease and it is easily passed through body fluids such as breast milk.

Is this what Malthus predicted when he said that one day the levels of human populations would be levelled by nature's intervention through 'sickly seasons, epidemics, pestilence, and plague', and 'gigantic inevitable famine'?

The most agonizing question of all belongs to the third and last witch. Her visions belong to the ghost of the man who many regard as the most influential scientist, naturalist and thinker of all time.

The reason Charles Darwin was so reluctant to publish the theories described in *On the Origin of Species* and *The Descent of Man* was simply that its conclusions led to the inevitable question of whether humans are fundamentally different from other animals. Many people today find Darwin's prophetic warning that 'Man still bears in his bodily frame the indelible stamp of his lowly origin,' hard to accept, either for religious reasons or simply because evidence from all around suggests that, unlike other creatures, humans are not susceptible to the same rules of survival and extinction.

Mankind's ability to sidestep nature's systems shows no sign of abating. Since the Second World War his artifice has developed systems that have radically changed the way we relate to each other and to the world around us. Televisions, computers, video games, mobile phones, text messaging and the internet have taken ordinary people into unnatural worlds where no wildlife can possibly get in their way. Seasons have been abolished as obstacles in the way of consumer choice, with the emergence of air-conditioned supermarkets which can source and deliver tens of thousands of different lines of refrigerated foodstuffs from around the world, twenty-four hours a day, 365 days a year.

Broadcast media, which emerged in a truly mass-consumer form only in the 1950s, transformed the ability of manufacturers to sell their products through advertising. Modern economic growth can now rely on marketing agencies developing elaborate strategies for convincing millions of consumers to buy products not found in nature that no one really needs. Fashions and fads are essential ingredients in modern man's 'virtual world' which further increases the distance between the human world, nature and other living things.

Western science is the self-appointed protector of man's artificial world. Synthesized drugs have engineered longer life spans, couples who can't have children naturally now have the chance through IVF, and slim-hipped women who in the past would probably have died during childbirth can now elect to have caesareans that minimize risks to their own health.

Such innovations tamper with the fundamental fabric of nature herself – the path of natural evolution through which species survive or fail based on each generation's natural adaptations to their surroundings. Artificial selection has been applied by humans to animals and plants since the advent of selective breeding and agriculture more than 10,000 years ago. But modern science, with its recent understanding of life's genetic code, DNA, aspires to giddy new heights. Initially, its aims are to help genetically engineer out nature's life-threatening diseases, or to develop drought-resistant crops. In the long term such 'solutions' compound the problems of ever-increasing human populations making demands on the same resource-depleted, environmentally wrecked planet. How far distant is this approach from Hitler's attempt to interrupt nature's flow to produce his master race, weeding out the weak from society, allowing the strong and wealthy who can afford expensive treatments to prosper and thrive?

Despite modern appearances, are humans really so different, so apart from nature? What will happen when the earth's oil runs out? Unless dramatic new levels of investment are made in nuclear energy and renewable energy sources, man's fossil-fuel-dependent virtual reality could be unplugged by global conflicts over increasingly scarce energy supplies. Financial markets may buckle under the weight of inflation as demand for food and energy soars.[19] Current estimates predict that oil supplies will 'peak' in the next few years (some think they already have), and that known reserves will have been exhausted by 2038. The global addiction to fossil fuels is likely to prove even harder to kick in the twenty-first century than imperial China's addiction to imported opium in the nineteenth.

The mantra of ever-increasing economic expansion assumes a world with limitless resources. As discovered by the Roman Empire (see page

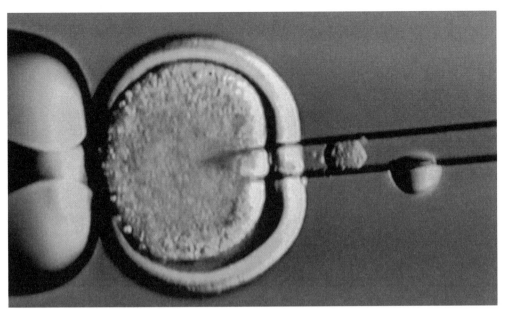

197), territorial and economic growth cannot be assured *ad infinitum*. Plundering the wealth of other continents – a Roman habit, inherited first by European explorers and later by Western governments and business corporations – has already been stretched to its limits following the take over of North America by European settlers and the scramble for Africa. For how much longer can cheap Asian labour subsidize living standards in the democracies of the West?

What about developing alternative lifestyles that are more sustainable in the long term? Darwin concluded that man has evolved, inescapably, as part of the natural world. Perhaps now is the time to relearn how to live within nature's means, as some, like Gandhi and his followers, have tried to demonstrate. Switch off the electricity, turn out the lights, sell the car, grow vegetables, walk to work, bring back the small local school, learn a craft, buy only what you need, make your neighbours your friends and have fun in simple, traditional ways such as playing cards, story-telling, drama, dancing and building dens outdoors.

But Darwin's conclusion that humans evolved in the same way as all other life forms suggests that people aren't naturally well adapted to such long-term, rational planning. It has always been 'the blind watchmaker' – nature – that determines,

however randomly, the long-term state of life on earth, while individual species either collaborate or battle in the here and now.[20] Evolution depends on the strongest survivors passing on useful traits to their successors, while the weakest fall into obscurity and eventually extinction.

Humans today appear to follow their natural instincts every bit as much as their ancestors did. Modern democracies, like hedge-fund managers, plan around the here and the soon-to-be. Their concern is not with making sacrifices in the present for the sake of alleviating possible risks in an uncertain future yet to come. As one twenty-first-century President of the United States famously declared, modern Western lifestyles are 'blessed'.[21] The pursuit of life, liberty and happiness in the present is what most often seems to count.

Those who hold this view believe humans should carry on as near as possible to the *status quo*, except perhaps for a few tweaks here and there. The consequences are predictable – at least until some unmanageable disaster strikes: lots more rich people, lots more poor, lots more in between, and lots more to come. These are the sceptics, like Pyrrho (see page 194), who believe that all the fuss about finite raw materials and overpopulation is more of a hoax propagated by fanatics and societies' envious have-nots.

Others believe humans, unlike other animals, do have within themselves the capacity for a rational escape from their evolutionary origins. Huge investment in a search for new sources of raw materials, for example by colonizing the moon, could be a stepping stone for exploration elsewhere. Techniques to capture and store carbon dioxide emissions before they leak out into the atmosphere could be perfected and made mandatory throughout the world. Proposals to limit carbon dioxide emissions could be driven by a comprehensive trading system in which governments, companies and individuals bid to purchase a fixed number of credits that cap the total amount produced. Consumers could make a start by making short-haul aeroplane flights morally unacceptable – a modern-day taboo, to take a leaf out of the Australian Aboriginals' book.

Such efforts would have to be applied globally, in a rational, consistent and universal manner. Governments would have to agree to caps on CO_2 emissions for their military operations too. Thousands of years of tribal conflict, more recently manifested in nationalistic pride and sporting contests, would have to be set aside for the sake of the greater global good, and for generations yet to come. The European Union and the United Nations are examples of modern attempts to end centuries of tribal rivalries that have got in the way of collective, thoughtful, long-term policy.

※※※※※

What if the depletion of the earth's finite natural resources does bring about the fall of global Capitalism? What if climate change really is the beginning of nature's check on the exponential rise of human populations? What if humanity's evolutionary instincts prevent it from collectively reaching beyond the short-term satisfaction of its immediate material desires? If these warnings of Marx, Malthus and Darwin do indeed come to pass, their prophecies will take centre-stage in the next act in the drama that is life on earth.

Now, at last, the clock finally strikes midnight on our twenty-four-hour history. What on earth happens next promises to be a lively beginning to the first one thousandth of a second in a brand new day.

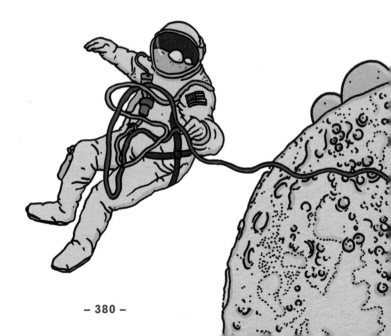

Epilogue

AN INFINITE BLAST OF ENERGY from the Big Bang pervades every aspect of life, animate and inanimate, be it past, present or future. How small and insignificant is all human history compared to the aeons that created the sun, moon and stars – a saga that began 13.7 billion years ago. Take a fresh look at our planet, the earth, with its extraordinary story of life. Look with wonder at creation all around without forgetting that more than 99 per cent of all species that have ever existed have since disappeared into extinction.

What is the meaning of human history, which at a generous pinch began just twenty seconds before midnight on our twenty-four-hour scale? What should we make of the birth of human civilizations, themselves a mere pimple on the end of humanity's tale, a tenth of a second away from midnight on the clock?

This is the true context of where we, *Homo sapiens*, fit in. History that's wrapped up into packets and projects is what disconnects us from our natural heritage. Stories that ignore the inextricable relationship between planet, life and people do so at the price of perpetuating a belief that humans are so special a species that everything non-human they touch pales into insignificance.

Now, as humanity's scale tips nature's ecological balance into chaotic disarray, we may find ourselves at our most exposed. Since the path to the future follows on directly from the paving stones of the past, isn't it time to revive our expectations of history and to keep asking: what on earth happened?

24:00:00

Time's Tables

WHO AND WHAT have had the biggest impacts on shaping the destiny of the planet, life and people? Is it the earth itself, other life or people who are most in control of events historically, today and in the future?

Peak events puncture history's horizon, each one making its own unique but interconnected contribution to the unfolding story of planet, life and people. The following tables feature only the most significant events as discovered on my own personal journey through the saga of *What on Earth Happened?* Please come and share your own thoughts at www.whatonearthhappened.com.

Top Ten Natural Events that Shaped the World

	Page	Event	Date (years ago)	Why?
1	17	Collision of the earth and Theia that created our moon	4.49 billion	The ultimate collision between two early planets that resulted in the formation of our moon, and the earth's magnetic shield that protects life against solar radiation.
2	52	Creation of Pangaea	300 million	This single super-continent challenged plants and animals to find ways of living inland – leading to the evolution of seed-bearing trees and hard-shelled eggs.
3	52	Volcano that erupted for over a million years creating the Siberian Traps	252 million	The Siberian Traps were formed by this huge volcanic event that also contributed to the Permian Mass Extinction – the largest ever in pre-history.
4	60	Indian volcano that created the Deccan Traps	65.5 million	Huge quantities of carbon dioxide from this eruption, possibly triggered by part of the meteorite impact that also struck Mexico at this time, caused climate temperatures to see-saw out of control, contributing to a devastating mass extinction.
5	59	Asteroid that smashed into Mexico 65.5 million years ago	65.5 million	A six-mile-wide asteroid smashed into the Yucatán Peninsula (Mexico) with a force 10,000 times greater than all nuclear warheads in existence today, triggering a mass extinction of species, including the dinosaurs.
6	84	Crashing of Indian plate into Asia that created the Himalayas	40 million	This mountain range is thought largely responsible for a dramatic cooling of the earth over the last forty million years, as water vapour cooled by the peaks created monsoon rains that dissolved large volumes of carbon dioxide from the air.
7	86	Great American Interchange that connected North and South America	3 million	As a result of this land-merger, animals from two continents were able to mingle, and a new ocean current, the Gulf Stream, spluttered into life, warming Europe and the Middle East, eventually helping the establishment of human civilizations.
8	98	Toba eruption in Indonesia	75,000	This volcano may have nearly wiped out the species Homo sapiens, but in the end reduced populations to between 1,000 and 10,000 individuals, creating unusual genetic similarity within our species.
9	114	Younger Dryas	12,700	Sudden cooling of the climate followed by a rapid warming led to the first human experiments with agriculture and animal domestication – the advent of farming.
10	146	Eruption of Thera	3,635	Tsunamis triggered by this volcanic eruption devastated several advanced early Mediterranean civilizations such as the Minoans in Crete and the Egyptians, giving violent horse and chariot invaders such as Hyksos and Mycenaeans a chance to establish themselves.

Top Ten People that Changed Human History

	Page	Person	Date	Why?
1	128	**Hammurabi**	1810–1750 BC	A King of Babylon who wrote one of the earliest comprehensive legal codes, establishing the principle that an accused is innocent until proven guilty.
2	169	**Ashoka**	304–232 BC	An Indian king who spread the ideals of Buddhism throughout Asia. He was the first ruler to put animal and human rights on an equal footing.
3	202	**Jesus Christ**	2 BC–36 AD	The pacifist son of a Jewish carpenter whose miraculous powers helped his followers believe he was the son of God. Jesus's death on a cross ultimately led to the establishment of what is now the world's biggest religion, Christianity, adopted as a state creed by the Roman Empire in 391 AD.
4	231	**Mohammed**	570–632 AD	An Arabic trader who revealed a series of divine messages that led to the establishment of Islam, now the world's second largest religion. The rise and spread of Islamic political and trading empires helped to connect Eastern and Western cultures.
5	284	**Hernán Cortés**	1485–1547	A Spanish mercenary and conquistador who masterminded the destruction of the Aztec empire of Mexico through a mixture of luck and cunning. Amongst his company was an African slave who brought deadly smallpox to the Americas.
6	338	**Richard Trevithick**	1771–1833	A Cornish inventor who, with the assistance of a neighbour (William Murdoch), was the first successfully to build a high-pressure steam engine that could provide power without the assistance of atmospheric pressure. This invention gave humanity its first mobile power source and ushered in a new age of industrialization.
7	341	**Friedrich Wöhler**	1800–82	A German chemist who artificially manufactured the first organic chemical, urea. Before this time people believed all chemicals produced by living things could not be made by man. His discoveries led to the birth of organic chemistry from which derive everything from plastics and synthesized drugs to explosives and artificial fertilizer.
8	32, 334	**Charles Darwin**	1809–82	An English naturalist who developed a new theory on the origins of life, published in 1859. Darwin's theory of evolution by natural selection suggested that species can become extinct. It provided powerful evidence to support the view that the earth was billions of years old and that all living creatures are ultimately related to a common ancestor, including humans, who, he claimed, were descended from apes.
9	359	**Karl Marx**	1818–83	A Jewish German political philosopher who predicted the eventual demise of Capitalism. His Communist Manifesto inspired a number of revolutionaries, leading to an epic ideological struggle between modern civilizations.
10	367	**Adolf Hitler**	1889–1945	An Austrian veteran of the First World War who so despised Western powers for their part in Germany's humiliating surrender in 1918 that he became obsessed with restoring all life on earth to a 'natural order'. Hitler attempted to re-create a master human race that could eventually breed 'impure' stock out of existence. Hitler's ideas led to a second global war that stimulated the invention of atomic energy and rocket technology.

	Page	Item	Date	Why?
1	155	Rice	7000 BC in China and Korea	Today's massive human population owes as much to the cultivation of rice as it does to artificial fertilizers. This plant can feed more humans per hectare than any other crop.
2	113	Wheat	7000 BC in Middle East	Today's artificially cultivated varieties only began to be farmed by humans following the cold Younger Dryas period c.12,700 years ago. Their labours produced a domesticated crop with large seeds that held firmly to the stalk (this made grinding and harvesting easier).
3	217	Maize	5000 BC onwards in Central America	Painstakingly bred from wild teosinte by early Central American farmers, maize eventually became the staple crop for all indigenous American peoples. By the sixteenth century European explorers had spread maize around the world.
4	301, 361	Poppies	4000 BC onwards in Asia, the Middle East and Europe	The medicinal use of opium goes back to the first early Neolithic farmers but by the nineteenth century opium extracted from poppy seeds had become a major international commodity. Morphine, another derivative, is still one of the world's most popular painkillers.
5	217, 308	Potatoes	3000 BC in South America but 1600 onwards in Europe	Hundreds of varieties of this highly nutritious vegetable were selectively cultivated by South American natives, although only four were exported by sixteenth-century European settlers to Europe. A lack of diversity led to devastating blights in nineteenth-century Europe and mass emigrations to the Americas and Australia.
6	239, 282	Sugarcane	3000 BC in South-East Asia	Deforestation to clear ground for sugar plantations in the New World led to dramatic changes in the landscape and ushered in a new era of slavery, which eventually spilled out into armed conflict in the American Civil War.
7	155, 245	Mulberry trees	2500 BC in China	Leizu, a Chinese queen, is said to have discovered how threads from the cocoon of a moth larva could be woven into silk. Later a new paper-making process was discovered that used the bark of mulberry trees. Paper is a major contributor to economic growth and global deforestation.
8	158, 301, 318	Tea	Cultivated in China from before c.1000 BC	This camellia leaf was used medicinally by Chinese rulers and by Buddhist monks to keep them awake for prayers. The British became addicted to tea by the nineteenth century – so much so that its supply from China was secured by the illegal exchange of opium, grown in Bengal, which provoked international conflict.
9	180	Olives	700 BC in Mediterranean	An energy-rich crop that grows in craggy soil and requires no hard labour to cultivate. Wealth from olives provided ancient Greek cities with enough time and leisure to pursue scientific investigations and new experimental societies including democracy and republicanism.
10	346	Quinine	c.1600 in South America	This extract from the bark of the cinchona tree provided European settlers with their first effective protection against the deadly disease malaria, spread by mosquitoes. It therefore became a passport to the successful European colonization of Africa, eventually cultivating the conditions for creating what is now known as the Third World.

	Page	Creature	Date (years ago)	Why?
1	37	**Corals**	540 million	These tiny sea creatures have built some of the sea's most vibrant habitats. Coral islands are the skeletal remains of countless numbers of these marine organisms which have substantially reduced atmospheric CO_2 levels with their reef-building antics.
2	37	**Jellyfish**	530 million	These were the first sea creatures to develop cell tissues with specialized functions that later evolved in other species into separate organs. The deadly sting of some jellyfish provoked other creatures to defend themselves by making large protective shells.
3	38	**Sea squirts**	500 million	These animals are rooted to the sea bed and feed off microscopic food by filtering water through their bodies. Baby sea squirt larvae swim using a primitive chord that beats the water. In their descendants these chords evolved into backbones, from which the family of vertebrates (called chordate) and eventually humans emerged.
4	40	**Lungfish**	420 million	The first vertebrate fish whose ancestors experimented with escaping the terror of the high seas by using a primitive air-breathing lung adapted from one of their gills. It also learned to walk across muddy river estuaries using its fins, precursors to the tetrapods, the world's first four-legged animals.
5	48	**Velvet worms**	420 million	Ancestors of today's velvet worms developed stubby legs that helped them move faster across the sea floor to seek food, making them ideally suited to experiment with life on land where they re-nourish the soil. Eventually the velvet worm's many legs and segmented bodies evolved into insects.
6	48	**Dragonflies**	350 million	These creatures adapted slits once used for breathing underwater into flaps that helped them fly from place to place. Dragonflies were also the first creatures to gain a view of life on land through an ingenious system of compound eyes. Increased levels of oxygen in the air meant they once grew as large as seagulls are today.
7	66	**Honeybees**	150 million	Descended from wasps, these flying insects developed a taste for nectar and pollen, a fancy that stimulated plants and trees all over the world to cover the land in blossom. Social behaviour that evolved in bees' nests signalled nature's ability to construct complex civilizations out of large communities of living creatures.
8	68	**Termites**	200 million	These insects evolved from cockroaches and beetles and were part of nature's proving ground for creatures whose lives depended on social behaviour (like humans). Their collective intelligence, exhibited through language, rulers, teamwork, agriculture, education and sacrifice, later became hallmarks of civilized human society.
9	79	**Gorillas**	7 million	Ancestors of today's gorillas were the first members of the great ape family to experiment with living on the ground instead of up in the trees. As the climate dried out during the Ice Ages and grasslands replaced much of the world's forests, it was from the mild-mannered, vegetarian gorilla that humanity's most direct ancestor evolved.
10	116	**Grey wolves**	300,000	Today's domestic dogs are directly descended (as a sub-species) from the grey wolf that evolved during the Ice Ages. Without the help of dogs, early farmers would have struggled to keep control of their flocks of animals on which their sustenance depended.

Top Ten **Extinct Creatures that Shaped Life on Earth**

	Page	Creature	Date (years ago)	Why?
1	31	**Trilobite**	530 million	These were the first living creatures thought to have been able to see using compound eyes (sight later re-evolved on land in dragonflies).
2	49	**Tiktaalik**	375 million	This four-legged animal adapted to living on land by being the first creature to use a neck to lift and swivel its head from side to side, helping it to hunt for food and better detect danger.
3	50	**Ichthyostega**	350 million	Young offspring of these sea creatures colonized the land to protect themselves from predators in the sea. Their descendants evolved into the world's first dedicated land animals, the amphibians.
4	51	**Hylonomus**	315 million	This was the first creature known to have developed a hard-shelled egg, allowing it to breed inland, beginning the domination of life on land by reptiles, the family to which all dinosaurs belonged.
5	51	**Dimetrodon**	260 million	A mammal-like reptile whose ingenious back-mounted sail allowed it to warm up its blood to hunt early in the day when other reptiles were still too cold. Warm-bloodedness later allowed mammals to hunt at night when they were safer from dinosaur attack.
6	54	**Lystrosaurus**	230 million	A warm-blooded mammal-like reptile that somehow survived the Permian Mass Extinction (252 million years ago) and therefore preserved a vital evolutionary link between mammal-like reptiles and their descendants, the mammals.
7	65	**Sinosauropteryx**	140 million	A dinosaur that developed the use of feathers as a means of insulation. These were later adapted by birds for flight, showing that birds are living descendants of the dinosaurs.
8	74	**Hyracotherium**	35 million	A dog-like creature that roamed the forests of North America which later evolved into the horse. Horses then migrated across the Alaskan land-bridge into Asia where they were eventually domesticated by man 7,000 years ago, dramatically altering the course of human history by providing a new source of transportation as well as military and agricultural power.
9	104	**Australopithecus**	3 million	The earliest known ape that walked on two feet, freeing its hands to be able to carry food and make tools. Motor skills to control intricate hand-movements stimulated the growth of brain size, which required more food, which required better tools for hunting – provoking an evolutionary spiral that led to the development of human brains which are four times larger than those of our closest genetic relatives, bonobos and chimpanzees.
10	351	**Variola Major (Smallpox)**	Declared extinct in 1979	A deadly virus that spread to humans following the domestication of animals after the last Ice Age melt 12,000 years ago. In the twentieth century alone the virus was responsible for as many as 500 million human deaths. Before then, it accounted for the decimation of indigenous American and Australian populations when European settlers spread the disease. The virus was the first to have been completely eradicated from nature by man.

Top Ten Threats to Life on Earth

	Page	Threat	Why?
1	61	**Meteorite strike**	That's what finished off the dinosaurs 65.5 million years ago and at some point experts say another boulder of a similar size or bigger is bound to collide into the fragile earth, causing another mass extinction of species. So dark are many of these objects in space that there may not be too much warning, either.
2	342, 376	**Man-made Pollution**	The entire manufacturing output of all plastics made since the early 1900s will exist for thousands of years to come because none of it biodegrades. CFCs from fridges and aerosols continue to wreak havoc with the ozone layer. Nuclear waste from the world's 441 nuclear power plants will stay highly toxic for tens of thousands of years.
3	342	**Climate change**	Melting ice-caps, shifting rainfall patterns, floods, droughts and extreme weather events that destroy harvests are some of the predicted consequences of the continuing increase in atmospheric CO_2 from 284 ppm (parts per million) in 1834 to 383 ppm today. The rise is down to modern humans pumping our fossil fuel pollution into the atmosphere, like a massive global volcano.
4	344	**Over-population**	For tens of thousands of years human populations stayed at a roughly stable level of c.five million. Then, with the birth of agriculture after the last Ice Age melt, settled civilizations piled on the population pressure. By the time of Jesus Christ numbers had soared to c.200 million. By 1804 the total rocketed passed the one billion mark to become nearly seven billion today, tipping the world's living systems into ecological imbalance.
5	222, 261, 266	**Deforestation**	The itch to settle led humans to abandon living in forests because farming required felling trees for fields, turning lands once lush with vegetation into desert and scrub. Deforestation is a big cause of the current mass extinction of species and is also damaging the earth's natural capacity to soak up climate-changing carbon dioxide.
6	310	**Shrinking biodiversity**	As living things get transported by humans from one habitat to another in the modern globalized world, species become exposed to new predators, and without time to adapt to survive, many become extinct. Loss of trees, sterilization of the soil through over-farming and the use of pesticides and over-fishing are giving rise to concerns that humans are increasingly vulnerable in a world that has lost its ecological balance.
7	133, 149, 374	**Inequality**	Farming gave rise to permanent human settlements in which some people controlled supplies of stored food while others became dependent. Domesticated horses gave others military superiority, reinforced by bronze weapons and chariots. Roughly a third of the modern human population is overfed whilst another third is malnourished, giving rise to social grievances and international terrorism.
8	307, 377	**Disease**	Living in close proximity to farm animals spawned diseases that have plagued humans ever since they began to live in towns and cities. Smallpox, a virus which originated c.12,000 years ago and jumped to humans from cows, is human history's biggest killer. Today, viruses such as HIV Aids (originated in monkeys), avian influenza (from birds) and ebola (from fruit bats) could be just as lethal.
9	344, 377	**Famine**	Genetically modified foods that are disease and drought resistant offer the hope of salvation from death by famine for some people. However, the deliberate manipulation of nature's operating systems is fiercely opposed by those who believe humans are not adequately equipped to take over from nature as custodians of the earth's living systems.
10	372, 374	**Global war**	The potential for humans to wipe out most living things as well as themselves has only existed since the dawn of the bio-engineering and nuclear ages in the mid-twentieth century. Proliferation has recently increased the risk that such an outcome could take place through a state-sponsored global war over natural resources (e.g. oil) or by maverick groups of disaffected individuals.

	Page	Item	Date	Why?
1	13	**The Big Bang**	13.7 billion years ago	Echoes from this monumental, universe shattering event were unexpectedly detected in 1964 by American scientists Arno Penzias and Robert Wilson using a home-made space radio-telescope.
2	54, 55	**Lystrosaurus fossils**	250 million years ago	This mammal-like reptile, looking like a cross between a hippopotamus and a pig, somehow survived the biggest mass extinction of species on earth 252 million years ago. Its fossils, found on every continent, help prove that Alfred Wegener's theory of plate tectonics is correct.
3	61	**Iridium layer**	65.5 million years ago	High concentrations of a rare type of the metal iridium, not naturally occurring on the earth, were found in a layer of clay just at the moment when dinosaur relics disappear from the fossil record. First discovered in the early 1970s, the same tell-tale sign of a massive meteorite impact has since been found in rocks all over the world.
4	104	**Altamira caves**	20,000 years ago	Eight-year-old Spanish child Maria Sautuola couldn't believe her eyes when she noticed paintings of huge bison on the ceiling of a cave in Spain. They turned out to be among the oldest expressions of human art ever known, painted by hunter-gathering people more than 20,000 years ago.
5	159	**Oracle bones**	c.1600 BC	Ancient writing that resembles Chinese today was etched on oracle bones and turtle shells discovered near the Royal Tombs of Yin where Shang rulers buried their dead. More than 20,000 have been discovered, revealing questions put by kings to the gods in their capacity as intermediaries between earth and heaven.
6	124	**Nineveh tablets**	c.600 BC	A staggering 20,000 tablets of cuneiform writing, unearthed in the 1840s, have transformed our historical understanding of ancient Sumerian, Assyrian and Babylonian civilizations. The Royal Library of King Ashurbanipal contained everything from king lists and epic histories (Gilgamesh) to complex mathematical treatises.
7	176	**Cyrus cylinder**	539 BC	Discovered in 1879 under the walls of Babylon, this inscribed cylinder details what is hailed as the first charter of human rights. Tolerant Persian King Cyrus abolished slavery and even paid for the reconstruction of the Jewish Temple in Jerusalem after it was destroyed by Babylonian King Nebuchadnezzar.
8	170	**Ashoka pillars**	c.264 BC	Indian King Ashoka was so dismayed at the horrific casualties of the Battle of Kalinga that he converted to Buddhism and vowed to dedicate his reign to restoring peace on earth for all living things. His achievements were recorded on a series of pillars, one of which still stands outside the city of Sarnath, near Varanasi, in India.
9	221	**Sacred Cenote of Chichen Itza**	300 BC – 800 AD	In 1897 American archaeologist Edward Thompson dredged a small lake that contained sacrificial knives, plates and other artefacts revealing the grim business of Central and South American human sacrifice. Desperate rulers believed these practises would encourage the gods to bring them good fortune and rain to water their crops.
10	247	**Caves of the Thousand Buddhas**	900 AD	Early twentieth-century Hungarian archaeologist Aurel Stein discovered a trove of more than 40,000 perfectly preserved Buddhist texts, many hand-written by monks, which included the oldest dated printed document ever found, the Diamond Sutra (868 AD).

	Page	Item	Why?
1	16	**What happened before the Big Bang?**	Is our universe just the next in a long line of bangs and crunches? Is it one of many parallel universes, each one spurred into existence by other big bangs with their own different laws of fundamental physics, creating a kind of multiverse? Or, is there some form of superior intelligence that created our universe and life on earth?
2	18	**What triggered life on Earth?**	Amino acids delivered on meteorites from outer space? A chemical soup concocted at the mouth of sea-floor volcanic vents? A chance spark of life accidentally triggered by a primordial lightning storm? An alien or divine architect?
3	62	**What caused the first flowers to bloom?**	Charles Darwin called it an 'abominable mystery', and even today experts have no clear idea. Was it the presence of pollinators such as bees, or did bees emerge only after flowers were there to provide a diet of pollen and nectar?
4	79	**Who was the missing common ancestor of humans and chimpanzees?**	Modern genetics show that the split between chimps and humans occurred no more than about four to seven million years ago. But what was this creature like?
5	90	**What caused Australopithecus to swivel on to two feet?**	Walking on two feet has as many advantages as disadvantages to apes who came down from trees to live in grassland savannahs. With brains no larger than those of a chimpanzee it wasn't simply intelligence that provoked human ancestors to rise up on two feet. So what was it?
6	96	**When did humans learn to talk?**	Guesstimates range from between 50,000 and 110,000 years ago, but no one can be sure. Language gave humans a big advantage in being able to organize themselves to hunt more efficiently.
7	108	**What caused the Pleistocene extinctions in the Americas and Australia?**	Large animals (megafauna) rapidly disappeared once Homo sapiens populated Australia 40,000 years ago and then the Americas 12,000 years ago. Was it overhunting, climate change, a mixture of the two, or something else, such as disease, that caused such devastation?
8	194	**What happened to the tomb of Alexander the Great?**	Once the most popular tourist attraction of the Roman world, all records of the location of Alexander's golden tomb have since disappeared from history, not yet recovered by archaeologists.
9	286	**How did the ancient Nazca people construct their giant geoglyphs?**	Fashioned sometime between 300 BC and 800 AD, these people made more than 70 geometric and natural shapes, some as long as 270 metres, by brushing arid desert grit to one side into a pattern of paths. But how could they have made such intricate shapes when the only way of seeing what they had done was from hundreds of feet up in the air?
10	106, 133, 158, 163	**Is there such a thing as life after death?**	From the animistic beliefs of hunter-gathering man to the monotheistic religions of Judaism, Christianity and Islam and from Hindu reincarnation to the Buddhist nirvana, human history is one long record of the belief in a force beyond earthly existence. But is it just a delusion as some modern scientists claim?

Further Reading

GENERAL HISTORIES

A World History by William McNeill
(Oxford University Press, 1967)

A New Green History of the World
by Clive Ponting (Penguin, 2007)

Civilizations by Felipe Fernandez-Armesto
(Macmillan, 2000)

The Universe Story by Thomas Berry
and Brian Swimme (HarperCollins, 1994)

The Ancestor's Tale by Richard Dawkins
(Houghton Mifflin, 2004)

Maps of Time by David Christian
(University of California Press, 2004)

Collapse by Jared Diamond
(Viking Penguin, 2005)

A Little History of the World
by E. H. Gombrich
(Yale University Press, 2005)

OTHER HELPFUL BOOKS

Wonderful Life by Stephen Jay Gould
(Hutchinson Radius, 1990)

The Secret Life of Trees by Colin Tudge
(Penguin, 2006)

Life by Richard Fortey
(Vintage, 1999)

Oxygen by Nick Lane
(Oxford University Press, 2002)

In the Beginning by John Gribbin
(Viking Penguin, 1993)

The Civilisation of the Goddess
by Marija Gimbutas
(Harper San Francisco, 1991)

Gunpowder by Jack Kelly
(Atlantic Books, 2004)

Deforesting the Earth by Michael Williams
(University of Chicago Press, 2003)

Science and Civilization in China
by Joseph Needham
(Cambridge University Press, 1986)

Plows, Plagues and Petroleum
by William Ruddiman
(Princeton University Press, 2005)

Climate History and the Modern World
by H.H. Lamb (Methuen, 1982)

The Ages of Gaia by James Lovelock
(Oxford University Press, 1988)

On the Origin of Species
by Charles Darwin (Harvard University Press,
1964 – facsimile of original edition of 1859)

The First Eden by Richard Attenborough
(William Collins, 1987)

Guns, Germs and Steel by Jared Diamond
(Norton, 1997)

Notes

Chapter 1: Crunch, Bang, Ouch!

1 This is called Cosmic Inflation Theory, and was first put forward by an American scientist, Alan Guth, in 1979.

2 The first man to suggest that the echo should still be possible to detect was George Gamov, a Russian scientist, who published his theory in 1948.

3 John Gribbin, *In the Beginning* (Penguin, 1993), page 18.

4 This estimate is based upon research performed in 2004 by a team of astronomers: Luca Pasquini, Piercarlo Bonifacio, Sofia Randich, Daniele Galli and Raffaele G. Gratton.

5 This echo is officially called Cosmic Microwave Background Radiation, or CMR.

6 Russian scientist Andrei Linde (b.1948), who collaborated with Alan Guth on developing the Cosmic Inflation Theory (see Note 1, above), was one of the first to propose a multiverse based on an infinite number of cosmic bubbles.

7 Gribbin, page 149.

8 Brian Swimme and Thomas Berry, *The Universe Story* (HarperCollins, 1994), page 65.

9 Dana Mackenzie, *The Big Splat* (John Wiley, 2003), page 189.

Chapter 2: First Twitches

1 During photosynthesis six molecules of water combine with six molecules of carbon dioxide to form a carbohydrate (sugar energy) and six molecules of oxygen. i.e $6H_2O + 6CO_2 = C_6H_{12}O_6 + 6O_2$.

2 Cyanobacteria still live in the oceans and in lakes, ditches and rivers in the form of blue-green algae. They produce the oxygen that maintains our supply in the air. Without them, it has been estimated that all the oxygen in the atmosphere would disappear completely within about 2,000 years. See 'The Natural History of Oxygen', by Malcolm Dole in *The Journal of General Physiology*, No. 49, page 9.

3 Iron ore was mined and the iron extracted by early man to make primitive weapons and tools. Today we use it to build everything from cars to stainless steel knives and from saucepans to skyscrapers.

4 Ozone is a special form of oxygen that combines three molecules, O_3.

5 This theory was developed by Lynn Margulis in the 1970s. See Lynn Margulis's *Origin of Eukaryotic Cells* (Yale University Press, 1970).

Chapter 3: Tectonic Teamwork

1 The gas is called dimethyl sulphide.

2 These creatures use carbon and calcium from the sea water to create their microscopic shells.

3 See James Lovelock, *Ages of Gaia: A Biography of our Living Earth* (Oxford University Press, 1998), pages 105–13 and Don Anderson, *The New Theory of the Earth* (Cambridge University Press, 2007), page 8.

4 For a full account of this episode in earth history, see Gabrielle Walker, *Snowball Earth* (Bloomsbury, 2003).

Chapter 4: Fossil Fuss

1 Some creatures have both male and female parts. They are called hermaphrodites.

2 The worm has been called Dickinsonia, while the beetle-like creature is known as Spriggina, named after Reg Sprigg.

Chapter 5: Davy Jones's Locker

1 Carolus Linnaeus, *Systema Naturae* (1767), page 29.

2 Richard Dawkins, *The Ancestor's Tale* (Marina Books, 2004), page 464.

3 The research, conducted by Philip Anderson of the University of Chicago, was published in November 2006 in the journal *Biology Letters*.

4 The tracks were found in East Lothian by Martin Whyte, a geologist from the University of Sheffield.

Chapter 6: Friends of the Earth

1 Colin Tudge, *The Secret Life of Trees* (Penguin, 2006), page 73.

2 Roy Watling, *Fungi* (Smithsonian Books, 2003), page 24.

3 *Ibid.*, page 21.

4 Nick Lane, *Oxygen* (Oxford University Press, 2002), page 86.

5 *Ibid.*, page 98.

Chapter 7: Great Egg Race

1 Based on estimates in 1973 by C.D. Bramwell and P.B. Fellgett that it would take the animal 205 minutes to heat up from twenty-six to thirty-two degrees without the sail, and only eighty-two minutes with it. See *Nature* (Vol. 242), 'Thermal Regulation in Sail Lizards'.

Chapter 8: Dino-Wars

1 We know their numbers got fewer because painstaking analysis of Cretaceous rocks in North America, such as in the Hell Creek area of Montana, shows that there are far fewer dinosaur fossils found in the rocks between sixty-five and seventy million years ago.

2 Another impact zone, called the Shiva Crater, has recently been discovered under the sea off the west coast of India, leading some scientists to think that this may have triggered the Deccan eruptions. Other craters have been found that date to the same time in the Ukraine, the North Sea, Canada and Brazil, indicating that perhaps the meteorite did split up just before impact.

3 See Richard Fortey, *Life* (Vintage, 1999), pages 242–45.

Chapter 9: Flowers, Birds & Bees

1 This theory is based on recent research by J. Michael Moldowan at Stanford University in which chemicals such as oleanane, a substance sometimes secreted by flowering plants to ward off insects, have been detected in old rocks. See *www.stanford.edu*

2 Researchers from the Floral Genome project at Penn State University in Pennsylvania believe that 'whole genome duplications' can drive sudden bursts of evolution.

3 Tudge, *Secret Life of Trees*, page 152.

4 This system is similar to what is called range voting, where each voter in an election can score each candidate within a range of, say, 0 to 100. At the end of the election the scores are added up and the winner is the one with the best average score. See Warren D. Smith, 'Ants, Bees and Computers Agree Range Voting Is Best Single-winner System' at *www.math.temple.edu/~wds/homepage*

Chapter 10: Prime Time

1 Dawkins, *The Ancestor's Tale*, page 197.

2 Geographical isolation provides a barrier to gene flow. Individual mutations will have a bigger and quicker effect in a smaller gene pool. Therefore species that are isolated are more likely to adapt faster to their new surroundings through natural selection, and to create new species.

3 Dawkins, *The Ancestor's Tale*, page 217.

4 One of the biggest and best fossil sites for mammals is in a pit near the German town of Messel. An early bat called palaeochiropteryx was found there.

5 Dawkins, *The Ancestor's Tale*, page 171.

6 Orang-utan behaviour has been extensively studied recently by an international group of scientists. See *Science*, Vol. 229.

7 This theory was put forward in 1998 by American scientists Caro-Beth Stewart and Todd Disotell.

Chapter 11: Ice Box

1 This theory is called the Uplift-Weathering Hypothesis, and was first put forward by Maureen Raymo, Flip Froelich and Bill Ruddiman in 1988.

2 See *New Scientist*, 3 July 1993.

3 As the Mediterranean Sea evaporated at a rate of thousands of cubic kilometres a year, a huge weight was lifted off the sea floor, destabilizing the region and triggering volcanic eruptions. New mountains formed by the volcanoes cut off the Mediterranean from the Atlantic. Isolation caused more evaporation, until more volcanic activity set in. This cycle of flooding and evaporation, like the emptying and filling up of an enormous bath, is known as the Messinian Salinity Crisis.
4 The Karoo Ice Age came and went over roughly a hundred million years, between 350 and 250 million years ago.
5 Grasslands need about 500 to 900 millimetres of rain per year; a rainforest needs more than 2,000. A desert may receive only 300.

Chapter 12: Food for Thought
1 The genes inside all living cells contain an exact set of instructions of how to build a complete version of the animal or creature to which they belong. By comparing the different genetic codes of one species with another, scientists can estimate roughly how long it took for the genes to evolve into different species. See *Nature*, Vol. 437, 1 September 2005, pages 69–87.
2 See 'Chimpanzee Locomotor Energetics and the Origin of Human Bipedalism' by Michael Sockol, David Raichlan and Herman Pontzer in *PNAS*, 16 July 2007.
3 See *Nature*, No. 443, pages 296–301.
4 Dawkins, *The Ancestor's Tale*, page 83.
5 David Christian, *Maps of Time* (University of California Press, 2005), pages 161–62.
6 Reported in *Smithsonian* magazine, November 2006.

Chapter 13: Humanity
1 Marco Polo (1254–1324) was a Venetian merchant and explorer who was one of the first Europeans to travel the Silk Road to China.
2 Analysis of Neanderthal fingers and thumbs was carried out on fossil finds at La Ferrassie in France in 2003.
3 These habits carried on into ancient human civilizations such as the Egyptians, the Sumerians and the Chinese.
4 Bob Fink, a Canadian musicologist, has extensively studied the Neanderthal flute. See his essay at *http://www.greenwych.ca/fl-compl.htm*

Chapter 14: The Great Leap Forward
1 The genes responsible for red hair, freckles and pale skin date back to before *Homo sapiens* evolved. According to experts from the Oxford Institute of Molecular Medicine, these genes could not have come from African humans, because they provided very little protection against the sun's ultraviolet rays, so in all likelihood they came from the Neanderthal line. Some modern populations have high numbers of people with these inherited genes, such as Scotland where 10 per cent of all people have red hair and 40 per cent carry the genes responsible for it.
2 The theory of a human genetic bottleneck caused by the Toba eruption was first proposed by Stanley H. Ambrose in 1998.
3 For example, a study tracing the genetic descent of 1042 individuals from their DNA was performed by Peter Underhill and his team at Stanford University in 2001.
4 The phrase was coined by Jared Diamond in his book *The Third Chimpanzee* (Harper Perennial, 1992).
5 The Ancient Human Occupations of Britain Project (AHOB) has completed its first phase of research, and discovered evidence of eight human settlements in all, stretching back 700,000 years. See *www.nhm.ac.uk*

Chapter 15: Hunter-Gatherer
1 One type is called the 'bottle gourd' to this day.
2 Estimates come from Michael Kramer, 'Population Growth and Technological Change: 1 million BC–1990' in *The Quarterly Journal of Economics* (1993), pages 681–716.

Chapter 16: Deadly Game
1 For a summary of this theory see Paul Martin and David Steadman, *Extinctions in Near Time* (New York Plenum Press, 1999), pages 17–55.

Chapter 17: Food Crops Up
1 This rapid melt is known in scientific circles as 'meltwater pulse 1A'. The ice shelf that collapsed is thought to have been either in Antarctica or the sheet covering northern Britain and Ireland.
2 Different types of oxygen, called isotopes, are trapped in the ice as snow falls to the ground. Worldwide temperatures can be deduced by determining the proportion of different oxygen isotopes in the ice.
3 Dr Douglas Kennett and Dr John Earlandson from the University of Oregon, believe a meteorite plunged into the North American ice sheets near Lake Agassiz, causing ice flows to rupture and triggering the Younger Dryas event. This is known as the YDB Comet Theory.
4 See Offer Bar-Yosef, 'The Natufian Culture in the Levant' in *Evolutionary Anthropology 6* (1998), pages 159–77 and Peter Bellwood, *The First Farmers* (Blackwell, 2007). For an alternative explanation, see the Oasis Hypothesis proposed by Vera Gordon Childe in 1928, in which he suggested that climate warming *after* the Younger Dryas period led to human populations crowding around oases, triggering the development of new agricultural methods for survival. See Vera Gordon Childe, *Man Makes Himself* (Oxford University Press, 1936).
5 Jared Diamond, *Guns, Germs and Steel* (Norton, 1997), pages 158–63.
6 One is in Ain Mallaha, another in Hayonim Terrace, both in northern Israel. See the *Journal of Achaeological Science*, Vol. 24, No. 1 (January 1997), pages 65–95.

Chapter 18: Written Evidence
1 Alexander Marshack (1918–2004), an American archaeologist, believed the earliest attempts at writing may have been made by *Homo sapiens* as far back as 20,000 years ago using markings or notches made on bones to record the cycles of the moon and other celestial patterns. See Marshack's *The Roots of Civilization* (Weidenfeld & Nicolson, 1972), pages 81–108.
2 A map of the world showing the earth as a flat disk was found on one of the clay tablets in the library of Ashurbanipal.
3 The tablets were rediscovered in 1901, and are now exhibited in the Louvre in Paris.

Chapter 19: Divine Humanity
1 There were a few female pharaohs, but they portrayed themselves as men, wearing beards.
2 Alan Weisman, *The World Without Us* (Virgin Books, 2007), pages 75–6.
3 The ancient Egyptians believed individual souls mirrored the progress of the sun god, Ra, who they thought descended into the underworld each night to be re-energized by Osiris, the god of the dead, ready to rise again full of energy for the next day.
4 See Janet Johnson, 'The Legal Status of Women in Ancient Egypt' in *Women in Ancient Egypt* (New York Hudson Hill, 1997), pages 175–86 and 215–18.
5 There were some periods of expansion during the 3,000-year-long Egyptian civilization, most notably with the conquest of what is now Israel, Lebanon and Syria by Thutmose III (ruled 1479–1425 BC).

Chapter 20: Mother Goddess
1 Masson described his travels in his book *Narrative of Various Journeys in Balochistan, Afghanistan and Punjab, 1826–1838* (Munshiram Manoharlal Publishers, 2001).
2 This was found in the Indus Valley city of Mohenjo-daro, eighty kilometres south-west of Sukkur in Pakistan.
3 Recent genetic studies show that crops and animals were first domesticated in the Near East, not in Europe, so they must have been taken across the continent by migrating people during the period 7000–3500 BC.
4 This migration route is supported by recent studies which show that a majority of the British population have genetic links with people living in northern Spain and Portugal. See Stephen Oppenheimer, *The Origins of the British, A Genetic Detective Story* (Constable, 2006), pages 375–78.
5 See Leonard Cottrell, *The Bull of Minos* (Bell and Hyman Ltd, 1984), page 146.

Chapter 21: Triple Trouble

1 A Hurrian myth, *The Songs of Ullikumi*, is very similar to the later work by the Greek writer Hesiod, whose *Theogony* tells the story of how Zeus became king of the gods.
2 See David Traill, *Schliemann of Troy* (John Murray, 1995), pages 304–305.
3 Biblical scholars have developed a 'Documentary Hypothesis' which estimates that these books were first written down in a series of different versions in about 500 BC. The theory was first formulated by German scholar Julius Wellhausen in his book *Prolegomena zur Geschichte Israels* published in 1878.

Chapter 22: Dragon's Lair

1 See Pliny's *Natural History*: 'The Seres [Chinese] are famous for the woollen substance obtained from their forests; after a soaking in water they comb off the white down of the leaves … So manifold is the labour employed, and so distant is the region of the globe drawn upon, to enable the Roman maiden to flaunt transparent clothing in public.'
2 See John Moorhead, *Justinian* (Longman, 1994), page 167 and Procopius, *History of the Wars*, Book 8.17, 1–8.

Chapter 23: Peace of Mind

1 The sun is estimated to have increased in luminosity by 30 per cent since it ignited, yet the temperature of the earth has cooled dramatically since life first emerged. See Lovelock, *The Ages of Gaia*, page 35.
2 These are: *bhakti* yoga (love and devotion), *karma* yoga (selfless action), *jnana* yoga (learning what is real and what is not) and *dhyana* yoga (stilling of the mind and body through meditation).
3 *Ahimsa* is the first of five vows (*yamas*) in *raja* yoga.
4 For a good story about how Buddha converted a terrorist see *Buddha and the Terrorist*, retold by Satish Kumar (Green Books, 2006).
5 Jains believe that the universe oscillates infinitely from birth to death in a continuous cycle called the *kalchakra*. The same idea is found among some modern scientists, who think the universe keeps oscillating between big crunches and big bangs (see page 16).
6 Pliny's *Natural History*, Book VI.
7 He first articulated this in 1972, and has reiterated it several times since, and hosted various international conferences on the concept.

Chapter 24: East–West Divide

1 In 1993 the Tel Dan Stele was found in northern Israel. It has the first inscription referring to a dynasty known as the House of David, corroborating the existence of the biblical King David. The Cyrus Cylinder found in 1879 under the walls of Babylon backs up the story in the Book of Ezra telling of the Jews' return from Babylonian exile and the building of the Second Temple in Jerusalem. And in 1947 a Bedouin animal-herder called Mohammed Ahmed el-Hamed discovered a series of priceless scrolls hidden in pottery jars stored deep inside a cave in Qumran, near the Dead Sea. These documents, known as the Dead Sea Scrolls, contain copies of many books from the Old Testament dating back as far as 200 BC – some 800 years older than other surviving texts. Experts agree that texts dating 500 years after the events they describe are significantly more likely to have a historical basis than those dating more than 1,500 years after.
2 The twelve tribes were: Reuben, Simeon, Levi, Dan, Naphtali, Gad, Asher, Issachar, Zebulun, Jopseh, Judah and Benjamin.
3 Herodotus says Astyages had a dream which his wise men interpreted as a sign that his grandson would eventually overthrow him. He ordered his steward, Harpagus, to kill the child. But Harpagus couldn't go thorough with it, and instead gave Cyrus to a herdsman to raise as his own. When Cyrus was ten he came across Astyages, who noticed a family resemblance and summoned Harpagus to explain why the boy had not been killed. After Harpagus had confessed to not killing the boy, Astyages tricked him into eating his own son, but allowed Cyrus to return to his parents. In revenge Harpagus helped Cyrus defeated the Medes.
4 It is estimated that there are about 200,000 Zoroastrians worldwide today, making it a comparatively minor religion.

5 See the Behistun Inscription (see page 124), which Darius I had carved into a mountainside. It testifies that Darius's enemies were defeated because they told lies: 'As to these provinces which revolted, lies made them revolt.'
6 Herodotus says Xerxes had 1.7 million infantry, 80,000 cavalry and over 500,000 crew on board his ships. Modern scholars, however, estimate that the real number of troops could not have been more than 250,000.
7 Historians who have emphasized the importance of the Greek victory at Salamis include Victor Hanson in *Carnage in Culture* (Doubleday, 2001), pages 55–59.

Chapter 25: Olympic Champions

1 The people of Athens were fortunate to have access to the nearby silver mines at Laurium, which provided them with the raw materials to mint their own currency.
2 Coins are thought to have originated in the nearby kingdom of Lydia (western Turkey) during the reign of King Croesus, around 600 BC. Rich deposits of soft but valuable metals such as gold, silver and copper could be panned from the River Pactolus, near Sardis, leading to the easy manufacture of coins, which became commonplace from then on across western Turkey, the Persian Empire and throughout the Greek city states.
3 Plutarch was a famous Greek historian who lived from 47 to 127 AD. His *Lives* chronicles the stories of many Greek and Roman heroes.
4 This phenomenon is known as retrograde motion. When the earth completes its orbit in a shorter period of time than the planets outside its orbit (e.g. Mars, Jupiter, Saturn, etc.), it periodically overtakes them, like a faster car on a multi-lane highway. When this occurs, the planet we are passing will first appear to stop its eastward drift, and then to drift back towards the west. Then, as the earth swings past the planet in its orbit, it appears to resume its normal west-to-east motion.
5 Anaximander (610–546 BC), another philosopher who lived in Miletus, was taught by Thales. Only one fragment of his work survives, but he is thought to have accurately described the mechanics of the planets, as well as to have produced one of the first ever maps of the world. The Anaximander crater on the moon was named in his honour.
6 Thales developed a theory that earthquakes were caused by violent underground tsunamis. This may have originated form the ancient Babylonian belief that the earth originally arose out of water, and that the land was surrounded by water on all sides and perhaps underneath.
7 Plato's *The Republic*, Chapter 32, Despotism.
8 *Ibid.*, Chapter 16, Marriage and Childbirth.
9 See *Dictionary of Greek and Roman Antiquities* edited by William Smith (1870).
10 To atone for killing his wife and children in a fit of madness, Heracles was ordered by the gods to carry out twelve labours set by his arch-enemy Eurystheus.

Chapter 26: Conquerors of the World

1 Another Greek thinker, called Anaxagoras, influenced Aristotle and brought the ideas of Thales to Athens. He developed theories to explain eclipses, meteors, rainbows and the sun, which he described as a mass of blazing metal. He was eventually arrested on charges of heresy, and forced to leave Athens. Democritus and Epicurus were two other Greek thinkers who further developed the idea of a mechanistic universe, and were the first known to propose that all matter in the universe is composed of invisible, irreducible little building blocks called atoms. This idea was rediscovered more than 2,000 years later, when John Dalton published his paper on the absorption of gases in 1805.
2 *On the Universe*, attributed to Aristotle, Chapter 6.
3 Pyrrho believed that wise men should prevent themselves from being identified with any specific claim, cause or belief, thus laying the foundations for agnosticism, which was later formalized by a new academy of thinkers. Scepticism has had a major influence on modern Western science, where concepts are often described as 'theories' and 'hypotheses', because many experts consider that no amount of experimental rigour can be regarded as perfect proof.

Chapter 27: Hurricane Force

1 Some of this wall, built after a visit to Britain by Emperor Hadrian in 122 AD, is still standing. It was 120 kilometres long, and stretched from Wallsend on the River Tyne to the Solway Firth.

2 The Romans were among the first people to discover how to make concrete, by mixing volcanic ash and pumice with quicklime. The knowledge then disappeared from Europe after the fall of the Roman Empire for more than 1,300 years before it was rediscovered by the British engineer John Smeaton in 1756.

3 According to an inscription found in the Colosseum, it was treasure stolen from Jews that financed the cost of this vast project. The inscription reads: 'The Emperor Vespasian ordered this new amphitheatre to be erected from his generals' share of the booty.' The Jews were defeated following a brutal seven-year war, during which Roman legions under Titus (who became Emperor in 80 AD) destroyed the city of Jerusalem once again (in 70 AD), including its sacred Second Temple – the one financed by the tolerant Persian Emperor Cyrus the Great (see page 176). According to the historian Josephus, more than a million Jews died during this war. Nearly 100,000 more were sold into slavery and dispersed throughout the Empire.

4 See 'Lead Exposure in Italy: 800 BC to 700 AD' in *International Journal of Anthropology*, Vol. 7, No. 2.

5 Simon Bar Kokhba, who also described himself as the son of God, persuaded the Jewish community in Judaea to revolt against Roman tyranny. The two-year war, beginning in 135 AD, was ordered by the Emperor Hadrian.

6 Galerius issued this decree ordering a general edict of toleration only on his deathbed. Formerly he had been a proponent of the persecution of Christians, and an architect of the decrees of Diocletian.

7 See Origen's *Contra Celsum*, translated by Henry Chadwick (Cambridge University Press, 1980), page 199.

8 See Eusebius of Caesarea, *The Church History and Life of Constantine: 275–339 AD* (Clarendon Press, 1999).

Chapter 28: Timbuk-taboo

1 William Ruddiman, *Plows, Plagues and Petroleum* (Princeton University Press, 2005), page 93.

2 This compares with a total Australian population of about twenty million today. According to the Australian Bureau of Statistics, the Aboriginal population since 1996 has grown at twice the rate of the overall population reaching about 458,000 in 2001 (2.4 per cent).

3 The expression 'Dreamtime' was coined in 1899 by two Victorian anthropologists, Walter Spencer and Francis Gillen. See *The Native Tribes of Central Australia* (1899).

4 Note how much longer the 40,000-year-old Australian Aboriginal culture has lasted than other surviving human civilizations – even Chinese civilization, for example, has lasted no more than 3,500 years.

5 Spencer Wells, *The Journey of Man* (Alan Lane, 2002), pages 56–57.

6 See *The Lost World of the Kalahari* (Hogarth, 1958), page 14. Some recent observers have questioned the authenticity of some of van der Post's writings. See J.D.F. Jones, *Storyteller: The Many Lives of Laurens van der Post* (John Murray, 2001). Others, such as Christopher Booker, have leaped to his defence. See 'Small Lies and the Greater Truth' published in *The Spectator* (20 October 2001).

7 See Felipe Fernandez-Armesto, *Civilizations* (Pan, 2001), page 47.

8 The nearest land to Easter Island is Pitcairn Island, 2,000 kilometres away, and the nearest mainland is Chile, 3,600 kilometres away.

9 See Laurens van der Post, *Lost World of the Kalahari*, Chapter 1.

10 According to Survival International about 10,000 Penan are still living in Borneo today, although the government there has forced most of them to abandon their traditional nomadic way of life and settle in towns. It is thought that only about 350 of them still live in the forest.

11 See Wade Davis, 'A World Made of Stories' in *Nature's Operating Instructions* (University of California Press, 2004).

Chapter 29: Amaizing Americas

1 The earliest attempts at domesticating *teosinte* into maize were uncovered by archaeologists at Guila Naquitz Cave in the Oaxaca Valley of central Mexico, and date back 6,250 years. However, modern genetic analysis indicates that the first genetic modifications as a result of human selection may have started even earlier.

2 Apart from sustaining billions of people around the world, the economic impact of these crops has become monumental. In 2004 more than thirty-three million hectares of maize were grown worldwide, more than rice or wheat, with a market value estimated at US$23 billion. See *www.commodityonline.com*

3 The Cascajal Block, as it is called, mixes these symbols with more abstract boxes and blobs. See the journal *Science*, 15 September 2006.

4 See Friar Diego de Landa, *Yuccatan Before and After the Conquest*, translated by William Gates (Dover Publications, 1978), page 82.

5 It is now in the Newberry Library, Chicago.

6 This plate is now in the Peabody Museum at Harvard University.

7 Analysis of the bones was carried out by Ernest Hooton in 1940. See Alfred Tozzer, *Chichen Itza and its Cenote of Sacrifice*, *Peabody Museum* (Peabody Museum, Cambridge, 1957), page 205.

8 This tale comes from the Franciscan missionary Bernardino de Sahagún (1499–1590), who compiled the *Florentine Codex*, twelve books about Aztec life based on interviews with the natives. See Christian Duverger, *La Flor Letal: Economía del Sacrificio Azteca* (Fondo de Cultura Económica, 2005), pages 128–29.

9 Diamond, *Collapse* (Penguin, 2007), pages 170–77.

10 Other Native American civilizations such as the Pueblo peoples from Chaco Canyon in New Mexico were severely affected by a drought beginning in 1130. It caused them to migrate further south in a desperate attempt to find water. See Diamond, *Collapse*, page 152.

11 Estimates of the number of sacrifices range from 3,000 to 84,000, although the latter figure is almost certainly an exaggeration.

12 See C.A. Burland, *Peoples of the Sun* (Book Club Associates, 1986), pages 225–28.

Chapter 30: What a Revelation!

1 See Ronald Segal, *Islam's Black Slaves* (Atlantic Books, 2001), page 86.

2 The Abbasids claimed direct descent from Mohammed's own clan, the Hashim, in order to secure the support of the Shia Muslims of Persia. Once in power, however, they changed their story and became Sunnis.

3 Stirrups were probably introduced to the Arabs by the professional Persian knights (*Azatan*) established by the Sassanids (226–651 AD), who took the idea from the invading Avars of central and eastern Asia, with whom they allied against the Byzantines in 626 AD.

4 On the eastern side were the Persian Sassanids, and on the west the remaining rump of the Roman Empire, the Byzantines, based in Constantinople.

5 See the *Sahih Bukhari*, a collection of prophetic traditions collected by Muslim scholar Muhammad ibn al-Bukhari (810–870 AD), Vol. 6, Book 60, No. 201, translated by M. Muhsin Khan courtesy of the University of Southern California. Online text at *www.usc.edu*

6 One of the most famous Sufi poets was an Afghan called Rumi (1207–73), who founded the Mevlevi Order, better known as the 'Whirling Dervishes', who perform their worship through a dance called the *sema*.

7 It was not until the twelfth century that a Latin version of this masterpiece appeared in the West, translated from an Arabic copy found by Christians in Spain.

8 Jonathan Bloom and Sheila Blair, *Islam* (TV Books, 2000), pages 106–107.

9 The English philosopher Roger Bacon (1214–94), mentions the Latin translation of *Kitab al-Manazir*, al-Haytham's work on optics. See David C. Lindberg, *Roger Bacon and the Origins of Perspectiva in the Middle Ages* (Clarendon Press, 1996), page 11.

10 The earliest known depiction of a stirrup is on a model of a rider found in a Chinese tomb dating to 322 AD.

Chapter 31: Paper, Printing & Powder

1 Bloom and Blair, *Islam*, page 228.

2 It is ironic that if it weren't for the ingenuity of Chinese

civilization, Columbus would probably never have read Polo's book, since the paper it was printed on was an entirely Chinese invention. See Bjorn Landstrom, *Columbus* (Alan & Unwin, 1967), page 27.

3 Paper-making was first known in Europe in Fabriano, Italy, in 1268. It had reached Troyes in France by 1348, Nuremberg in Germany by 1390, Basel in Switzerland by 1433, Hertfordshire in England by 1495 and Dortrecht in the Netherlands by 1586. See J. Needham, *Science and Civilization in China* (Cambridge University Press, 1986), Vol. 1, pages 299–302.

4 Needham, *Science and Civilisation in China*, Vol. 5, page 123.

5 The imperial modifications to ancient Shinto never fully took hold. Today there are an estimated 100,000 traditional Shinto shrines still used in Japan. Followers of the religion worship spirits called *kami* that they believe are found in all natural forms, animate and inanimate, from animals to rocks and pools.

6 In 988 AD the Chinese National Academy contained an archive of 4,000 carved wood blocks for printing various Confucian classics. By 1005 AD this had risen to 400,000 – a hundred-fold increase in seventeen years. See Needham, *Science and Civilisation in China*, Vol. 5, page 370.

7 Patricia Ebrey, Anne Walthall and James Palais, *East Asia* (Houghton Mifflin, 2006), page 156.

8 From *Classified Essentials of the Mysterious Tao of the True Origin of Things*, a mid-ninth-century book on Chinese alchemy quoted in Needham, *Science and Civilization in China*, Vol. 5, page 11. Realgar is a naturally occurring mineral of arsenic sulphide.

9 Needham, *Science and Civilisation in China*, Vol. 5, page 223.

10 Shiba Yoshinobu, *Commerce and Society in Sung China*, translated by Mark Elvin (University of Michigan Press, 1970), page 33.

11 The technique is mentioned in a Chinese work composed between 20 and 100 AD called the *Louen-heng*. See Li Shu-hua, 'Origine de la Boussole 11: Aimant et Boussole', in *Isis*, Vol. 45, No. 2 (July 1954), page 176.

12 Described in his book *Dream Pool Essays*, written in c.1088 AD.

13 Zhu Yu, an official who governed the province that included the port of Guangzhou, mentions it in his maritime history: 'According to government regulations concerning seagoing ships, the larger ones can carry several hundred men, and the smaller ones may have more than a hundred men on board. The ships' pilots are acquainted with the configuration of the coasts; at night they steer by the stars, and in the daytime by the sun. In dark weather they look at the south-pointing needle.' See Needham, *Science and Civilisation in China*, Vol. 4, page 279.

14 The first mention of the compass in Europe is in Alexander Neckham's *On the Nature of Things*, written in Paris in 1190, and the first recipe for gunpowder appeared in Roger Bacon's *Opus Majus* in 1267.

15 Archaeological evidence comes from increased water levels in the Caspian Sea. See H.H. Lamb, *Climate History and the Modern World* (Routledge, 1995), pages 175–76.

16 Robert Claiborne, *Climate, Man and History* (Angus & Robertson, 1973), page 346.

17 See Paul Chevedden, 'The Invention of the Counterweight Trebuchet' in *Dumbarton Oak Papers*, No. 54 (2000).

18 These empires were initially called the Blue Horde and the White Horde. In 1266 they were consolidated by another of Genghis's sons, Berke (died 1255), into a single empire called the Golden Horde, which then became an Islamic state and was eventually absorbed into the Ottoman Empire in 1502.

19 The Mongols are known to have used gunpowder and firearms in Europe as early as 1241 at the Battle of Mohi in Hungary. See Jacques Gernet, *A History of Chinese Civilisation* (Cambridge University Press, 1982), page 379.

20 According to the *History of Song* (1345). Full text available at *www.gutenberg.org/etext/24183*

Chapter 32: Medieval Misery

1 From Jordanes' *The Origin and Deeds of the Goths*, written in c.551 AD. See translation by Charles Mierow (Princeton, 1908), page 57.

2 See Alessandro Barbero and John Cullen, *The Day of the Barbarians: The Battle that Led to the Fall of the Roman Empire*, translated by John Cullen (Walker & Co, 2007).

3 This Chronicle, written by Michael of Syria (d.1199), tells the story of all history from Creation up until his own day. Only one original copy survives, kept in a locked box in a church in Aleppo. There is a French translation by J.B. Chabot, *Chronique de Michel le Syrien, Patriarche Jacobite d'Antioche (1166–1199),* Five volumes: 1899, 1901, 1905, 1910, 1924.

4 Historians from Edward Gibbon in *Decline and Fall of the Roman Empire* (1781–89) to Paul K. Davis in *100 Battles from Ancient Times to the Present Day* (1999) have cited certain battles as key moments that have changed the course of human history.

5 Josiah C. Russell, 'Population in Europe' in *The Fontana Economic History of Europe*, Vol. 1: The Middle Ages (The Harvester Press, 1976), page 36.

6 There had been a previous outbreak in Athens between 430 and 427 BC when it seemed the Athenians may yet win the Peloponnesian War. See Thucydides account in *The Greek Historians* edited by M.I. Finley (Penguin, 1980), pages 274–75.

7 Arianism was the accepted version of Christianity among many of the Germanic tribes such as the Goths, Visigoths, Ostrogoths, Vandals and Lombards who invaded Europe as the Roman Empire declined.

8 David Keys, *Catastrophe: An Investigation into the Origins of the Modern World* (Century, 1999).

9 Lynn White, *Medieval Technology and Social Change*, (Oxford University Press, 1966), page 2.

10 See accounts from the geographical compendium of Ibn Rustah, a tenth-century Persian explorer who described his experiences of the Rus in Novgorod.

11 The ritual is described in *The Orkneyinga Saga: The History of the Earls of Orkney* translated by Hermann Pálsson and Paul Edwards (Hogarth Press, 1978), page 33.

12 Michael Williams, *Deforesting the Earth* (University of Chicago Press, 2003), pages 106, 118.

13 Needham, *Science and Civilisation in China*, Vol. 2, page 326.

14 Williams, *Deforesting the Earth*, page 104.

15 *Ibid.*, page 113.

16 *Ibid.*, page 116.

17 Russell, 'Population in Europe', pages 34–36.

18 *Translations and Reprints from the Original Sources of European History*, Series 1, Vol. 3:1 (University of Pennsylvania, 1896), pages 15–16.

19 Modern agricultural methods, transformed by fertilizers, machinery and the mass-production techniques of agribusiness, are 300 times more efficient, yet the number of people living in Europe now is only four times greater than in 1315 when the continent's first great famine struck.

20 M.A. Jonker, 'Estimation of Life Expectancy in the Middle Ages' in *Journal of the Royal Statistical Society*, 2003, Vol. 166, No. 1, pages 105–17.

21 This theory was first put forward in 2001 by Susan Scott and Christopher Duncan, researchers from Liverpool University. They have since claimed that thanks to the Black Death approximately ten per cent of all Europeans are genetically resistant to HIV Aids. See *Journal of Medical Genetics*, No. 42 (2005), pages 205–208.

22 Ruddiman, *Plows, Plagues and Petroleum*, pages 120–23.

23 *The Opus Majus of Roger Bacon*, translated by Robert Belle Burke, Vol. 2 (University of Pennsylvania, 1928), pages 629–30.

24 Juan Marian, *Historia general de Espana* (1608), Vol. 2, page 27. English translation by Capt. John Stephens, *The General History of Spain* (1699), Vol. 1, page 264.

25 Jack Kelly, *Gunpowder* (Atlantic Books, 2004), pages 49–53.

Chapter 33: Treasure Hunt

1 Ross Dunn, *The Adventures of Ibn Battuta* (Croom Helm, 1998), pages 45–6.

2 Elikia M'Bokolo, *Afrique noire: Histoire et Civilisations*, Vol. I (Haiter-Aupelf, Paris, 1995), page 264. Also see Alan Weisman, *The World Without Us* (Virgin Books, 2007), page 83.

3 Ross Dunn, *The Adventures of Ibn Battuta* (Croom Helm, 1986), page 290.

4 Bloom and Blair, *Islam*, page 228.

5 See the writings of Ruy Gonzáles de Clavigo. See *Embassy to Tamerlane 1403–1406*, translated by Guy Le Strange (Routledge, 1928).

6 There has been recent speculation that on the sixth voyage, in 1421, Zheng He travelled all the way around the world, sailing past the Cape of Good Hope and discovering Greenland, Iceland, Australia, Antarctica and America. This theory, put forward by Gavin Menzies in *1421: The Year the Chinese Discovered the World* (Bantam Books, 2003) has not been accepted by many historians.

7 See *La Escatologia Musalmana en la Divine Comedia* (1919), by the Catholic scholar Miguel Asin, which charts the similarities between Dante's poem and the *Hadith*, the oral teachings of the prophet Mohammed. Controversy over how far Dante borrowed from Islamic thought for his depiction of heaven, purgatory and hell is still ongoing.

8 Diamond, *Collapse*, pages 266–76.

9 The Pope's right to appoint bishops regardless of the will of secular authorities was the subject of fierce rivalry in the eleventh century between the Holy Roman Emperor Henry IV and Pope Gregory VII. The Investiture Contest, as the episode is known, led to a civil war in Germany. As a result dukes and abbots gained control over Holy Roman Imperial authority in Germany, leading to its fragmentation into many small states until it was eventually reunited in the nineteenth century by Bismarck (see page 354).

10 See Johann Burchard, *Pope Alexander VI and his Court: Extracts from the Latin Diary of the Papal Master of Ceremonies, 1484–1506*, edited by F. L. Glaser (Nicholas Brown, New York, 1921), page 194.

11 Patrick Macey, *Bonfire Songs: Savonarola's Musical Legacy* (Clarendon Press, 1998), pages 30–31.

Chapter 34: Moules Marinière

1 Slaves were snatched from as far away as Easter Island, discovered by Europeans in 1722. Although some commentators have speculated that environmental catastrophe exacerbated by human deforestation accounted for the ruin of the island's population (see Diamond, *Collapse*), the effects of enforced slavery are an equally likely explanation, since many natives, including King Kamakoi and his son, were kidnapped and taken as captives from the early 1800s to work in Peruvian mines. See Benny Peiser, 'From Genocide to Ecocide: The Rape of Rapa Nui' in *Energy and Environment*, Vol.16, No. 3 and 4 (2005).

2 The estimate comes from Amigos de Potosí, a charity established in the Netherlands to help the surviving mining families of Potosí. See *www.amigosdepotosi.com*

3 Clive Ponting, *A Green History of the World* (Sinclair Stevenson, 1991), page 131.

4 Columbus studied the works of Marinus of Tyre (70–130 AD) and Alfraganus, who lived in the ninth century AD. His mistake was partly because Alfraganus's calculations were based on Arabic miles (1,800 metres) not Italian miles, as Columbus was assuming (1,238 metres). See Bjorn Landstrom, *Columbus* (Allen & Unwin, 1967), page 30.

5 *Ibid.*, page 131.

6 *Ibid.*, page 93.

7 An engraving in a journal purportedly by Vespucci published in 1505 shows naked women in feathered head-dresses distributing human limbs. One of them is gnawing on a human arm. See *Amerigo* by Felipe Fernandez-Armesto (Weidenfeld & Nicolson, 2006), page 162.

8 *Ibid.*, page 52.

9 Ponting, *A Green History of the World*, pages 196–7.

10 *Ibid.*, page 130.

11 Smallpox is first thought to have appeared several thousand years ago in ancient Egypt and India and then in Europe by the fourth century BC. See Donald Hopkins, *Princes and Peasants: Smallpox in History* (Chicago University Press, 1983), pages 13–21.

12 Stuart Stirling, *The Last Conquistador* (Sutton, 1999), page 44.

13 *Ibid.*, pages 140–41.

14 There is some dispute about who was the first person to circumnavigate the globe, because on an earlier trip to the Spice Islands (sailing west) Magellan had enslaved a Malaccan interpreter named Enrique, and known as Henry the Black. Enrique accompanied Magellan on this voyage east, although he mysteriously left the expedition in the Philippines, shortly after Magellan's death. If he made it back to Malacca, and there is some evidence to suggest that he did, then technically he would have been the first man to circumnavigate the globe, albeit in two journeys.

Chapter 35: Fancy a Beer?

1 Geoffrey Parker, *The Military Revolution* (Cambridge University Press, 1996), pages 90–91.

2 *Ibid.*, page 94.

3 Troy Beckham, 'Eating the Empire' in *Past and Present*, No. 198, February 2008.

4 David Christian, *Maps of Time* (University of California Press, 2004), pages 344–45.

5 Ponting, *A Green History of the World*, page 145.

6 See *De Sapientia Veterum* in *The Works of Francis Bacon*, edited by James Spedding (Longman, 1858), Vol. 6, page 747.

7 See Descartes, *Principles of Philosophy IV* (1644), Article 203.

8 Blenheim Palace, built in England between 1705 and 1722, was modelled on Versailles, as were numerous palaces in Spain, Italy, Sweden and Germany.

Chapter 36: New Pangaea

1 Williams, *Deforesting the Earth*, page 180.

2 NRS, *Naval Tracts of Sir William Monson*, V, page 268.

3 By 1760, 84 per cent of all European masts were made of Baltic fir. See Williams, *Deforesting the Earth*, page 200.

4 Ponting, *A Green History of the World*, page 279.

5 *Ibid.*, page 278.

6 Horses had become extinct in the Americas at the end of the last glaciation, 12,000 years previously.

7 Ponting, *A Green History of the World*, page 231.

8 *Ibid.*, page 173.

9 Sugarcane, originally brought to Europe from south Asia by Muslim traders, took so much energy from the soil that a typical plantation became infertile after about twelve to fifteen years. By then fresh areas had to be cleared to make way for new farms, deforesting more areas and expanding coastal settlements further inland. Thanks to slave labour, it was usually easier to clear more forest than to ship in supplies of manure to regenerate the land. See Williams, *Deforesting the Earth*, page 217.

10 By 1671 timber from mainland America had to be imported to Barbados to fuel the sugar industry. It was even reported by one visitor that 'all the trees are destroyed so that wanting wood to boyle their sugar, they are forced to send for coales from England'.

11 See Williams, *Deforesting the Earth*, page 233, and Ping-ti Ho, *Studies on the Population of China 1368–1933* (Harvard University Press, 1959), pages 183–9.

12 See Cha'ao-Ting Chi, *Key Economic Areas in Chinese History as Revealed in the Development of Public Works for Water Control* (Allen & Unwin, 1936), page 22.

13 See J.S. and W.S. Hampl, 'Pellagra and the Origin of a Myth: Evidence from European Literature and Folklore' in *Journal of the Royal Society of Medicine*, Vol. 90 (1997), pages 636–39.

14 Brian Fagan, *The Little Ice Age: How Climate Made History 1300–1850* (Basicbooks, 2000), page 189.

15 See Redcliff Nathan Salaman, *The History and Social Influence of the Potato* (Cambridge University Press, 1985), page 308.

16 See Ponting, *A Green History of the World*, page 181.

Chapter 37: Mixed Response

1 See Robert Davis, *Christian Slaves, Muslim Masters* (Palgrave Macmillan, 2003), page 23.

2 To the Shia shrine of Ardabil, Abbas gave jewellery, weapons, horses, sheep and goats, as well as fine manuscripts and 1,162 ceramics – many of them rare Chinese blue-and-white porcelains. See Bloom and Blair, *Islam*, page 203.

3 Bloom and Blair, page 216

4 The letter, dated 1617 appears in *Outlines of European History: Volume 2*, edited by James Harvey Robinson (Boston: Ginn & Co, 1914–1927), pages 333–35.

5 The Peacock Throne was the ultimate symbol of Mughal wealth. It had feet of solid gold and was studded with more than 108 rubies, 116 emeralds and countless diamonds and pearls. This masterpiece of vanity was destroyed in 1747, when Nadir Shah was assassinated in a coup by officers of his own guard.

6 Diamond, *Guns, Germs and Steel*, page 257.

7 For a contemporary account of the encounter from an American perspective see John S. Sewall, 'The Invincible Armada in Japan' in *New Englander and Yale Review* (1890), pages 201–12.

Chapter 38: Free Reign

1 This phrase was first used in Edward Bulwer-Lytton's play *Richelieu: On the Conspiracy* (1839): 'Beneath the rule of men truly great, the pen is mightier than the sword.'

2 Newton's classic *Philosophiae Naturalis Principia Mathematica* (1687), which states his three universal laws of motion, was not improved upon by scientists for more than 200 years.

3 This important limitation on state power reflects the Deist belief of the Founding Fathers that religion was a matter of personal choice, not state establishment.

4 Note the difference here with the English Bill of Rights, which allowed firearms only for the purpose of self-defence (see page 325).

5 These were probably caused by the effects of El Niño, a climatic event off the coast of Peru. See Fagan, *The Little Ice Age*, pages 165–6, and Richard H. Grove, 'Global Impact of the 1789–93 El Niño', in *Nature*, No. 393 (1998), pages 318–19.

6 Dr Antoine Louis (1723–92), physician to the King, actually took Guillotin's ideas and designed the first machine.

7 Kelly, *Gunpowder*, pages 164–66.

8 Nelson and Wellington were buried in St Paul's Cathedral, London.

9 Diamond, *Collapse*, page 388.

10 See Chasing Freedom Information Sheet at *www.royalnavalmuseum.org*

Chapter 39: Monkey Business

1 Francis of Assisi (1181–1226), founder of the Catholic Franciscan monks, and Maimonides (1135–1204), a Jewish rabbi and philosopher, were notable exceptions who championed an equal regard for all living things.

2 The Qing rose to power following a peasants' revolt against the last Ming Emperor, led by former shepherd Li Zicheng.

3 In 1550, annual English coal production was about 210,000 tonnes. By 1650 this had risen to 1.5 million tonnes, and by 1790 to over six million. The big increase took place from 1790 onwards, however, reaching sixteen million tonnes by 1815. Coal didn't start to replace wood-burning in the USA until about 1850, thanks to its still abundant supplies of timber. See Ponting, *A Green History of the World*, page 281.

4 Patents were first used by the republican government of Venice as early as 1474. However, Britain was the first country to adopt the system. It soon caught on in the newly independent United States of America, with the passing of a Patent Act in 1790.

5 Ponting, *A Green History of the World*, page 277.

6 In a world where independently available power sources are everywhere, as ushered in by high-pressure steam from c.1850, Smith's capitalist premise that some places are more naturally suited for manufacturing than others is weakened.

7 Low-pressure steam engines push a piston out of a cylinder to create a partial vacuum, leaving atmospheric pressure to push the piston back again.

8 One of the ships used to lay the cable was HMS *Agamemnon*, built in 1852, the first British battleship to be designed from the keel up with steam power installed.

9 See *www.worldmapper.org*

10 Strictly speaking, this was the first controlled, heavier-than-air powered flight since Zeppelin balloons fill with hydrogen gas, which is lighter than air, had been flown previously.

11 See Army Airforce Statistics Digest (WWII), table 79 at *www.afhra.maxwell.af.mil*

12 When crude oil is distilled in a modern refinery, roughly eighty-six per cent goes into making energy-rich fuels such as petrol or kerosene. The other fourteen per cent provides the compounds, called polymers, which man uses for making artificial substances such as plastic and nature uses for making everything from cells, blood and skin to petals, leaves and bark.

13 See John Postgate, 'Fixing the Nitrogen Fixers' in *New Scientist*, 3 February 1990, and Vaclav Smil, *Enriching the Earth* (MIT Press, 2004), page 204.

14 Controversy still surrounds the widespread ban on DDT, with some arguing that its lack of use in certain African countries has led to many more human deaths through diseases such as malaria, carried by mosquitoes. One reason for the lack of a worldwide DDT ban is the lucrative export market for cut flowers, which relies on chemicals like DDT to produce cheap but perfect-looking bouquets. See Weisman, *The World Without Us*, page 74.

15 See 'Leaded Gasoline, Safe Refrigeration, and Thomas Midgley, Jr', in S. Bertsch McGrayne, *Prometheans in the Lab* (McGraw-Hill, 2002).

16 The actual annual growth in the number of humans being added to the world population is now in decline, from eighty-seven million per year in the 1980s to seventy-five million in 2006. However, even at this reduced rate, the world's population is expected to exceed nine billion by 2050. See 'World Population Prospects (2006)' published by the United Nations at *http://esa.un.org/unpp/*

17 See *The World Factbook* (CIA, 2007).

Chapter 40: White Man's Race

1 The Union of South Africa was granted semi-independent status as a dominion of the British Empire on 31 May 1910, and became a fully independent republic in 1961.

2 See Paul Broca, 'Sur les cranes de la caverne de l'homme mort (Loere)' in *Revue d'anthropologie*, No. 2 (1873), pages 1–53. In this article Broca claimed that his experiments showed that on average whites had heavier brains than blacks with more complex convolutions and larger frontal lobes. Such differences have also been claimed subsequently by other researchers See J. Phillipe Rushton, *Brain Size, IQ, and Racial Group Differences: Evidence from Musculoskeletal Traits* (Intelligence 31, 2003), pages 139–55. Also the American palaeontologist, evolutionary biologist and scientific historian Stephen Jay Gould (1941–2002) studied this research in *The Mismeasure of Man* (Penguin, 1997), pages 124–39. He suggested that Morton had fudged data and 'overpacked' skulls with filler in order to justify his opinions.

3 See Madison Grant, *Passing of the Great Race* (Bell & Sons, 1917), page 46.

4 See James McPherson, *Battle Cry of Freedom: The Civil War Era* (Oxford University Press, 1988), page 244.

5 Ernest William Winkler (ed.), *Journal of the Secession Convention of Texas 1861, Edited from the Original in the Department of State* (Texas Library and Historical Commission, 1912), pages 61–65. Some southerners believed that God's curse on Noah's son Ham and his descendants in Africa was a biblical justification for slavery.

6 See Alexander Stephens's Cornerstone Speech, Savannah, Georgia, 21 March 1861.

7 See *Civil Rights Cases, 109 US 3* (1883). In 1879 as many as 40,000 black Africans are thought to have fled to Kansas by riverboat or on foot, because it was regarded as a more tolerant state. The migrations continued in the 1890s, to Nebraska, Colorado and Oklahoma.

8 'Bringing them Home: Report of the National Inquiry into the Separation of Aboriginal and Torres Strait Islander Children from their Families'. Since the report was issued every Australian state has made a formal apology for the episode, while the Federal government finally made an official apology, following the election of Prime Minister Kevin Rudd, on 13 February 2008.

9 Australian Bureau of Statistics, *Population Distribution, Aboriginal and Torres Strait Islander Australians*, 15 August 2007.

10 Letter, dated 15 March 1843, by Lieutenant-Colonel de Montagnac, in *Lettres d'un soldat*, (Christian Destremeau, 1998), page 153.

11 Peter Forbath, *The River Congo* (Secker & Warburg, 1978), page 370.

12 *Ibid.*, page 375.

13 The episode has been described by a United Nations report as one of the earliest attempts at genocide in the twentieth century. See *Whitaker Report* (1985).

14 So vital was this waterway that in1882 British forces occupied all Egypt just to protect it. They stayed there until 1954 but re-invaded two years later (aided by French and Israeli forces in what has come to be known as the Suez Crisis).

15 The de Beers diamond company, founded by Rhodes, still controls approximately half the world's trade in rough-cut diamonds.

16 Attributed to Rhodes in R. Dumont and N. Cohen, *The Growth of Hunger: A New Politics of Agriculture* (Marion Boyars, 1980), page 29.

17 This is the first verse of 'The White Man's Burden', written by Kipling in 1899 in support of the United States' invasion of the Philippines.

Chapter 41: Back to the Future

1 In July 2005 Marx was voted by BBC Radio 4 listeners as the greatest philosopher of all time.

2 See E. Backhouse and J.O.P. Bland, *Annals and the Memoirs of the Court of Peking* (Heinemann, 1914), pages 325–34.

3 See Fu, Lo-shu, *A Documentary Chronicle of Sino-Western relations* (University of Arizona Press, 1966), Vol. 1, page 380.

4 Ernst Johann (ed.), *Reden des Kaisers: Anspruchen, Predigten und Trinkspruche Willems II* (1912), page 402. Translated by Richard S. Levy.

5 The Tsar's failure to deliver on his promise of a democratic parliament following the massacre by his troops of hundreds of unarmed protesting factory workers during the Bloody Sunday riots of 1905 inflamed the situation further.

6 Lenin, *Collected Works* (Moscow: Foreign Languages Publishing House, 1960–80), Vol. 30, page 335.

7 The Sykes–Picot agreement was negotiated by French and British diplomats in November 1915, despite promises to Arab nations by British officials, including T.E. Lawrence, that their independence from Western rule would be granted in return for support against the Ottomans.

8 Adolf Hitler, *Mein Kampf* (Jaico edition, 1988), page 187.

9 *Ibid.*, page 69.

10 *Ibid.*, page 129.

11 The term 'Aryan' comes from the Sanskrit word meaning noble. Nazi scientists and historians claimed that all Indo-European-speaking peoples originated from a European master-race they called Aryan.

Chapter 42: Witch Way?

1 'Weird' in this context means the same as the Anglo-Saxon *wyrd*, or fate (see page 215).

2 The Algerian War of Independence took place between 1954 and 1962. Peace was eventually secured with the Evian Accords, which granted Algeria independence and France the rights to certain military bases, which it used as nuclear test facilities until 1966.

3 Saudi Aramco is still the largest oil corporation in the world, but is now wholly owned by the Saudi government.

4 See, for example, Milton Friedman, *Capitalism and Freedom* (University of Chicago Press, 1982).

5 The long-term consequences of the credit crunch that affected financial markets in the summer of 2007 and the ability of co-ordinated central bank activities to stabilize the financial system remains to be seen.

6 See World Institute for Development Economics Research, 'World Distribution of Household Wealth' (December 2006). The study likened present-day global inequality to one person in a group of ten taking ninety-nine per cent of a pie, leaving the other nine to share the remaining one per cent.

7 Examples include Zimbabwe, Somalia, the Democratic Republic of Congo, Nigeria, Chad, Angola, Algeria, Sudan, Congo, Liberia, Zambia and Sierra Leone. See The World Bank, *The Worldwide Governance Indicators Project 2006*.

8 Life expectancy in sub-Saharan Africa stood at 46.1 years in 2005, according to a report by the Global Commission of the Societal Determinants of Health, established by the World Health Organization.

9 See the United Nations report 'Global Biodiversity Outlook 2' (20 March 2006). See also World Conservation Union Red List of Threatened Species, 2007.

10 Eight hundred million hectares is eight million square kilometres. The area of the United States is 9.6 million square kilometres. See Williams, *Deforesting the Earth*, page 396.

11 *Ibid.*, page 456.

12 Two men from Brazil, Rayfran das Neves Sales and Clodoaldo Carlos Batista, were convicted of Sister Stang's murder on 10 December 2005. In May 2007 wood rancher Vitalmiro Bastos Moura was convicted of paying them to kill her after her letters to the authorities led to him receiving a substantial fine. He was sentenced to thirty years in prison. See 'Brazil: Rancher Guilty in Killing of US Nun' in *New York Times*, 16 May 2007.

13 See 'A Global Map of Human Impact on Marine Ecosystems' in *Science*, 15 February 2008.

14 See W. Rees and M. Wackernagel, 'Ecological Footprints and Appropriated Carrying Capacity: Measuring the Natural Capital Requirements of the Human Economy', in A.-M. Jansson and M. Hammer (eds.), *Investing in Natural Capital: The Ecological Economics Approach to Sustainability* (Island Press, 1994).

15 See Diamond, *Collapse*, page 360.

16 According to US Energy Information Administration, see *http://www.eia.doe.gov/neic/quickfacts/quickoil.html*

17 According to the Intergovernmental Panel on Climate Change report published on 16 November 2007, 'Warming of the climate system is unequivocal as is now evident from observations in global average air and ocean temperatures, widespread melting of snow and ice, and rising global average sea levels.' The same report estimates that global greenhouse emissions caused by human activity rose by seventy per cent between 1970 and 2004, and that at 379 parts per million, levels of carbon dioxide in the atmosphere 'exceed by far the natural range of the last 650,000 years'. It concludes: 'There is very high confidence that the net effect of human activities since 1750 has been one of warming.'

18 Writing in the *Washington Post* on 15 June 2007, United Nations Secretary General Ban Ki-Moon said the conflict in Darfur began as 'an ecological crisis, arising at least in part from climate change'.

19 There is little prospect of renewable technologies such as solar, wind and tidal power being able to meet the world's demand for energy in 2038, when it is predicted that oil will run out. And the amount of uranium required to power the number of power stations needed to convert wholesale to nuclear energy would far outstrip global supplies; current estimates predict the supply can last for only about eighty-five years. See OECD Nuclear Energy Agency and the International Atomic Energy Agency, 'Uranium 2005: Resources, Production and Demand'.

20 The phrase was first used by evolutionary scientist Richard Dawkins in his book *The Blind Watchmaker* (Longman, 1986): '[Natural selection] has no vision, no foresight, no sight at all. If it can be said to play the role of watchmaker, it is the "blind" watchmaker.'

21 In answer to a question about whether Americans should adjust their lifestyles to address the energy problem, President George W. Bush's press representative, Ari Fleischer, said at a press briefing in 2001: 'That's a big no. The President believes that it's an American way of life and that it should be the goal of policymakers to protect the American way of life. The American way of life is a blessed one.' See the White House website at *http://www.whitehouse.gov/news/briefings/20010507.html*

Index

Page numbers in *italics* refer to illustrations.

A

M

Picture Credits

Published by Bloomsbury USA, New York

Design by willwebb.co.uk; Illustrations by andyforshaw.com;
Picture research by Anne-Marie Ehrlich; Index by Vicki Robinson

Every effort has been made to contact and clear permissions with the relevant copyright holders. In the event of any omissions, please contact the Publishers with any queries.

All papers used by Bloomsbury USA are natural, recyclable products made from wood grown in well-managed forests. The manufacturing processes conform to the environmental regulations of the country of origin.

LIBRARY OF CONGRESS CATALOGING-IN-PUBLICATION DATA
Lloyd, Christopher, 1968–
What on Earth happened? : the complete story of the planet, life, and people from the big bang to the present day / written by Christopher Lloyd.
p. cm.
Includes bibliographical references and index.
ISBN-13: 978-1-59691-583-1 (hardcover : alk. paper) ISBN-10: 1-59691-583-8 (hardcover : alk. paper)
1. Earth sciences—History. 2. Civilization—History. 3. Earth—History. I. Title.
QE11.L56 2008
900—dc22
2008015348

First U.S. Edition 2008

1 3 5 7 9 10 8 6 4 2

Printed in China by RR Donnelley, South China Printing Co. Ltd.